More Vertical Antenna Classics

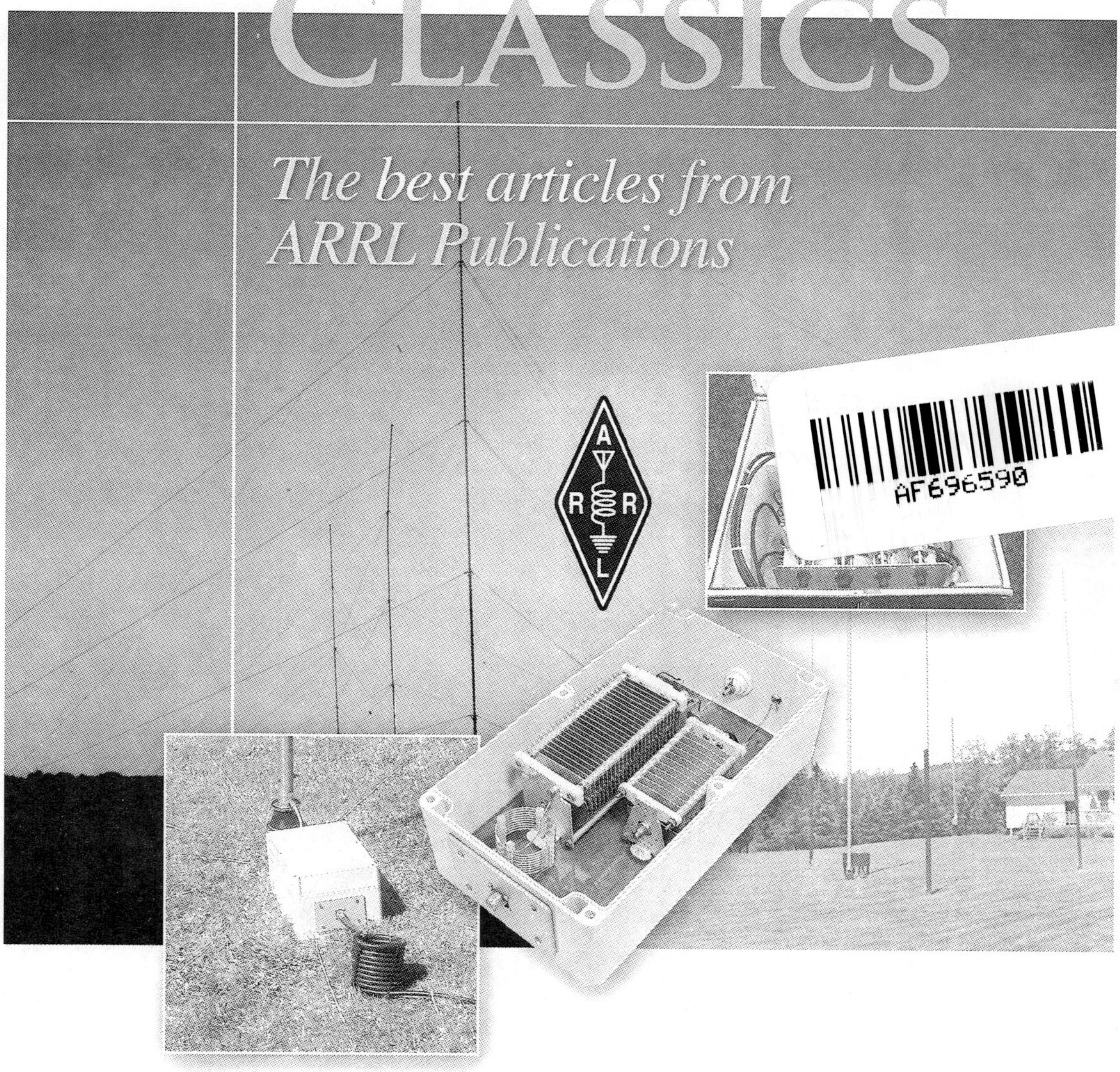

About the Covers

Front cover: The foreground image captures a sunset silhouette of the 160-meter vertical array at W7RH. Inset photos, clockwise from upper left: (1) the L-network of K1FK's 75-foot top loaded vertical; (2) the 75-foot top loaded vertical itself and (3)-(4) the L-network tuner described in "A 20 and 40 Meter Vertical on 'Autopilot'" by K6MHE.

Back cover: Two views of the "Ground Coupled Portable Antenna" by W3JIP.

Compiled by: Steve Ford, WB8IMY

Production: Shelly Bloom, WB1ENT

Cover Design: Sue Fagan

Published by:

ARRL *The national association for* AMATEUR RADIO

225 Main Street • Newington, CT 06111-1494

Printed in the USA

ISBN: 0-87259-979-5

1st Edition
1st Printing

Contents

Single-Band Verticals

Vertical Anenna Design and Use

Foreword

Vertical antenna designs have been Amateur Radio favorites for decades, and it is easy to understand why. Many designs are relatively easy to build and occupy very little space. Verticals can also be "stealthy," meaning that they are less visible to neighbors. Verticals stand up to weather much better than other antennas and they are quite easy to service should something go wrong.

When installed properly, vertical antennas can also be quite effective. In fact, some amateurs have created powerful vertical arrays with two or more antennas. These arrays function much like traditional directional antennas in their ability to concentrate signals.

In ARRL's *More Vertical Antenna Classics*, we've collected a diverse selection of articles from magazines such as *QST*, *QEX* and the *National Contest Journal*, as well as several *ARRL Antenna Compendium* publications. You'll find nuts-and-bolds construction articles that guide you through the process of building a particular antenna. You will also find more abstract theoretical articles that describe how vertical antennas work—and how they can be made to work better.

I hope you'll enjoy this book and find it to be a valuable reference.

David Sumner, K1ZZ
Chief Executive Officer
Newington, CT
September 2006

About the ARRL
The national association for Amateur Radio

The seed for Amateur Radio was planted in the 1890s, when Guglielmo Marconi began his experiments in wireless telegraphy. Soon he was joined by dozens, then hundreds, of others who were enthusiastic about sending and receiving messages through the air—some with a commercial interest, but others solely out of a love for this new communications medium. The United States government began licensing Amateur Radio operators in 1912.

By 1914, there were thousands of Amateur Radio operators—hams—in the United States. Hiram Percy Maxim, a leading Hartford, Connecticut inventor and industrialist, saw the need for an organization to band together this fledgling group of radio experimenters. In May 1914 he founded the American Radio Relay League (ARRL) to meet that need.

Today ARRL, with approximately 170,000 members, is the largest organization of radio amateurs in the United States. The ARRL is a not-for-profit organization that:

- promotes interest in Amateur Radio communications and experimentation
- represents US radio amateurs in legislative matters, and
- maintains fraternalism and a high standard of conduct among Amateur Radio operators.

At ARRL headquarters in the Hartford suburb of Newington, the staff helps serve the needs of members. ARRL is also International Secretariat for the International Amateur Radio Union, which is made up of similar societies in 150 countries around the world.

ARRL publishes the monthly journal *QST*, as well as newsletters and many publications covering all aspects of Amateur Radio. Its headquarters station, W1AW, transmits bulletins of interest to radio amateurs and Morse code practice sessions. The ARRL also coordinates an extensive field organization, which includes volunteers who provide technical information and other support services for radio amateurs as well as communications for public-service activities. In addition, ARRL represents US amateurs with the Federal Communications Commission and other government agencies in the US and abroad.

Membership in ARRL means much more than receiving *QST* each month. In addition to the services already described, ARRL offers membership services on a personal level, such as the ARRL Volunteer Examiner Coordinator Program and a QSL bureau.

Full ARRL membership (available only to licensed radio amateurs) gives you a voice in how the affairs of the organization are governed. ARRL policy is set by a Board of Directors (one from each of 15 Divisions). Each year, one-third of the ARRL Board of Directors stands for election by the full members they represent. The day-to-day operation of ARRL HQ is managed by an Executive Vice President and his staff.

No matter what aspect of Amateur Radio attracts you, ARRL membership is relevant and important. There would be no Amateur Radio as we know it today were it not for the ARRL. We would be happy to welcome you as a member! (An Amateur Radio license is not required for Associate Membership.) For more information about ARRL and answers to any questions you may have about Amateur Radio, write or call:

ARRL—The national association for Amateur Radio
225 Main Street
Newington CT 06111-1494
Voice: 860-594-0200
Fax: 860-594-0259
E-mail: **hq@arrl.org**
Internet: **www.arrl.org/**

Prospective new amateurs call (toll-free):
800-32-NEW HAM (800-326-3942)
You can also contact us via e-mail at **newham@arrl.org**
or check out *ARRLWeb* at **http://www.arrl.org/**

From *QST*, September 2005

A 20 and 40 Meter Vertical on 'Autopilot'

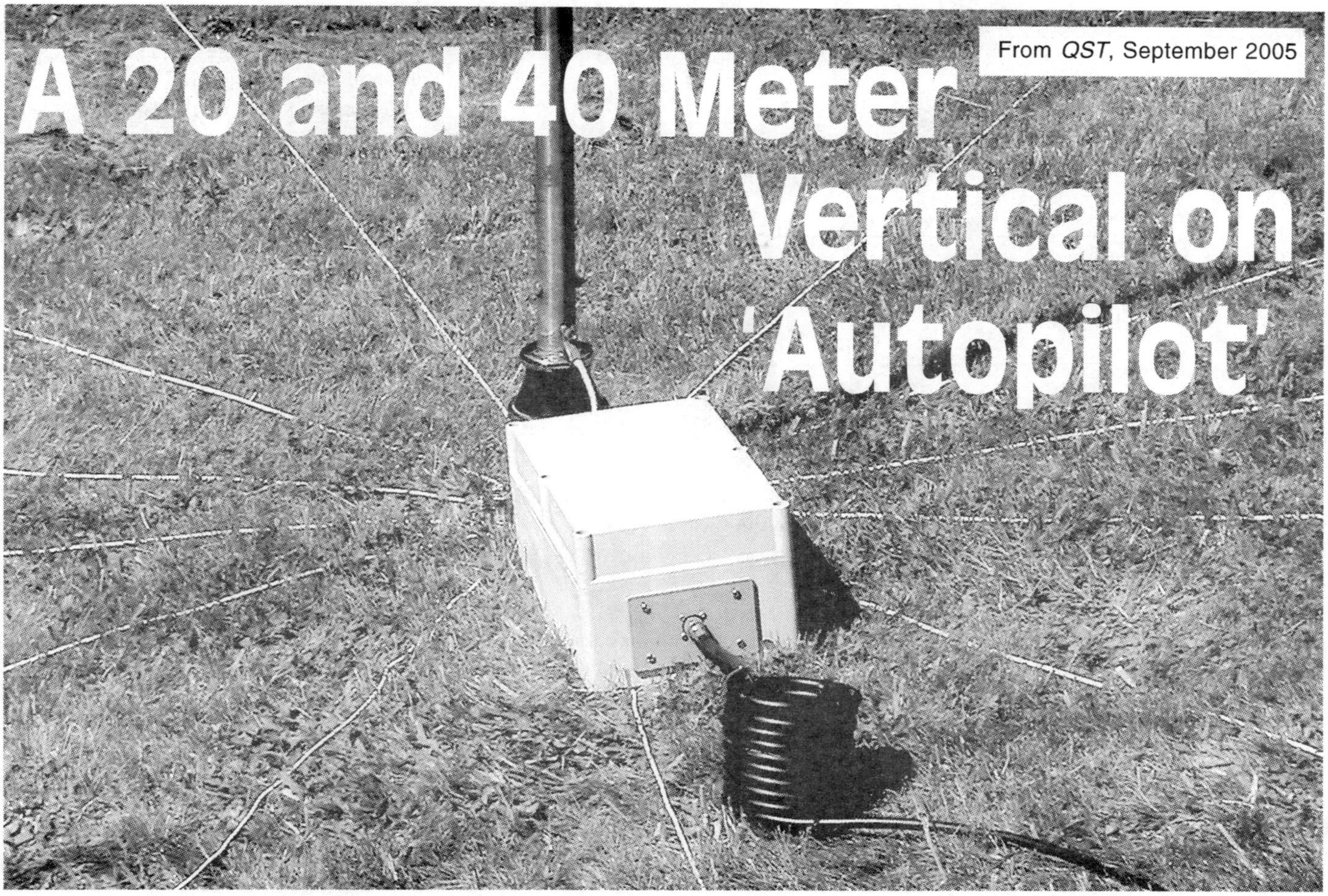

An old "Hints & Kinks" item results in an excellent Field Day antenna.

Dan Richardson, K6MHE

With its size, ease of installation, and reliable performance, the 40 meter quarter wavelength vertical antenna is a very popular HF antenna. What's more, with proper matching, it also performs very well on the 20 meter band. Being a half-wavelength on that band, the antenna provides an efficient low-angle radiation pattern, with roughly a dB of gain over the shorter standard quarter-wave vertical.

This article describes a dual-band vertical antenna system for 20 and 40 meters that I built for portable Field Day use. It uses a somewhat innovative tuning method that, after the initial setup adjustments are made, permits operation on ether band with no further tuning required—and it works great!

I discovered this scheme while perusing some past issues of *QST*. I had a very nice telescoping 35 foot marine vertical whip that I wanted to use for an upcoming Field Day

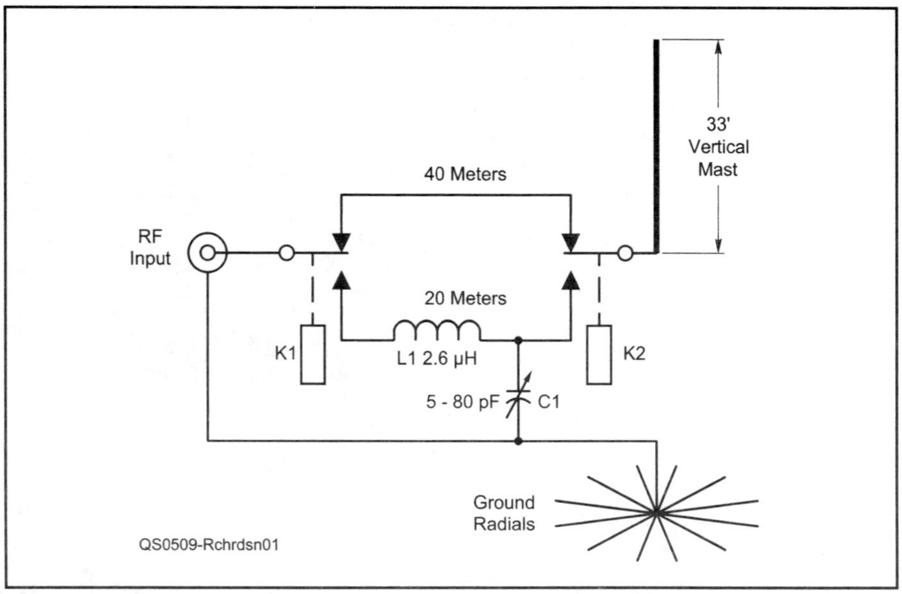

Figure 1—Using remote relays (K1, K2) for dual-band operation. One DPDT relay could be substituted for the two relays.

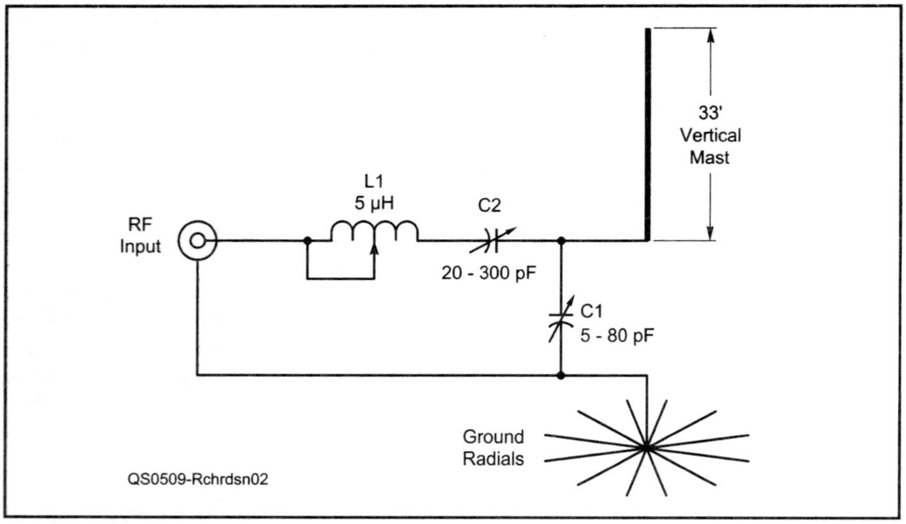

Figure 2—The modified L-network for dual-band operation—without relays.

and, while looking for some ideas, I came across this novel method for dual band operation. It was described by Wes Hayward, W7ZOI, in an old *Hints & Kinks* column.[1]

Typically, a quarter-wave vertical is fed by directly connecting the coaxial feed line to the base of the antenna. As the base impedance is low (typically 40-70 Ω, depending upon the ground/radial system utilized) this technique provides an acceptable SWR and works well. However, the base impedance of a half-wave (20 meters, in this case) vertical is much higher (around 1100 Ω); therefore a matching network is required to successfully feed the antenna with coaxial transmission line. Prior to discovering this method, I had only seen this type of dual-band operation performed using remote relays (K1, K2) in conjunction with an L network (L1, C1), as shown in Figure 1.

This innovative approach involves placing an additional capacitor in series with the inductor of an L network so that, when properly adjusted, a conventional 40 meter, 1/4 wavelength, vertical will operate on both the 40 and 20 meter bands with an acceptable SWR with no need for switching schemes or any additional adjustments.

How it Works

W7ZOI's modified network, shown in Figure 2, is easily understood when compared side-by-side with the simple L network of Figure 1. L1 and C2 form a series resonant circuit that behaves—on 40 meters—as a short circuit. The result is the same as having the relay contacts in the 40 meter position, as shown in Figure 1. The shunt capacitor (C1) reactance on 40 meters is roughly ten times that of the antenna's base (feed-point) impedance and therefore has a negligible effect on matching. The net result is the antenna's low base impedance essentially remains unchanged to the coaxial transmission line on 40 meters.

Twenty meter operation is accomplished by adjusting L1 and C2 so that the net inductive reactance is the same as it would be in the basic L network. The value of C1 remains the same in both networks.

So, we end up with a matching system that does double duty. The modified L network is adjusted so that L1 and C2 provide the proper amount of inductive reactance such that, when used with C1, matches the high base (feed point) impedance of the antenna on 20 meters. And, simultaneously, it maintains the series resonance that is required for matching the antenna's low base impedance on 40 meters. A pretty neat idea!

Construction

I built the entire network solely from junk-box parts. As is usually the case when building a project like this, my treasure trove did not contain capacitors of the exact values recommended by W7ZOI. I substituted a 100 pF capacitor for C1 and a 500 pF capacitor for C2. Originally, I was concerned that, by using larger values than those recommended, particularly for C2, the minimum capacitance values might be too large for proper tuning; however, this turned out not to be the case. The inductor was made from a section of coil stock (2 inch diameter; 14 gauge wire; 10 t at 8 tpi). All components were mounted on a 1/8 inch thick plastic panel, which, in turn, was installed in a watertight housing that had seen previous life as part of a shipboard antenna system. The completed network is shown in Figures 3 and 4.

If you choose to build this project take care to use capacitors that have a sufficiently high voltage rating. Mine are rated at 4.5 kV and should easily handle 500 W or more. [Vacuum capacitors would be ideal for this application, but they are expensive.—*Ed.*]

Tuning

As I stated earlier, my antenna's height is about 35 feet; therefore, its resonant frequency falls below the 40 meter band. This caused me no problems when tuning; however, I did have problems trying to adjust the network with an antenna height that was slightly less than 1/4 wavelength. I will discuss this in a moment.

Tuning was done empirically. Using an MFJ-259B antenna analyzer, the tap on the inductor was set near maximum inductance and capacitor C2 was adjusted for the best indicated SWR on 40 meters. Next, the analyzer's frequency was changed to the 20 meter band and C1 was adjusted for the best SWR, but the match obtained was not good. I next moved the tap on the coil in the direction of less inductance and C1 and C2 were adjusted again for their respec-

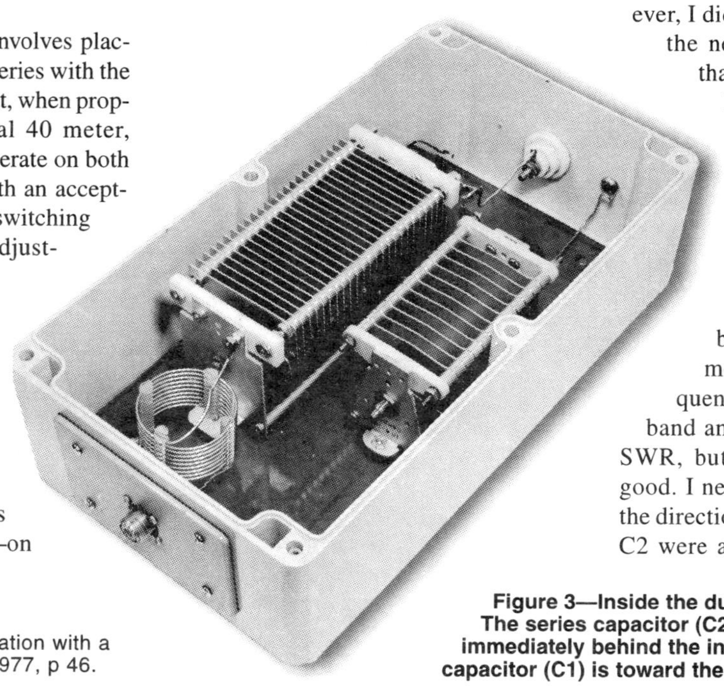

Figure 3—Inside the dual-band modified L-network. The series capacitor (C2) is at the top left, immediately behind the inductor (L1). The shunt capacitor (C1) is toward the right.

1W. Hayward, "Dual-Band Operation with a 33-Foot Vertical," *QST*, Jun 1977, p 46.

Figure 4—Tuning the network with an SWR analyzer. See text for tuning suggestions.

tive bands. This process was repeated until satisfactory SWR readings were obtained. (My requirement was to have the SWR fall below 2:1 on both bands. Many solid-state rigs start reducing their power or, worse yet, shut down at an SWR of 2:1.) The resulting SWR curves are shown in Figures 5(A) and (B). I should point out that this system was tuned using a marginal radial system (only twelve radials) and the SWR curves are therefore rather broad. Those curves would be more pronounced had the antenna been adjusted utilizing a more efficient ground system.

My telescoping mast comprises six sections that, when extended, reaches about 35 feet. As a matter of interest, I tried tuning the antenna with one of the sections retracted, resulting in an overall antenna height of about 30 feet, which is, of course, self-resonant above the 40 meter band (about 7.8 MHz). Using the shorter length antenna, I could achieve a good match on one band or the other but not on both simultaneously. It turned out that the additional inductance required to obtain a good match on 40 meters with the shorter antenna height resulted in having too much net inductive reactance in the L1/C2 series combination for a satisfactory match on 20 meters. Conversely, reducing the amount of inductance that provided a good match on 20 meters did not afford enough inductance for an acceptable match on the lower frequency. Hence, for this system to work properly, the antenna height should be no shorter than 1/4 wavelength for the lowest frequency of operation. I found no detrimental consequences, however, when it was slightly longer.

Conclusion

After the initial adjustments, the network performed flawlessly. Alternating back and forth between 20 and 40 meter operation required nothing more than changing the frequency/band selection on the radio, with no concern whatsoever about the antenna system.

I am surprised I haven't seen this technique in wider use. Given that more than 25 years has passed since it was first published, I felt it worthwhile to present it again for those who may not be aware of this practical and versatile design.

I wish to thank to Wes Hayward, W7ZOI, for his review and suggestions in the preparation of this article.

First licensed in 1955, Dan Richardson, K6MHE, enjoys experimenting with antennas. Dan was an electronic technician in the US Coast Guard during the 1950s and '60s; he then joined the family foundry business, retiring about 10 years ago. After he had moved to rural Northern California, Dan's devotion to antennas flourished, and he's published several articles in leading amateur journals. Dan can be reached at PO Box 2644, Fort Bragg, CA 95437 or **k6mhe@arrl.net**.

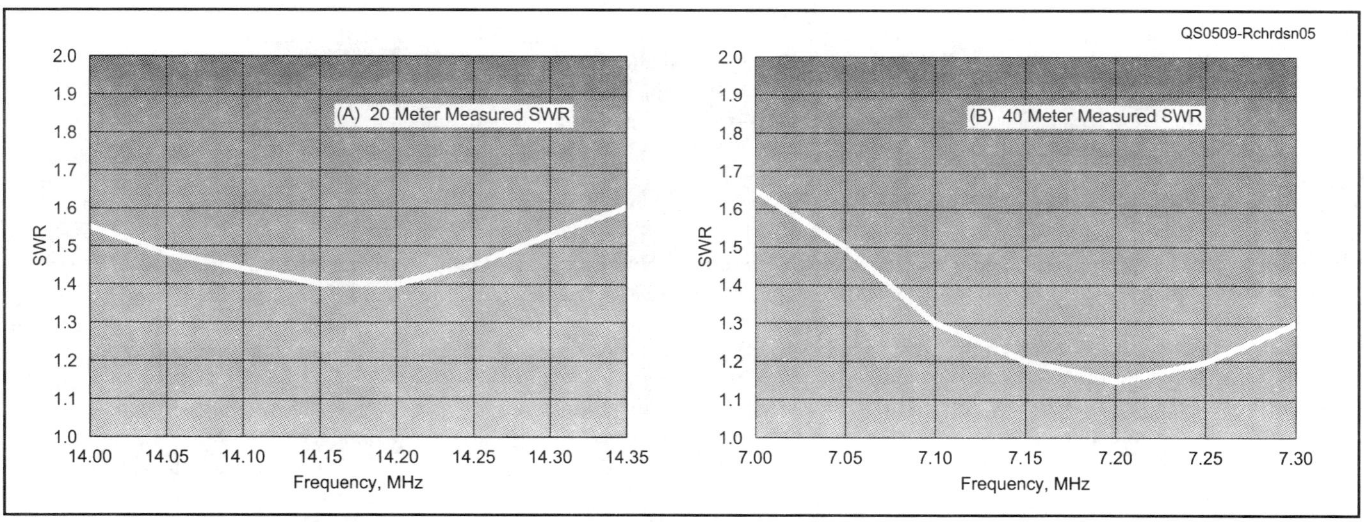

Figure 5—The SWR at 20 meters (A) and 40 meters (B).

By Dave Bowker, K1FK

A 75 foot Top Loaded Vertical Antenna

Some ideas and techniques for constructing and installing a large vertical element antenna system for 160 and 80 meters.

I have used this 75 foot top-loaded vertical ground plane antenna for several years now. The construction and installation techniques described can be adapted to similar designs.

Main Element Construction

The main element is constructed of 5 inch diameter (0.08 inch wall thickness) aluminum irrigation pipe, an 80 meter trap, four sloping top loading wires for 160 meters, and a four-point guy system that enables the element to be erected safely by a single person. Four 30 foot lengths of pipe are required—three to construct the main element and splices, and one for use in erecting the completed assembly. Two 30 foot lengths and a 15 foot length of pipe comprise the main element structure for an overall height of 75 feet. Top and side views are shown in Figure 1.

The lengths of irrigation pipe are modified for splicing sections together by removing the end couplings from each end with a hack saw.

The splice couplers, consisting of 48 inch long doublers made from the same pipe, are prepared by removing a ½ inch wide linear strip along the length of the section. The coupler is compressed to a smaller diameter by hose clamps, lubricated liberally with WD-40 and driven 24 inch into one end of the main element using a sledge and a piece of wood to protect the driven end of the doublers as shown in Figure 2. The clamps are progressively loosened and moved towards the driving end as the coupler is inserted into the main element.

After insertion into the main element, the coupler is riveted into place using eight ⅛ inch pop rivets spaced equally around the circumference of the element and spaced ½ inch from the end, as shown in Figure 3. Another set of eight rivets is placed 22 inches from the end of the main element to secure the lower end of the coupler to the main element. Finally eight rivets are placed linearly along the coupler seam, four equally and alternately spaced along each side of the seam.

The other end of the coupler is compressed with the hose clamps and the next main element section is driven onto it until the two main element sections mate. The riveting procedure described above is repeated to finish the spliced joint. Details of the rivets placed along the coupler seam can be seen in Figure 4.

Trap Construction

The trap serves to decouple the 160 meter top loading wires from the main element on 80 meters and allows the element to function as a 75 foot vertical on that band. The trap is constructed of a 12 inch length of 4 inch Schedule 80 CPVC pipe (outside diameter = 4½ inch) wound with 7¾ turns of PVC jacketed RG-213 coax.[1] Schedule 80 CPVC pipe is required for its mechanical properties (greater wall thickness) to support the top loading wires which also function as top guys. If only Schedule 40 PVC pipe is available, two pieces may be used, one slit and formed into a doubler similar to the doublers used to splice the main element sections, and bonded into place with CPVC cement.

[1]Notes appear on page 14.

A ½ inch diameter hole is drilled into the top of the form approximately 1½ inches down from the top to anchor the top end of the coax winding. Starting with an 11 foot length of coax, wind 7¾ turns onto the form and mark the position where the bottom of the winding will be terminated, remove the coax and drill another ½ inch hole to terminate the bottom of the winding. Prepare one end of the coax by stripping 4 inches of insulation from it and extracting the inner conductor from the braid as shown in Figure 5. Insert this end into the bottom hole of the form so ½ inch of insulation protrudes inside the form and tightly wind the 7¾ turns toward the top of the form. Insert the top end of the coax into the terminating hole and pull taut. Cut the coax inside the top of the form so 4½ inches remains and again remove the winding from the form and prepare the remaining end of the coax as described above.

The coax is now ready to be permanently wound onto the form using CPVC cement to bond each turn to the form as it is applied. Insert one end of the winding into the bottom termination hole with ½ inch insulation protruding inside the form and apply a liberal amount of cement in a strip about 1 inch wide around the circumference of the form and wind two turns tightly, holding in place while the cement sets. Continue this process until the top of the form is reached but do not cement the top turn to the form. Insert the top end of the winding into the top termination hole.

Temporarily connect the coax leads inside the form to effect a parallel tuned trap and check for resonance around

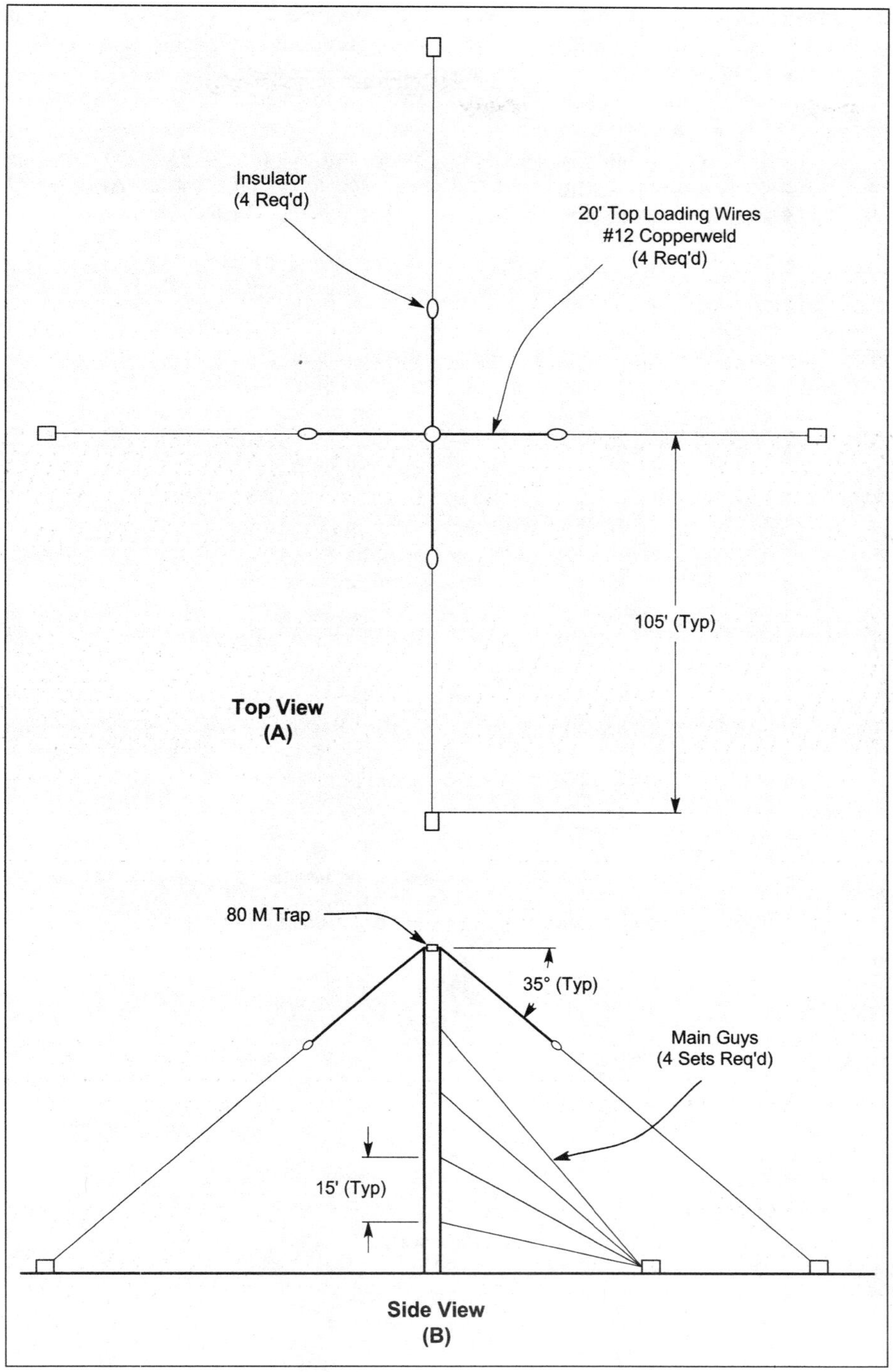

Figure 1—Top view (A) and side view (B) of the vertical antenna.

Figure 2—The doubler ready for insertion into the main section.

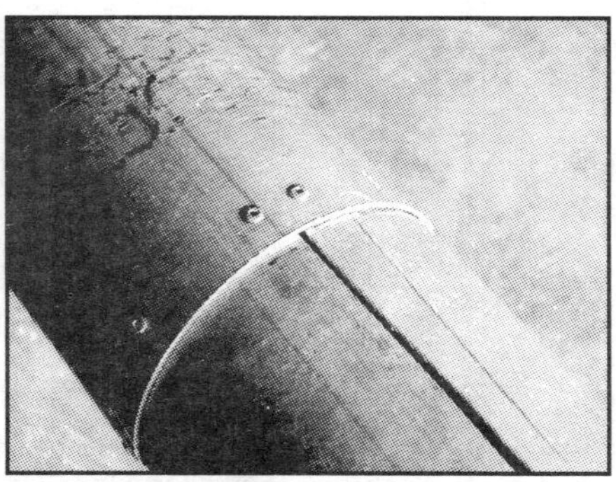

Figure 3—The doubler riveted into place.

3.6 MHz using a grid drip meter, adjusting the spacing of the top turn to achieve resonance in this frequency range. Once resonance is achieved, the top turn may be cemented in place. Check for final resonance then apply several liberal coats of CPVC cement over the entire coax winding and seal the termination holes in the form as shown in Figure 6.

The trap leads are permanently connected and soldered inside the form, wrapped with PVC electrical tape and given several coats of CPVC cement. The

Figure 4—The guy clamp attached to the main element.

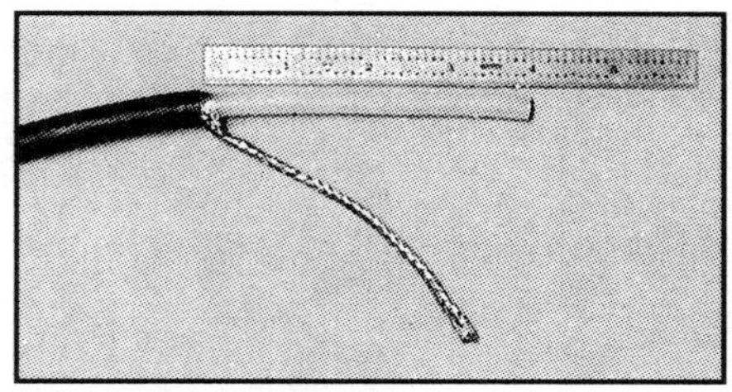

Figure 5—Preparation of the coax trap ends.

Figure 6—Completed 80 meter trap assembly ready to install.

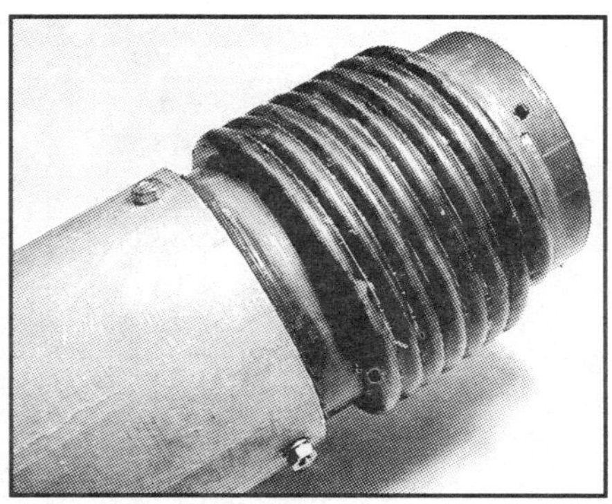

Figure 7—The 80 meter trap installed on the main element.

bottom of the trap is connected to one of the three bolts which will attach the trap to the inside of the main element. Four leads of flexible braid are soldered to the top of the trap. Four 1/4 inch diameter holes are drilled 3/4 inch from the top of the form to four eyebolts that secure the top loading wires. Another hole is drilled next to each eyebolt to pass a lead from the top of the trap and connect to each loading wire.

The trap is bolted into the top of the main element, leaving a spacing of approximately 1 inch between the bottom of the trap winding and the top of the element, as shown in Figure 7. An aluminum cover is installed over the top end of the trap to prevent rain, snow and bird droppings from entering into the element connections. The trap-to-element joint is sealed with RTV adhesive.

Guy Brackets

Four sets of guy brackets are constructed from 0.040 inch 6061-T6 or equivalent aluminum sheet and stainless hardware as shown in Figure 8. Each bracket consists of four sections configured to support the cable thimbles to prevent chafing of the 3/16 inch double braided black Dacron rope guys. The brackets are attached to the main element at 15 foot intervals. After aligning and bolting into place with the bracket bolts, eight rivets are installed to prevent slippage of the bracket—shown in Figures 4 and 8.

Top Loading Wires

The top loading for 160 meters is constructed from four thimbles inserted into the eyebolts, four lengths of #12 copper weld wire and four end insulators. The trap leads are silver soldered to the loading wires, leaving sufficient slack in the flexible leads to facilitate movement without breaking during installation. The wires are cut to length and insulators attached. The exact length of the loading wires depends upon several factors, including their slope angle as shown in Figure 1, the surge impedance of the main element and ground system parameters. The lengths may need adjustment during installation and testing to bring the element into resonance at or near the 160 meter band.

Information for calculating top loading and loading wire length for various configurations is available in the amateur antenna literature.[2] The exact frequency of resonance is not critical. The base impedance of the final installation will be measured. An L network is designed and adjusted to match the final base impedance to the 50 Ω transmission line.

Base Insulator, Guy Anchors and Guy Lines

The vertical requires four guy anchor points arranged in a square and four sets of main guys constructed of 3/16 inch double braided black Dacron rope. Each of the guys must be adjusted to precise lengths to facilitate raising the element to the vertical position. The top set of guys holding the 160 meter top loading wires should be anchored at a distance from the center of the square which will maintain the proper angle. For a 35° angle this distance will be 105 feet, as shown in Figure 1.

The base insulator pier must be located in the center of a square, the corners being the guy anchors. The location of the guy anchors relative to the insulator pier can

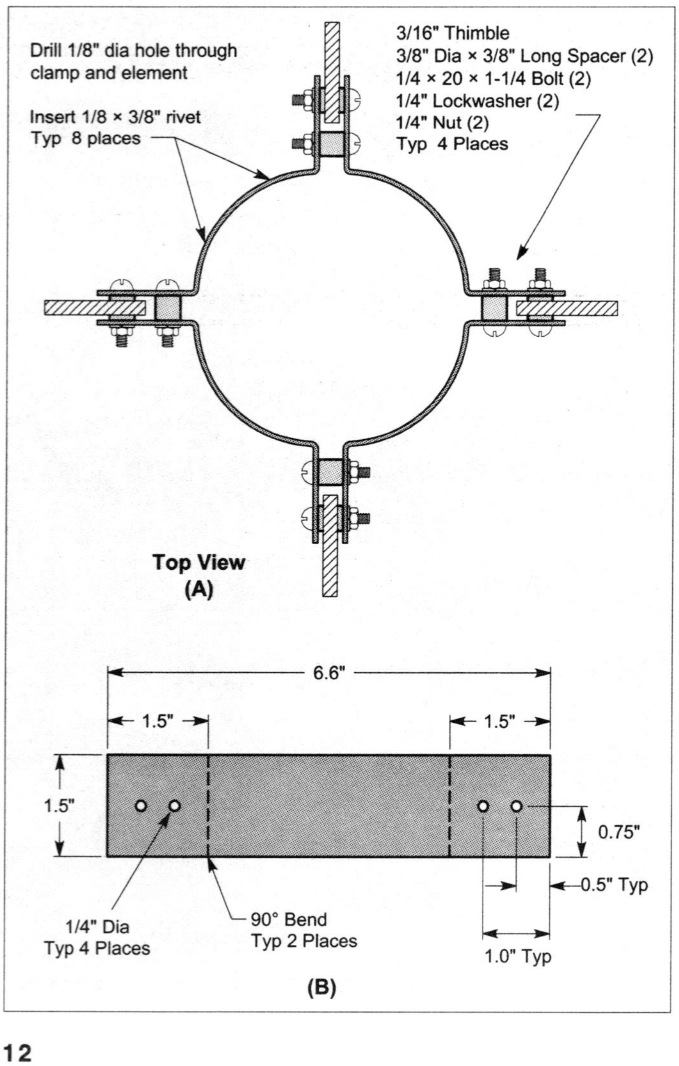

Figure 8—Guy clamp assembly details. At A, top view of complete guy clamp assembly. At B, guy clamp section flat pattern layout (four required per clamp assembly).

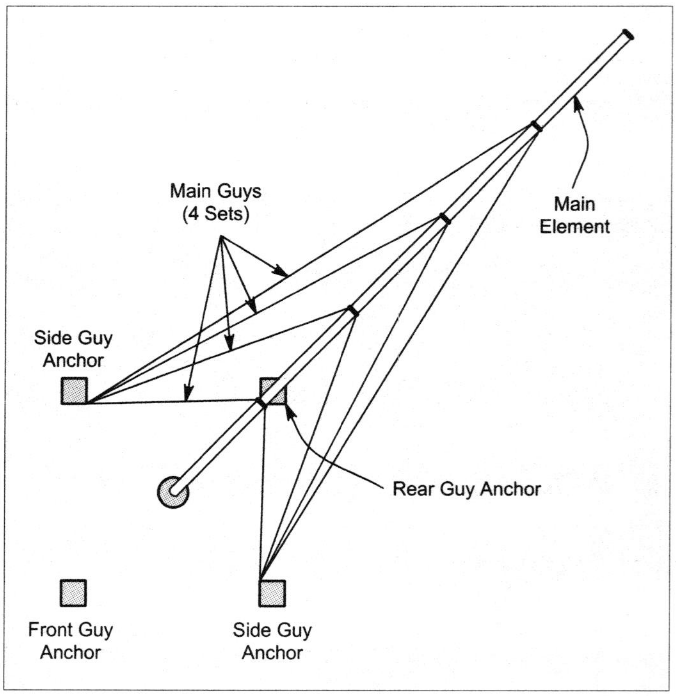

Figure 9—Layout for cutting and attaching guys. Top loading wires and guys are attached, but are not used for raising the element. They are not shown here. Only the side guys are shown to enhance clarity. The front and rear guys are the same length as the side guys.

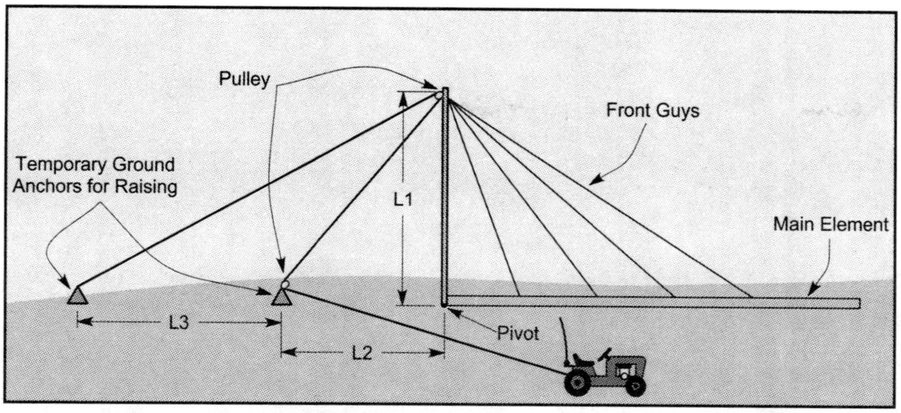

Figure 10—Layout of derrick for raising the element. L1=the length of the derrick boom. L2=L1. L3=any length greater than L2.

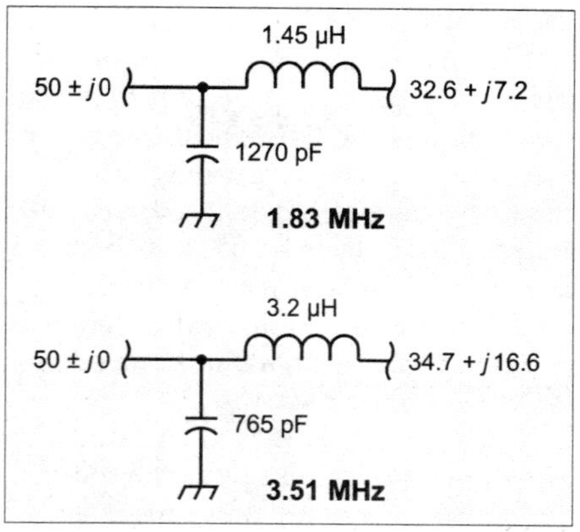

Figure 13—Typical L network values for matching.

Figure 11—Raising the element.

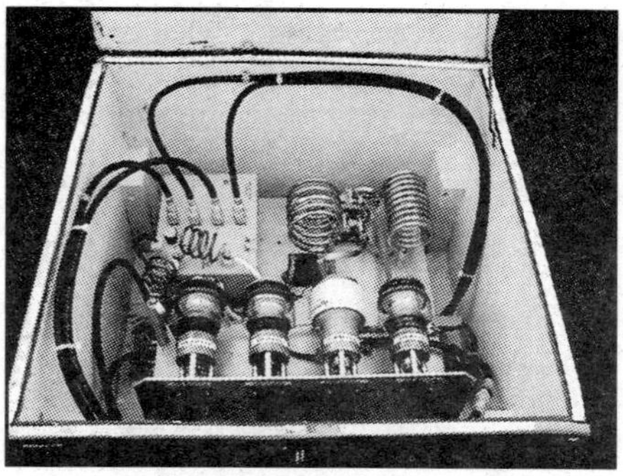

Figure 14—L-network for 160 meters (right of center) and 80 meters (far right of center).

Figure 12—Two views of the pivot fixture for positioning the base insulator.

be any convenient distance. The length of each guy may be calculated—alternatively, the base of the main element may be positioned over the insulator pier with the element laid across one of the guy anchors. Each guy rope is attached to the element and stretched taut to the side anchors as shown in Figure 9. The front and rear guys are cut to the same lengths as the side guys and the rear set of guys is attached to its respective anchor. The front set of guys will be attached to the derrick boom at a distance equal to the distance from the pier to one anchor. The top loading wires and guys are attached to the trap but are not used for raising the element to the vertical position.

The base insulator I used can be seen just below the pivot fixture in the right side photo of Figure 12. It consists of a surplus porcelain insulator and a machined aluminum plate upon which the main element rests. A substitute insulator can be made from a 4 inch length of 2 inch diameter (or larger) Schedule 80 CPVC pipe and two end caps.

Installation

The element is raised using a derrick, a 2:1 mechanical advantage block and tackle, a pivot fixture and a garden tractor. The derrick boom length should be at least equal to half the length of the main element and arranged to pivot with the element at the base insulator pier. Figure 10 illustrates the mechanical arrangement of the derrick boom, block and tackle, and garden tractor.

The derrick boom is guyed to the two side guy anchors for lateral stability during erection. The front guys are attached to the derrick boom at a distance up the boom equal to that from the base insulator pier to the front guy anchor. Figure 11 shows the actual raising of the element in progress.

Once the element is vertical (and the derrick boom horizontal), the front guys are removed from the derrick boom one at a time, starting with the lower guy, and attached to the forward guy anchor. After all guys are secured to their respective anchors, two hydraulic jacks are placed under the pivot pin, the pivot fixture is unbolted from the base insulator pier, and the main element is raised with the jacks sufficiently to permit the pivot fixture to swing out to the side while the base insulator is bolted into place. The jacks are then lowered to position the element onto the insulator and the pivot pin is removed. See the photos in Figure 12.

Matching Network

After the main element and ground system installation is completed, the feed point impedance is measured at the center of the operating frequency range on both bands. An L-network is designed, fabricated and installed at the base of the element for each band. Design of the L-networks may be accomplished with the aid of formulas in *The ARRL Antenna Book*[3] or readily available computer software.

The ground system consists of 120 $^1/_4$-wavelength radials laid in the sod. My ground scheme also takes advantage of the additional ground system of my 40 meter phased array, that consists of an additional 120 $^1/_2$-wavelength radials under each element. The two ground systems are bonded together wherever they cross and form a very low loss system. The measured feed point impedance of my vertical working against this ground system is 32.6 + j7.2 at 1.83 MHz and 34.7 + j16.6 at 3.51 MHz. The networks required to match these impedances to the characteristic 50 ± j0 impedance of the feed line are shown in Figure 13.

Completing the installation, the matching networks are contained in the same doghouse as the phasing and matching networks for my 40 meter phased array and Beverage switching components. They are shown on the right side of the doghouse in Figure 14. Vacuum variable capacitors and vacuum relays are used for reliability under extreme environmental conditions.

Acknowledgments

Thanks to Tony Trice, WØTY, for his suggestions in fabricating the splice couplers and review and critique of the manuscript.

Notes

[1]Belden 8267 or equivalent. Coaxial trap construction details can be found in Chapter 7 of the latest edition (20th) of *The ARRL Antenna Book*, p 7-15.

[2]*ON4UN's Low-Band DXing*, Third Edition, The American Radio Relay League, 1999, §3.6.3. Available from the ARRL Bookstore for $28 plus shipping. Order no. 7040. Telephone toll-free in the US 888-277-5289, or 860-594-0355, fax 860-594-0303; **www.arrl.org/shop/**; **pubsales@arrl.org**.

[3]*The ARRL Antenna Book* is available from the ARRL Bookstore for $39.95 plus shipping. Order no. 9043. Telephone toll-free in the US 888-277-5289, or 860-594-0355, fax 860-594-0303; **www.arrl.org/shop/**; **pubsales@arrl.org**.

Dave Bowker, K1FK, has been a licensed and active ham for half a century and is an ARRL member. A retired aerospace electronics engineer and manager, he is an exclusive CW operator and enjoys low-band DXing, contesting, designing/building homebrew equipment, and consulting. You can reach the author at 119 Bradbury Rd, Fort Kent, ME 04743; **k1fk@arrl.net**.

By Robert Johns, W3JIP

From *QST*, January 2001

A Ground-Coupled Portable Antenna

As the saying goes, "Imitation is the sincerest form of flattery." Here's a home-brewed antenna that proves the point.

This homebrew portable antenna for 40 through 6 meters is patterned after the ground-coupled design pioneered by Alpha Delta Communications, Inc.[1] Instead of using radials, this antenna employs a simple and very small grounding system that needs no tuning.

The antenna described here is a quarter-wave vertical sitting on a tripod base. The vertical mast and the tripod are each made of 2-foot-long telescoping sections of 3/4- and 5/8-inch-diameter aluminum tubing.[2] The mast itself resonates on 10 meters; lightweight aluminum tubing sections are added to the top of the mast to tune the antenna to 12, 15 and 17 meters.[3] These added tubing lengths can be installed vertically or horizontally. The antenna is fed at the top of the tripod, making the base a part of the radiating system. A bungee cord stretched from the top of the tripod to a stake in the ground keeps the structure stable.

Beneath the foot of each tripod leg is a grounding strip 2 1/2 inches wide and about 3 1/2 feet long, made of aluminum tape.[4] These strips are simply laid on the ground and form one plate of a capacitor coupling RF from the antenna to the ground. That's the whole grounding system! When I read about this in *QST* (see Note 1), I was skeptical, but intrigued. The arrangement is similar to that of a mobile antenna system in which the car body acts as one plate of a capacitor coupling RF to the road and ground. This grounding system works: The antenna radiates well and the SWR is reasonably low on all bands. (The tripod and grounding strips can also be used with any vertical element or mobile whip you have.) A loading coil added between the aluminum tubing mast and the flattop permits operation on 20, 30 and 40 meters. With the coil positioned this far up the antenna, the entire 10 feet of tripod and mast are unloaded radiators on all HF bands.

[1]Notes appear on page 19.

Building the Tripod

The top of the tripod, Figures 1 and 2, makes it easy to set up. The three 5/8-inch-diameter × 0.058-inch-wall aluminum tubes extending from the 1 1/2-inch PVC cap are permanently attached to it. To assemble the tripod, the legs slide over these tubes. A 3-inch-long, 3/8-inch carriage bolt passes through a hole in the top of the PVC cap to support the vertical element. This bolt also grips the 4-inch-long aluminum tubes inside the cap to form the three sloping legs of the tripod. See Figure 2 and its caption for details on how to make this top cap.

A 50-Ω coaxial feed line attaches to the antenna via an SO-239 chassis connector mounted on an aluminum angle bracket at the top of the PVC cap (see Figure 1). Make a 5/8-inch-diameter hole in the bracket to accept the coax connector body; you'll also need to drill four small holes for the connector's mounting hardware. The 3/8-inch bolt through the top of the cap keeps the aluminum bracket in place.

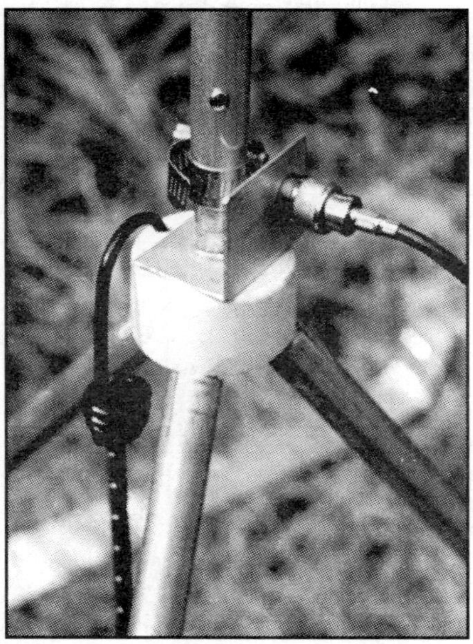

Figure 1—The top of the tripod with the bottom section of the mast connected to it. A bolt holds the three leg supports in the PVC cap slots. This bolt also passes through the 1 1/2 × 1 1/2-inch aluminum-angle piece that supports an SO-239 chassis connector for feed-line connection. A 1/4-inch hole in the cap top accepts the bungee-cord hook.

To assemble the tripod top, invert the cap so that you are looking down at the open end. Insert the carriage bolt through the 3/8-inch hole in the cap, through the mounting hole in the aluminum angle and add a lock washer and nut to the bolt. Initially, thread the nut about an inch onto the bolt so that the bolt is still loose and its head is out of the cap. Insert the three aluminum tubes into the slots in the wall of the cap and down against the carriage bolt where it passes through the hole in the cap. Tighten the nut so that the carriage-bolt head squeezes the tubes outward and into the slots. Once the nut is hand tight, wriggle each tube to seat it snugly with its tip into the countersunk hole with the bolt. Tighten the nut until the round wall of the cap is slightly deformed into a triangular shape.

Each tripod leg consists of a 0.058-inch-wall, 3/4-inch-diameter tube and a 0.058-inch-wall, 5/8-inch-diameter tube that fits inside the 3/4-inch tube. Each tube is two feet long; three can be made from 6-foot tubing lengths. Dimple each 3/4-inch tube about one inch from each end. The dimple acts as a stop and prevents the smaller tube from penetrating any farther. Form the dimples using a couple of firm hammer taps on a center punch placed against the tube. When joining the tubes, push a bit when inserting the smaller tube so that the side of the dimple holds the smaller tube in place.

Ground Strips

Although mating two strips of aluminum tape with their sticky sides together might seem like a routine job, it's probably the most difficult part of building this antenna! The adhesive is quite sticky and unforgiving, and handling the long strips can be messy. Get an assistant to help you with this task. You'll need three strips.

See Figure 3. Cut a 7-foot length of tape from the roll and lay it down, sticky side up, on the floor or a large table. Have your helper press a piece of heavy (#12) solid wire or a thin dowel across the width of the strip at the 3 1/2-foot midpoint and hold it in place. Pick up one end of the strip and carry it over the midpoint, keeping it tight so that it doesn't sag and touch the lower half. Keep both ends of the strip aligned while your helper at the midpoint presses the top piece of tape against the lower, working their way toward you. Trim (or remove) the excess wire or rod and the ground strip is done. Don't worry if the strips aren't aligned perfectly.

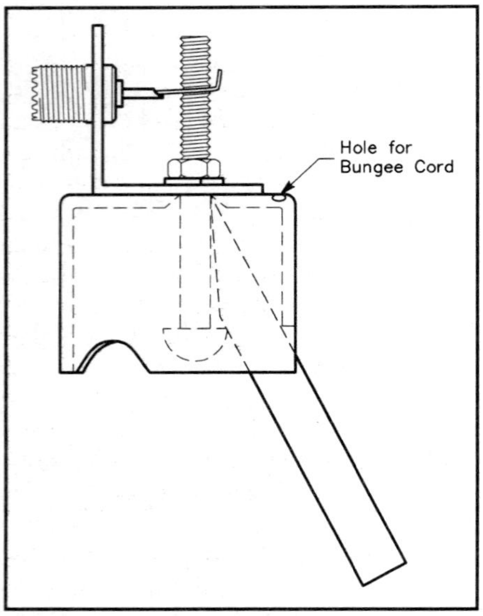

Figure 2—The tripod top cap. The three 5/8-inch-diameter aluminum tubes are 4 inches long, cut at a 30° angle at the end within the cap. From inside the cap, countersink a 3/8-inch hole in the cap top. This forms a trap that holds the ends of the aluminum tubes. Although only one of these leg supports is shown, all three are held between the bolt head, the three slots (either carved or filed in the wall of the cap) and the countersunk hole in the cap top. The slots in the cap are about 3/8-inch deep and wide enough to receive the aluminum tubes. An easy way to lay out the slots is to use the fluted handle from an outdoor water faucet as a template. The handle fits nicely against the cap and has six flutes about the circumference allowing you to mark three equally spaced locations.

The Mast

The 8-foot mast is made from two telescoping 3/4- and two 5/8-diameter × 0.058-inch-wall aluminum-tubing sections. Slot the ends of the 3/4-inch tubes so that they can be tightened around the smaller tubes with hose clamps.[5] To insulate the bottom 3/4-inch section from the 3/8-inch bolt in the tripod that supports the mast, its lower end is equipped with a plastic insulator. As shown in Figure 4, the insulator is a 2-inch length of acrylic tubing. The lower end of the acrylic tube extends about a quarter inch below the aluminum tube and is slotted so that the mast can be tightened around the bolt. Drill a 5/32-inch hole through the upper end of this insulator and the aluminum tube to pass a #6-32 bolt and nut to hold the insulator in place.

After mounting the SO-239 coax connector on the aluminum angle strip, solder a 2-inch length of #14 bare solid copper wire to the connector's center terminal and bend it close to the 3/8-inch bolt in the tripod top. Then bend the wire up and parallel to the bolt and about a quarter inch from it. When the bottom section of the mast is placed over the bolt, place this wire between the aluminum tube and a hose clamp. As you tighten the clamp, it makes the electrical connection from the coax to the mast and squeezes the slotted aluminum tube and insulator tightly against the bolt.

With the mast on the tripod, an easy way to make frequency adjustments is to separate the mast from its bottom section and lower it to the ground. You can then reach the flattop and coil without tilting the mast. For this reason, I don't tighten this joint. I place a #6-32 bolt through the 5/8-inch tube which is the second section of the mast, about one inch from its lower end so that it doesn't slide very far in. I still use a hose clamp over the 3/4-inch tube, adjusting it to make a snug sliding fit for the upper mast.

For the top antenna sections, I use 5/8-inch-diameter thin-walled aluminum tubing used for aluminum clothes poles. This material is lighter and cheaper than the 0.058-inch-wall tubing used for the tripod and mast, but is strong enough. Short tubing sections can be joined together using 2-inch-long sleeves made from the 3/4-inch-diameter × 0.058-inch-wall aluminum tubing. You need two 2-foot, two 1 1/2-foot and two 1-foot lengths of the 5/8-inch thin-walled tubing, three couplings and a **T** joint to connect the

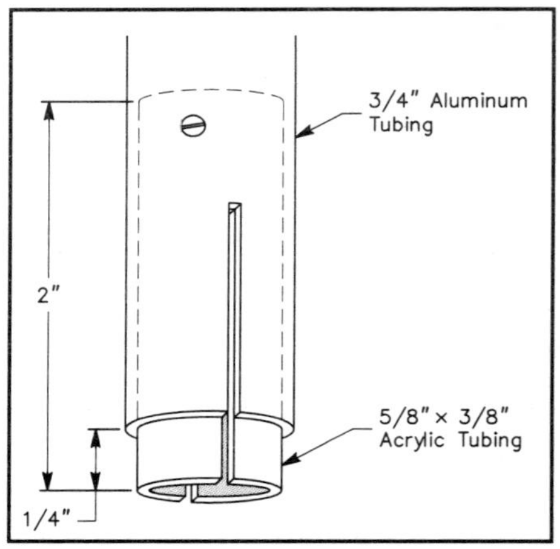

Figure 4—An acrylic (Plexiglas) tube insulates the antenna mast from the 3/8×3-inch bolt that supports it. The tube has a 3/8-inch ID and 5/8-inch OD so that it slips over the supporting bolt and telescopes inside the lower 3/4-inch mast section. Both the aluminum tube and the insulator tube are slotted using a hacksaw so they can be tightened around the bolt with a hose clamp. To mount a mobile antenna on the tripod, cut a 2-inch length of 1-inch-diameter acrylic rod and drill and tap one end to accept the 3/8×16 coarse-thread bolt of the tripod and 3/8×24 fine threads at the other end for the base of a mobile whip.

Figure 3—Making the ground strips is easier if you have some assistance. Refer to the text for an explanation of this sticky situation.

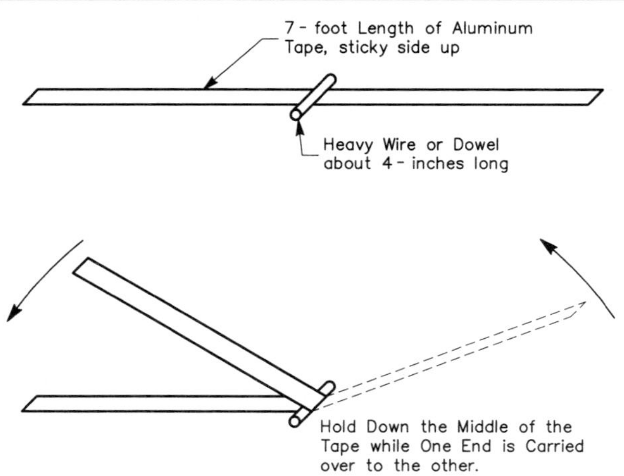

flattop to the mast or the top of the coil. See Figure 5.

Bungee Tie-Down

The antenna is quite light, and even with the wide base of the tripod it needs to be stabilized against wind gusts or someone tripping over the coax feed line. A bungee cord and a ground stake do an excellent job. The top of the tripod is about 3 feet high, so a 1/2- to 3/8-inch-diameter, 24-inch-long bungee cord works well. Any tent stake will do; drive it into the ground at an angle so it doesn't pull out easily. A special stake shaped like a large screw is ideal for this application.[6] It threads into the ground by hand and has a very low profile. (I leave the stake in the ground and my lawn mower doesn't even come close to striking it.) The stake won't go into hard, baked soil, however. For stability in such locales, or on pavement, hang some bricks, a rock or a jug of water beneath the tripod on the bungee cord.

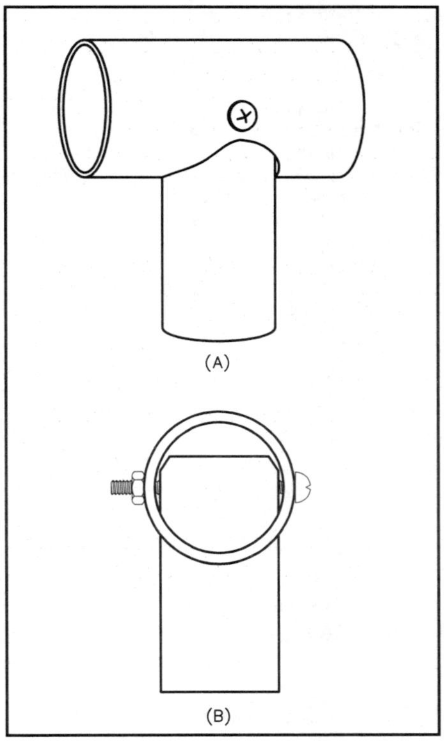

Figure 5—A T is needed to make the flattop. The 3/4-inch-diameter horizontal tubing has a 0.058-inch wall and accepts the 5/8-inch-diameter thin-wall tubes. The 5/8-inch-diameter vertical piece has a 0.058-inch wall and fits into the 3/4-inch tube at the top of the coil form. To make a 5/8-inch hole in the 3/4-inch tube, drill a hole then expand it with a 5/8-inch-diameter or larger countersink. (This process is heavy work for a countersink, so use a little lubricating oil.) Before drilling a 7/64-inch hole for a #6-32 bolt through the T, assemble the two pieces and squeeze them together tightly in a vise, making them perpendicular. To mount a flattop on the mast without the coil, first place a 3/4-inch-diameter coupling sleeve over the 5/8-inch-diameter top of the mast and fit the T into that coupling.

Table 1
Length of thin-wall tubes needed for operation on 10 through 17 meters.

Band (Meters)	*Length of Flattop (ft) and Number of Sections*	*Length of Vertical Top (ft)*
10	0	0
12	1 × 2	1.5
15	2.5 × 2	3.5
17	3.5 × 2	5.5

Antenna Operation on 10 through 17 Meters

For 10-meter operation, set up the tripod and place one end of a ground strip under each tripod foot. The ground strips may be laid in any direction. Adjust the mast to a length of about 7.4 feet. This length is quite a bit less than a quarter wavelength and I believe it's because of its closeness to the ground and the thickness of the tripod. No top hat is used on 10 meters. Adjust the mast length to resonate the antenna at your desired 10-meter frequency.

Table 1 provides lengths for the thin-wall tubes that you add to the antenna, either as a flattop or a vertical, for operation on 12, 15 or 17 meters. No change to the ground system is needed when changing bands. Table 1 assumes that you will leave the mast set for 10-meter operation. This simplifies band changing, such as moving from 10 meters to 15 meters and returning to 10 meters. These changes are quickly made by just adding the tubing lengths for 15 meters and removing them to return to 10 meters—no measurements, no tools.

6-Meter Antenna Operation

For 6-meter operation, the tripod must be insulated from ground and the mast reduced to a length of 52 inches from tripod to tip; see Figure 6. No ground-coupling strips are needed. Simple insulators can be made from 1/2-inch CPVC pipe and couplings. Cut three lengths of pipe about 4 to 6 inches long and hammer each into a coupling. Cementing them isn't necessary; they will be a tight fit. The other side of the coupling fits well over the 5/8-inch-diameter aluminum-tubing leg. Adding these insulators to the tripod resonates the antenna in the 6-meter band with good SWR. You can change the operating frequency by adjusting the length of the mast only—you don't need to adjust the size of the tripod.

Building the Loading Coil

For operation on the 20, 30 and 40-meter bands, a loading coil must be added to the antenna. A large tapped coil is shown in Figure 7; it tunes the antenna to 20, 30, or 40 meters and permits you to tune to the higher-frequency bands without changing the lengths of the top hat. The coil has 13 turns of #8 aluminum wire wound on a 4-inch styrene pipe coupling.[7,8] This coil form is secured to a 7-inch-long, 1/2-inch-diameter CPVC pipe using 1 1/4-inch-long, #6-32 brass or stainless steel machine screws and nuts. I like to reinforce the 1/2-inch pipe by hammering a 2-inch length of 1/2-inch wood dowel into each end. This allows me to tighten the nuts and bolts without flattening the pipe. These bolts also secure the ends of the 13-turn coil. Using a marking pen, I made black marks on the coil to identify the fifth and tenth turns. The marks serve to locate the proper tap points without having to count coil turns each time.

Figure 6—Here, the antenna is set up for use on 6 meters. The tripod construction remains the same, using legs approximately 4 feet long, but the mast has been shortened. No ground strips are needed and the legs are insulated from ground by 1/2-inch CPVC pipe extensions at their feet.

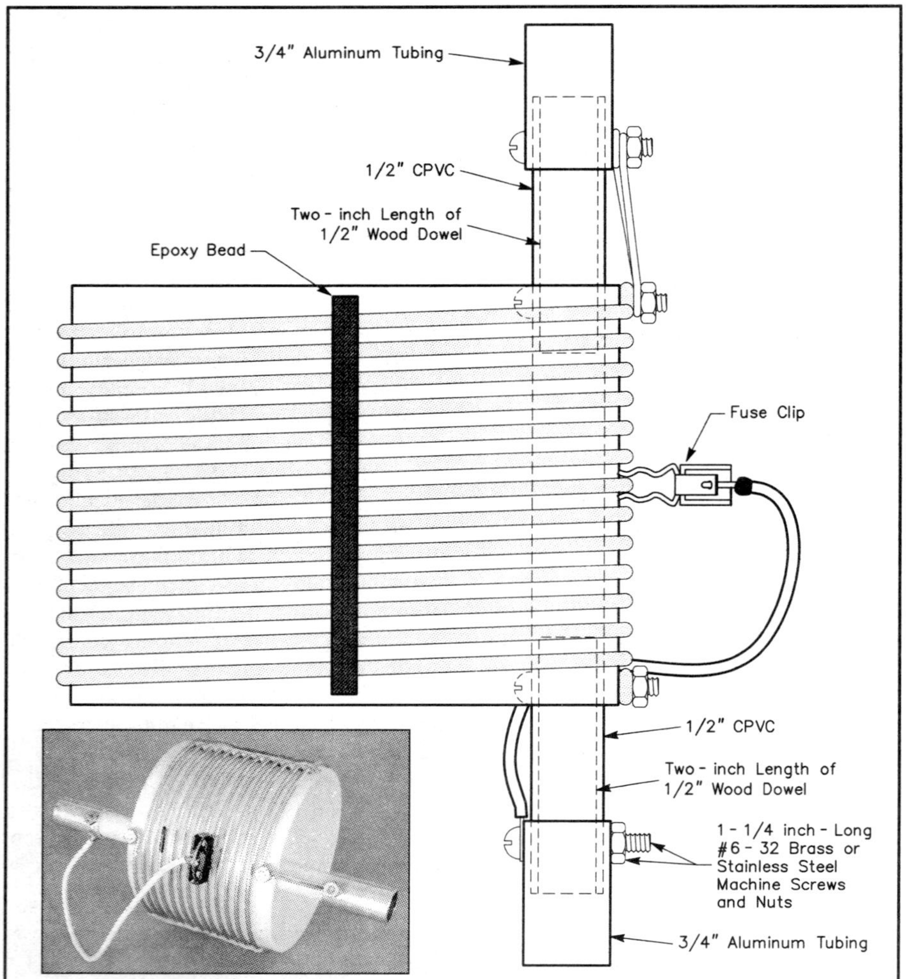

A close up of the completed coil.

Figure 7—To make the loading coil, 13 turns of heavy aluminum wire are spaced to fill the form. Secure the coil ends using the same bolts that hold the plastic pipe inside the 4-inch styrene coil form. Mount the coil on the mast with a 3/4-inch-diameter aluminum sleeve at the bottom of the plastic pipe; the tap wire is also connected here. An identical sleeve at the top of this plastic pipe connects to the thin tubing for the top vertical section, or to an aluminum T to hold the flattop.

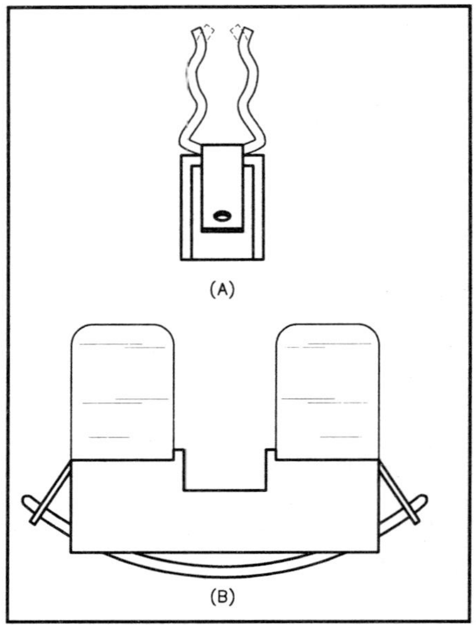

Figure 8—Making the tap connection to the coil. At A, the ends of the jaws of a 5-mm cartridge-fuse holder are bent inward (dotted lines) to grip the heavy wire of the coil. A side view of the fuse holder is shown at B. Bend the solder lugs at the ends of the fuse holder to accept a wire passing through them and beneath the fuse-holder base. When this wire is in place, bend the lugs farther up against the ends of the holder and solder them. Strip a 1/4 inch of insulation from the tap wire and solder it to the wire joining the lugs beneath the fuse holder. Round off any sharp points or rough edges with a file, because you'll be gripping this connector tightly for attachment to and removal from the coil.

Inside the styrene coil form is a ridge. Use a chisel or file to remove about a 1-inch-long section of this ridge to allow the CPVC pipe to lie flat against the inside of the form. Drill 7/64-inch holes at the ends of the styrene coil form and through the 1/2-inch pipe, then bolt them together as shown in Figure 7. Take a 16-foot length of aluminum wire, bend a loop at one end of it, attach the loop to one of the bolts and wrap the form as neatly as possible with 13 turns of wire, without bends, spacing the turns to fill the form. Wrap the end of the 13th turn around the bolt at the other end of the coil form and cut off the excess wire.

To tighten the wire on the form, clamp the form in a vise, grab the coil turns between both hands and progressively rotate the coil from one end to the other several times. This makes the turns tight enough to stay in place as you even out their spacing. To hold the turns in place permanently, run three ribs of epoxy the length of the coil. Use metal/concrete epoxy which has black resin and white hardener, making a dark gray mix that is easy to see against the white background of the coil form. One of these ribs is visible in Figure 7. To make nice straight ribs, first place strips of tape on each side of an intended rib location, apply the epoxy and remove the tape before the epoxy hardens.

Several types of alligator-clips will fit between the coil turns without touching neighboring turns, but I prefer to use a tap connection made from a fuse holder; see Figure 8.[9] After bending the fuse-holder-jaw tips, bend the jaws themselves to make them fit the wire tightly, but remain easy to attach and remove. Suit yourself as to how tight a grip they should have. Join the tap connector to the sleeve at the bottom of the coil form using a 9-inch length of stranded, insulated #14 copper wire, with a solder lug at the end. Use a similar piece of wire to join the top of the coil to the sleeve at the top of the coil form.

The sleeve at the coil bottom joins the coil to the mast. It is a 1 1/2-inch-long, 3/4-inch-diameter, 0.058-inch-wall aluminum-tubing piece. Insert the bottom of the 1/2-inch CPVC pipe halfway into this sleeve and drill a 7/64-inch hole through the sleeve and pipe. Fasten them together with a 1-inch-long, #6-32 brass or stainless steel machine screw and nut. The wire to the tap connector is attached with this same screw.

Antenna Operation with the Coil

To use the antenna on 40 through 10 meters, shorten the mast to 6 feet 2 inches and connect the coil to the mast. Atop the coil, add an element consisting of two horizontal 3 1/2-foot lengths of 5/8-inch-diameter tubing, or a single 7-foot vertical piece of tubing. With the full 13 turns of the coil, and part of an extra turn supplied by the tap wire, the antenna will likely resonate in the middle of the 40-meter band. To operate at the low end of the band, add a 1-foot length of tubing to one side of the flattop or the vertical tubing section. See Figure 9 for approximate dimensions of the assembled antenna.

It may seem as though Table 2 has some errors because it lists a greater

Table 2

This table identifies the number of coil turns (counted from the top of the coil) required to resonate the antenna on the 40- through 10-meter bands. These coil-tap settings are provided as a starting point only because installation conditions vary. To raise the antenna's operating frequency, reduce the number of turns used; to lower the operating frequency, increase the number of turns.

Band (Meters)	*Number of Coil Turns*
40	13
30	7.1
20	3.1
17	2
15	5
12	7
10	13

number of coil turns for operation on 15, 12 and 10 meters than for 17 meters! You're right—something strange is going on. It's because there are *two resonant frequencies* for each setting of the coil tap. Figure 10 shows the two paths that RF can take in the antenna. The upper part of the coil and the top hat provide the lower frequencies; the lower half of the coil provides the higher frequencies. A coil this large has considerable capacitance to free space, so it's not just an end-loading inductor at the higher frequencies. The antenna bandwidth is good, the SWR low and the antenna performs well on these bands. The charm of this coil system is that you can change bands by just moving the tap on the coil, without any adjustments to the mast length or the flattop. And a bonus: With 13 turns on the coil, the antenna works on 40 and 10 meters simultaneously.

The coil settings of Table 2 may need some minor adjustments if a vertical top section is used instead of the flattop. In general, the SWR is lower with the flattop and the antenna is easier to handle.

Power-Handling Capability and Safety

Because of the large coil and tubing used, you might be tempted to run high power with this antenna. I suggest you don't. The antenna may take it, but people can't. At high-power levels, dangerous RF voltages on the antenna are within range of physical contact. I have used the antenna at a 100-W level, but even that requires care and supervision.

Other Possibilities

With the tapped coil, this antenna can be tuned to any frequency from 7 to 40 MHz when operated on the ground-coupled tripod, and up to 110 MHz with the tripod insulated from ground.

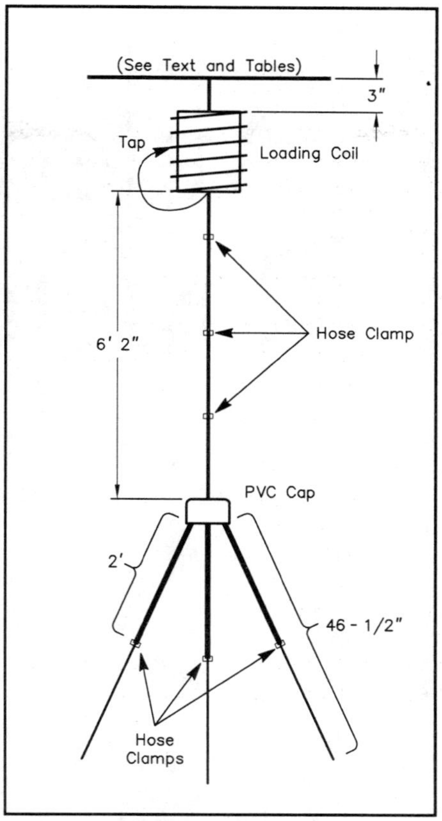

Figure 9—Approximate dimensions of the assembled antenna with the tripod, mast, loading coil and top hat.

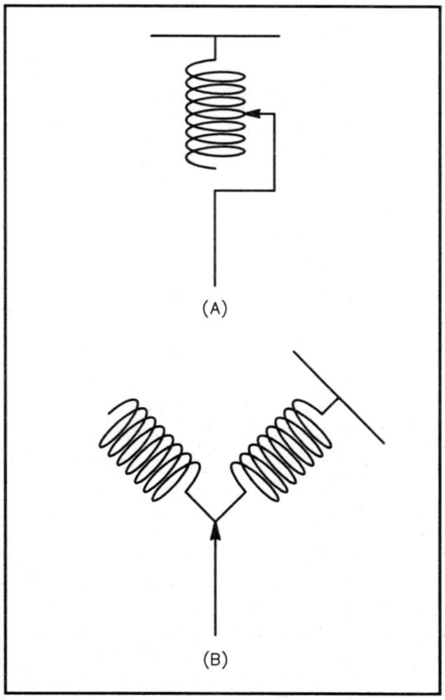

Figure 10—At A, the upper part of the antenna includes the coil, the adjustable tap and the top hat. The bottom of the coil is free and not connected to anything else. At B, this has been redrawn to show the two antenna circuits with the two resonant frequencies that are present. The upper half of the coil has a lower resonant frequency because of the length of the top hat above it.

The antenna also may be used with a longer mast for greater efficiency, or with a shorter mast when space is restricted. Even though the short version is only about 6 feet high, you can't use it indoors because it must be coupled to earth ground. The taller antenna gets out better, but band changing is more complicated. If operation on 75 and/or 80 meters is a must, you can add another coil to the antenna just below the 40-meter coil and change antenna frequencies with the 40-meter tap. Adding a coil made of 20 close-wound turns of #12 enameled wire wound on a 4-inch styrene form similar to the one in Figure 7 will allow you to tune the antenna from about 3.5 to 3.8 MHz, and from about 3.8 to 4.1 MHz with the top hat reduced to one 3.5-foot section and one 2-foot section. Six ground-coupling strips will provide a lower SWR on 80. A small vertical like this is not very effective for short-skip ragchewing, however. A λ/4-wire draped over bushes, flower beds or low tree branches offers more high-angle radiation.

Notes

[1]Rick Lindquist, N1RL, and Steve Ford, WB8IMY, "Compact and Portable Antennas Roundup," 'Alpha Delta Outreach/Outpost System,' Product Review, *QST,* Mar 1998, pp 72-73.

[2]Twelve feet of each tubing size is needed. The aluminum tubing is available from Texas Towers and Metal and Cable Corp. See their ads elsewhere in this issue.

[3]The thin-walled 5/8-inch-diameter aluminum tubing is available from Home Depot and hardware stores as aluminum clothes poles, each about seven feet long.

[4]Adhesive-backed aluminum tape 2 1/2 inches wide is available from Home Depot stores in the heating-vent section.

[5]You may want to consider using an antioxidant at the tubing joints. Antioxidant compounds available from electrical wholesale supply houses, Home Depot and hardware stores include Noalox (Ideal Industries Inc, Becker Pl, Sycamore, IL 60178; tel 800-435-0705, 815-895-5181, fax 800-533-4483) and OX-GARD (GB Electrical, 6101 N Baker Rd, Milwaukee, WI 53209; tel 800-558-4311). Use either sparingly; a thin coat is sufficient.—*Ed.*

[6]Aluminum angle 1 1/2 × 1 1/2 × 1/16-inch thick is available from hardware and Home Depot stores. The green plastic ground stake that threads into the ground has the name "Twizelpeg" stamped into it, and is available at camping supply stores.

[7]The #8 aluminum wire is RadioShack #15-035.

[8]The coupling is available from Home Depot in the drainage pipe section, and also from large plumbing or swimming pool distributors. The couplings are actually 4 1/2 inches in diameter and made from polystyrene, a very low-loss insulator.

[9]RadioShack #270-738.

Bob Johns, W3JIP, is an old gadgeteer who likes to play with antennas and coils. You can contact Bob at PO Box 662, Bryn Athyn, PA 19009; **ksjohns@email.msn.com**.

Photos by Joe Bottiglieri

By Dennis Kennedy, N8GGI

From *QST*, November 1997

A Packable Antenna for 80 through 2 Meters

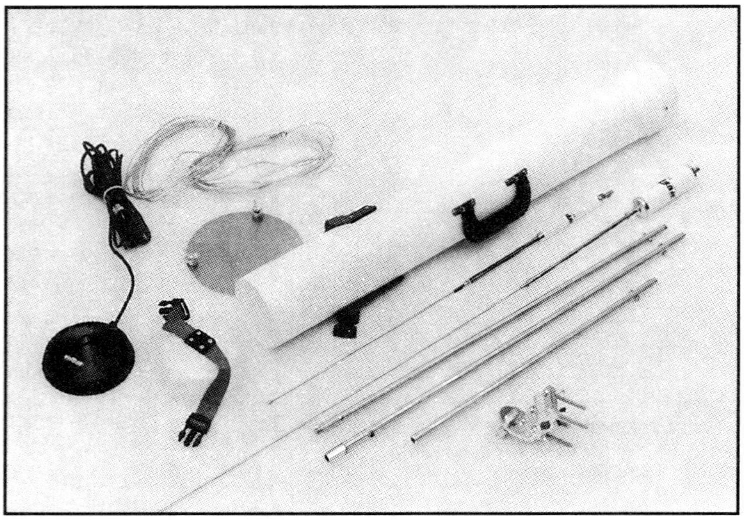

N8GGI converts a Hustler mobile antenna into a versatile airline-shippable antenna for mobile, marine or portable operation.

Here are 40, 20, 6 and 2 meter antennas, two radials for 20 meters, mounts and case.

The growing popularity of pint-sized HF and HF/VHF transceivers convinced me that I need one. I use it as a mobile rig, on business trips, vacations and to activate some US Islands. While the radio and power supply now travel easily, the antenna was the limiting factor for travel.

I bought a Hustler mobile antenna mast and 20 meter resonator and began using them with good success on the car. The fold-over 54 inch mast doesn't suit all the possible operating situations my wife might allow because it only folds to 90°. I wanted a versatile, compact antenna system that could pack compactly enough to ship as airline baggage. It would have a segmented mast and a variety of antenna bases for use on rental cars, sailboat stern rails, condominium balcony railings or anyplace else I might want to use the radio.

Splicing the Mast

First, I had difficulty locating 3/8-24 (UNF) hardware to fit most HF mobile antenna bases and the Hustler resonators. Common bolts in hardware stores are 3/8-16 (UNC). I needed a way to fit 3/8-24 threaded studs onto aluminum tubing and join the sections. My solution is to use a piece of 1/2-inch aluminum tubing with a 0.065 inch wall thickness. This leaves an inside diameter of 0.370 inch, five thousandths under 3/8 inch. I bought some 3/8-24 stainless steel bolts, cut off the heads and then ran a 3/8 inch drill a few inches into each end of each tube. The drilling is fairly easy because it only removes a few thousandths of an inch.

The unthreaded ends of the headless bolts slide snugly into the tubes, with about 3/8 inch of threads protruding as a stud. To secure each assembly, I mounted it in a vise and cross drilled (a drill press is helpful, but not necessary) the tube and bolt for #8-32 hardware (see Figure 1) that locks each shaft in place. Some antioxidant paste in the joints prevents corrosion and ensures conductivity. Combine the mast sections by screwing one end of each into a 3/8-24 stainless steel coupling nut.

The aluminum tubing comes in 72 inch lengths, and I discovered that I had exactly enough to make two mast sections for the HF antenna and a piece 19 1/8 inches long for 2 meters. To suit my ICOM 706 that covers HF, 6 and 2 meters, I have an antenna for any HF band (with suitable resonators), a 6 meter antenna (the 54 inch mast alone) and a $\lambda/4$ antenna for 2 meters (the 19 inch tube). Next, I needed antenna bases and mounting options.

A Mixture of Mounts

A Hustler MBM magnet mount works fine on the trunk lid of a rental car. (I wouldn't try driving at freeway velocities with the HF antenna on this mount, but it's okay for the 2 meter radiator or at low speeds.) I also bought a Radio Shack 21-937 mount, which works well on small horizontal pipe rails such as those found on boats. A small C-clamp or a couple of wood screws can easily attach the RS mount to a wooden deck railing. Attach $\lambda/4$ wire radials to the clamping bolts.

To make a free-standing antenna for a deck, dock or lawn, I cut a circle of thin steel plate about eight inches in diameter and put three bolts through it to attach radials. The ground connection couples capacitively to the steel plate through the mag mount, and the plate must be large enough to accomplish this. For my first test of this arrangement, I used one of my wife's steel cookie sheets with the radials secured to a bolt through the hole in the handle. (Thankfully, the neighbors didn't see me sitting on the deck working Houston on a cookie sheet!) Use monofilament fishing line as

[1]Notes appear on page 21.

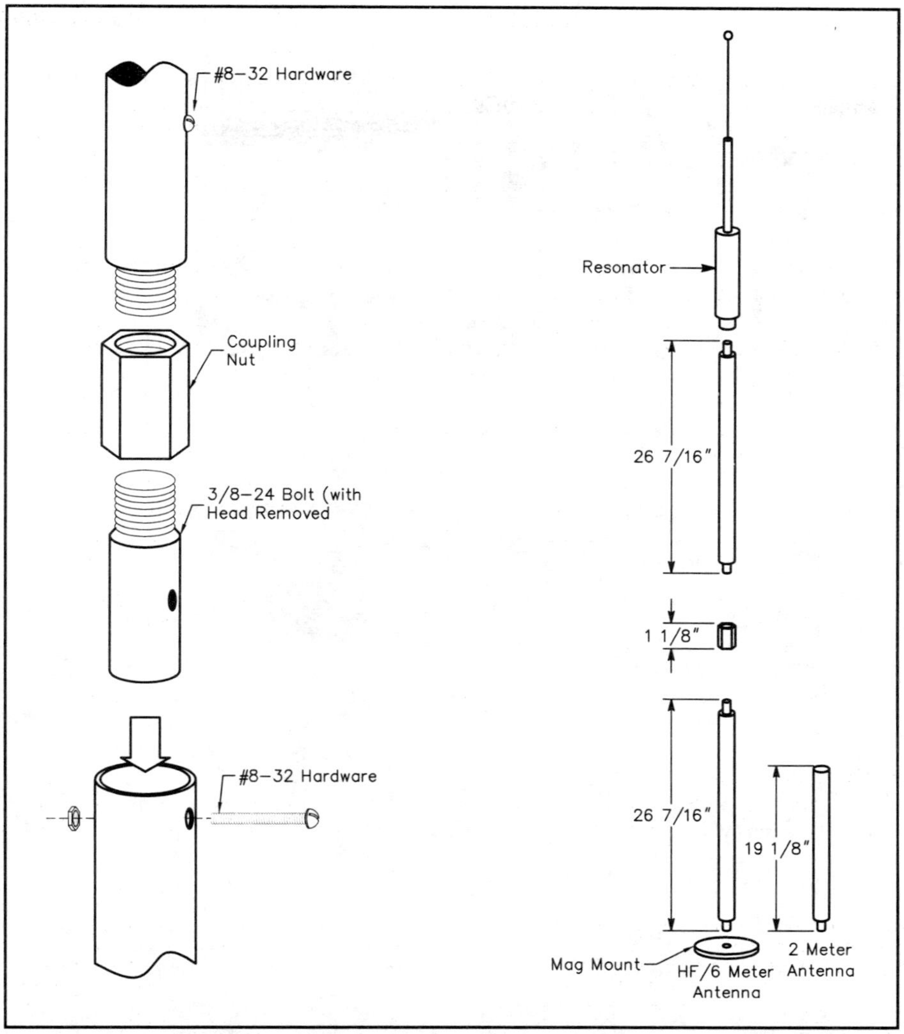

Figure 1—Construction details of N8GGI's portable antenna. Numbers in parentheses are McMaster-Carr catalog numbers (see Note 2).

Materials list:
Tubing—1/2 inch OD, 0.065 inch wall, 6061-T6 aluminum (#89965K54)
Bolts—3/8-24×3 inch long, stainless steel, hex head (#92198AQ362)
Coupling nuts—3/8-24 stainless steel, (#91811A031)

guys in windy situations.

By replacing the untuned car body with $\lambda/4$ wire radials, the bandwidths of the resonators seem to be greater than twice what Hustler claims for mobile operation.[1] Using only two radials, I get about 125 kHz on 40 meters and 250 kHz on 20 meters at an SWR of less than 2:1. Antenna resonance changes with environment and the particular mount used. You can regain resonance by adjusting the resonator tip rod according to the manufacturer's recommendations.

A Suitable Case

My final design task was to create a carrying case that prevents airline baggage "gorillas" from killing the antenna. I use a piece of 4 inch Schedule 40 PVC pipe and two end caps to make a shipping tube. Use glue or screws to secure one end cap. Nylon straps (with plastic catches like those used on fanny packs and life jackets) hold the removable cap. Pop rivet the nylon straps to the pipe and cap. Rivet on a strap handle for convenience. The only items that don't fit into the shipping tube are the magnetic mount and the steel radial plate. A tube large enough to accommodate these pieces would be too heavy and unwieldy.

All the parts for this project come from local fastener distributors and metal dealers. You can also buy them from a commercial supplier, such as McMaster-Carr Supply Company in Chicago.[2]

This antenna obviously won't compete with a full-size antenna, but I've had plenty of good signal reports from the US and Europe while running 100 W. Variations of this theme should allow some pretty stealthy antennas for those in apartments, condos or under antenna restrictions. Run the wire radials through a flower bed or along a deck railing to hide them. You can disassemble the antenna in seconds when the "antenna police" are patrolling.

Notes

[1]This may seem an improvement, but increased bandwidth for simple antennas usually indicates increased losses. More radials should improve the performance.—*Ed.*

[2]McMaster-Carr Supply Co, PO Box 4355, Chicago, IL 60680-0330; orders tel 630-833-0330, fax 630-834-9427. Their current home page is at **http://www.mcmaster.com/.** They sell to commercial customers and won't ship catalogs to individuals, but they're planning a Web catalog for this fall. Individuals may be able to place orders there.

Dennis was first licensed in 1961 as K8UKU and got back into hamming in 1984 with his present call. He is a graduate industrial designer and owns a custom exhibit and manufacturing firm in Columbus employing 25 people. His last QST *article was "Build it Yourself with Plastics,"* QST, *Oct 1987. You can reach Dennis at 133 N Lyndale Dr, Westerville, OH 43081.*

By Thomas Kuehl, AC7A

From *QST*, October 1998

An Efficient Multiband Vertical for 160 through 20 Meters

Build this low-loss, low-cost antenna for the low bands.

I have had the pleasure of hearing the well-known antenna master, Lew McCoy, W1ICP, half jokingly malign the reputation of the vertical antenna. Although there is some uncertainty about who actually was the first to state that "vertical antennas radiate equally poorly in all directions," Lew did make this statement in his antenna presentation at the 1996 ARRL Southwestern Division Convention. Well, despite this folklore, I have always found vertical antennas to be surprisingly good performers. Considering their simplicity, it is hard to beat them as low-cost DX antennas. In my personal experience, they often defy their reputation for poor performance.

The antenna described here is a versatile, multiband vertical designed for use on 160, 80, 40 and 20 meters. My goals were to produce a reduced-size design that would fit on a modest lot, could be erected by just a few people, has maximized efficiency and costs no more than $200. Even if you decide not to build the antenna as described, you can apply the principles to single-band antennas or scale the antenna for the 80 through 10-meter bands.

Figure 1 shows the major points of the design. The main support is a Rohn 50 telescoping mast, commonly used for fringe TV installations. Fully extended, the mast height is a little over 44 feet; it serves as the vertical element on both 160 and 80 meters. With the help of two short cross arms, the mast also supports vertical wires for 40 and 20 meters. The remainder of the antenna consists of two top-hat wires that add capacitive loading on 160 and 80 meters. On 80 meters, a pair of parallel-resonant traps disconnect some of the 160-meter top hat to provide appropriate loading. This is the only use of traps in the antenna, and I've made every effort to minimize their loss.

Alternatively, the mast can be made from many different materials, such as aluminum tubing, military-surplus cast-aluminum mast sections, tower sections or a wire hung from a conveniently placed tree. The primary effects of such substitutions will be variations of the wire lengths shown in Figure 1.

The operation on each band may not be completely evident, so a band-by-band explanation may help:

160 Meters

The 44-foot mast is well short of the 132 feet required for $\lambda/4$ resonance on 160 meters; therefore, loading is required to establish resonance. Rather than resort to inductive loading, with its inherently high losses, I incorporated a low-loss capacitance hat. A "folded top hat" design provides the additional capacitance necessary for resonance and requires less horizontal space than a flat top hat. The top hat approach also increases the radiation resistance, which improves the antenna's efficiency.[1]

With a vertical length less than 0.1 λ on 160 meters, this antenna system's total resistance (R_A) will be low—usually in the range of 5 to 25 Ω. R_A is the combination of radiation resistance (R_R) and ground-return resistance (R_G). The final R_A value depends heavily on the physical installation of the top hat and the value of R_G (see Note 1). On this band, a lower R_A indicates that system losses are low, and one can expect reasonable efficiency.

In order to achieve a match to a 50-Ω feed line, a matching transformer with some latitude is required. Figure 2 details a broadband transformer that allows a number of impedance-matching ratios; one of them should satisfy the R_A at your particular installation. Should a higher impedance ratio be required (high R_A), make taps in the primary winding.The figure also shows how an SPDT relay can switch out the transformer secondary on 80 through 20 meters. R_A is already close to 50 Ω on these bands, so a transformer usually isn't required. Leaving the transformer primary in the circuit on other bands has no adverse affect, and provides a direct path to ground for static discharge. The dc relay voltage can be supplied by a separate cable, or through the feed line if isolation chokes and capacitors are used.

[1]Notes appear on page 26.

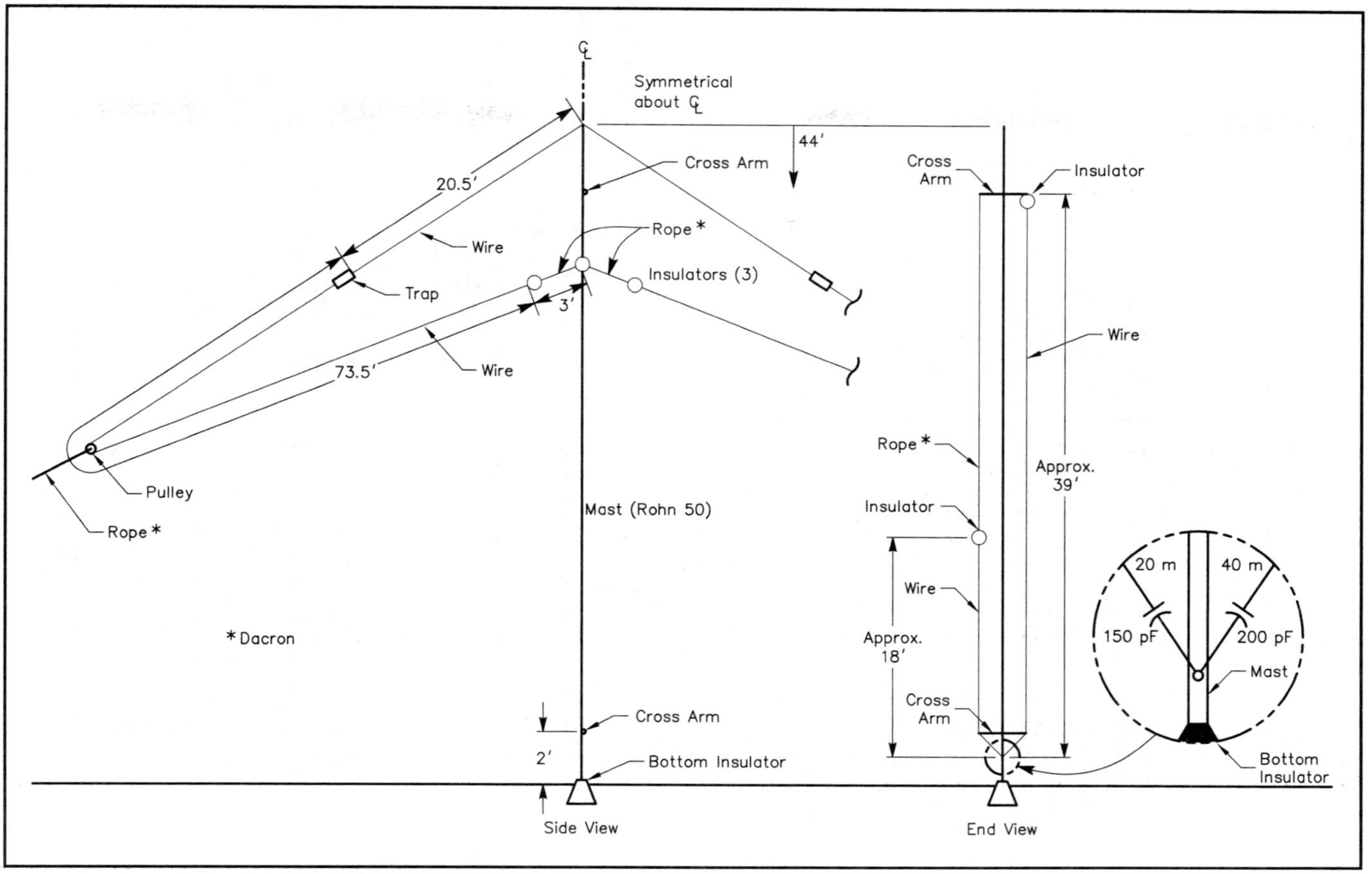

Figure 1—A diagram of AC7A's multiband vertical antenna system.

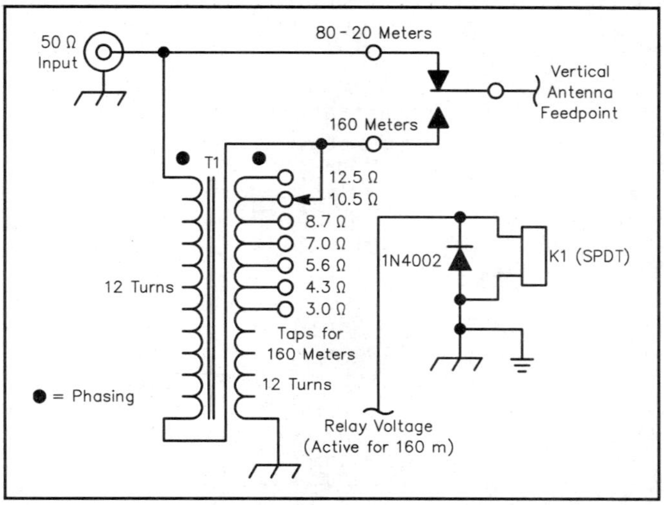

Figure 2—A circuit to switch in a multitap-matching transformer for 160-meter operation. T1 is an FT-140-61 core with 12 bifilar turns of #15 AWG (see text) Polythermaleze wire. Arrange the bifilar wires side-by-side; do *not* twist them. Elevate turns 6 through 11 of the secondary above the toroid's outer face to facilitate taps. K1 is a surplus relay (12 V, with 15 A contacts). Mouser stocks a suitable relay (Magnacraft & Struthers-Dunn; W78RCSX-97, Mouser #528-7880-97) You can contact Mouser at 958 N Main St, Mansfield, TX 76063-4827; tel 800-346-6873; e-mail **sales@mouser.com**, **http://www.mouser.com**. [The author has used this relay and wire with 150-W output.–Ed.]

Figure 3—A homebrew spacer relieves strain from the series capacitor at the bottom of the 40 or 20-meter wire elements. For permanent installation, this must be weatherproofed.

80 Meters

A λ/4 vertical on 80 meters ought to be about 66 feet in length. Therefore, the mast's 44-foot length fulfills two-thirds of the antenna's requirement. By trapping off most of the top hat, the remainder is used to establish resonance on 80 meters. The traps were designed as high-Q circuits to minimize losses.

By inclusion of the traps on 80 meters, the small, finite loss resistance of the traps is transformed to a larger value at the feed point. This small resistance, in conjunction with R_R and R_G, produces a good match to a 50-Ω feed line.

40 and 20 Meters

Short cross arms at 2 and 39 feet above

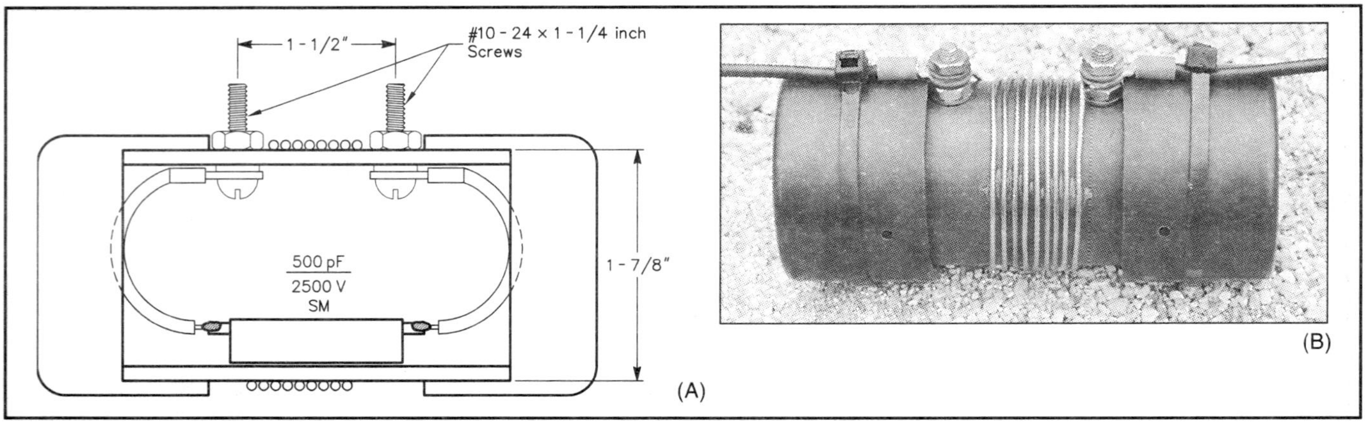

Figure 4—A is a cross section of the traps. B is a photo of the finished trap. Notice the ring terminals on the top-hat wires and the cable ties used for strain relief.

the antenna base support the 40 and 20-meter antenna wires and space them away from the mast. The physical spacing reduces tuning interaction between bands and helps keep the wires out of the way when the mast is being extended or collapsed. They are joined at the bottom of the mast, which serves as a common feed point (see detail in Figure 1).

The 40 and 20-meter antenna wires were intentionally made long. This adds inductive reactance that can be resonated by a series capacitor. This not only allows the antenna to be tuned at the base, but also eliminates the need to raise and lower the mast while adjusting the antenna for these bands. It also raises the R_A over that of a $\lambda/4$ vertical, which improves efficiency and provides a good match to a 50-Ω feed line.

The 40-meter vertical wire is 39 feet long, and the 20-meter vertical is 18.5 feet long. Both are opened about six inches from the base connection point to insert the series capacitors. When tuned, they provide an excellent match to a 50-Ω feed line. Systems with very low ground losses ($R_G < 5\ \Omega$) can achieve a radiation efficiency on the order of 90%.

Construction

There are no steadfast rules in the construction of a "homebrew" antenna; not everyone has access to the same materials and tools. Therefore, I encourage you to be creative in making do whenever necessary.

As a safety consideration, assess your physical ability to erect the telescoping mast before attempting to do so. When in place, the top end of the bottom section will be more than eight feet above the ground. Each mast section must be pushed up and locked into position while working from this height. This is best accomplished by two people, working from two tall ladders. New mast sections slide easily, but with time, corrosion and dirt buildup make the task more difficult. Keep the mast sections clean and lubricated to help with future disassembly. Aside from extending and collapsing the mast—a physically tiring act—nothing extraordinary is required to erect this antenna.

The Rohn 50 telescoping mast can be mail ordered from many Amateur Radio outlets, but be prepared for shipping fees that nearly equal the mast cost. With a little searching, I was surprised to find that they are stocked by a local electronic supplier—at a cost comparable with the mail-order distributors.

The two cross arms were made from a single five-foot piece of antenna mast, the thin-wall type used to support TV antennas. The swaged end was removed with a hacksaw and discarded. The remainder was cut into two equal-length pieces, about 27 inches long. A widely spaced **U** bolt (2 11/16 inches) fits around the lower mast section and through holes drilled for it at the center of the cross arm.

If they're available, use **U**-bolt clamps to help mount the upper cross arm. They also help keep the cross arms from windmilling when the wires are pulled. Since the lower cross arm does not carry a load like the upper cross arm, I didn't use a clamp, and it hasn't presented a problem.

Holes were drilled 1.5 inches from the cross arm ends to facilitate installing eye bolts through the top cross arm and to pass the 40 and 20-meter wires through on the lower cross arm. Remember that these holes are drilled at 90° to the **U**-bolt holes.

The 40 and 20-meter wire elements fasten to the upper-cross-arm eyebolts with Dacron rope. An egg insulator is used between the wire and rope. A short piece of rope is sufficient for the 40-meter wire; about 20 feet supports the 20-meter wire. Lay the extended mast on the ground before installation to set up the cross-arm positions and measure the exact rope lengths required to support the wires.

I mentioned that the 40 and 20-meter wires are spaced about six inches from the mast at the antenna base for placement of the series tuning capacitors. The capacitor leads will not withstand the wire tension, but mounting the capacitor across an egg insulator solves the problem. Alternatively, make an appropriate insulator from a piece of acrylic plastic mounted between two studs. Some form of weatherproofing is necessary. See Figure 3 for details of a homemade insulator.

The wire used for the top hat and vertical wires was purchased from Antennas West.[2] It is sold under the name "Quiet-Flex." It is a durable, 40-strand, #14, 3500-V-insulated copper wire, with a black, UV-resistant jacket. Other #14 insulated wire will work, but Quietflex will stand up to most any kind of sun or weather condition for a *long* time. It is flexible at most any temperature and is a pleasure to use. To protect it as it passes through the lower cross arm, I slipped a four-inch piece of RG-58 outer jacket over the wire to protect it from chafing—3/16-inch-diameter heat-shrink tubing should work as well. Just remember to slip it over the wire before making any permanent connections.

All antenna-wire connections were made using crimped and soldered ring terminals that fit #10-24 screws. Each terminal ring is sandwiched between a pair of washers and nuts at each connection. I used heavyweight ring terminals and find that they take much more punishment than their lightweight counterparts.

The top-hat wires can be connected to the top of the mast in several ways. You could drill two opposing holes about one-half inch from the mast end and install a #10-24×1 1/2-inch-long machine screw in each hole, with the threaded end pointing outward. Secure the ends of the top-hat wires to these studs using #10-24 hardware.

I mounted the studs in a piece of heavy-gauge PVC. It is slit at one end, slipped over the end of the mast and held in place by a stainless-steel hose clamp tightened over the slit end. The studs connect to the mast via short jumper wires fabricated from antenna wire and ring terminals. A self-tapping screw in a drilled pilot hole holds each terminal to the mast.

The top hat is folded back upon itself by running the wire through a small pulley at the fold point. This allows the fold-back point to move as necessary for each unique installation. Since the wire is insulated, either a metal or plastic pulley can be used. The free end of each wire terminates at an insulator. The wire insulators are connected, one to each end, of a six-foot rope that slides through a third insulator. That third insulator is attached to the guy ring at the top of the second highest mast section (see Figure 1).

Place the top-hat fold points as high as possible; a horizontal installation is optimum. Higher top hat positions increase the R_R portion of R_A on 160 and 80 meters, which helps maximize efficiency. Should tall trees or other structures be available, they will work as end supports. There are few tall trees here in the desert, so I use some thin-wall mast sections. In my installation, a 10-foot mast was used at each end. Since my lot is long, the masts are spaced about 30 feet from the fold points. This positions the top hat ends about 18 feet above ground.

It requires some ingenuity to hold the mast in place and insulate it from the ground. I buried the bottom half of a champagne bottle in the ground, centered the mast's bottom section over the bottle and tightly guyed the mast with galvanized steel wire. Egg insulators isolate the mast and guy wires. The galvanized wire has less give than rope and provides for more rigid mast placement.

Dacron rope was used for all the mast sections above the bottom section. Although you could guy the mast at each telescoping section, I installed guys only at the bottom, center and top sections. The mast bears no load at the top, and the top hat serves to guy the antenna. This approach has worked well, and it's much less confusing when the guys and various wires are being pulled into position. The antenna has withstood strong winds and stayed up without any problems.

The ARRL Antenna Book shows many suitable guy-anchor systems.[3] I use three-foot construction stakes obtained at a local home-improvement center. They are strong steel stakes that taper to a point at one end and have three convenient holes through them right near the top. Rather than threading the Dacron rope directly through the stake, I slip a piece of galvanized guy wire through the hole and twist it on itself. The guy then terminates in an egg insulator that provides a good union for the Dacron rope. The guy wire is several feet long to keep the Dacron rope above ground and away from rabbits that might chew through it.

One last point: As with any vertical antenna, the performance of this antenna is highly dependent on the ground conductivity near the antenna. The usual method of improving the ground conductivity is to use many radial ground wires around the antenna base. This and other possibilities for developing a ground system are covered in *The ARRL Antenna Book*.[4] Also, refer to the sources described in end notes of the author's previous articles on building efficient, short vertical antennas.

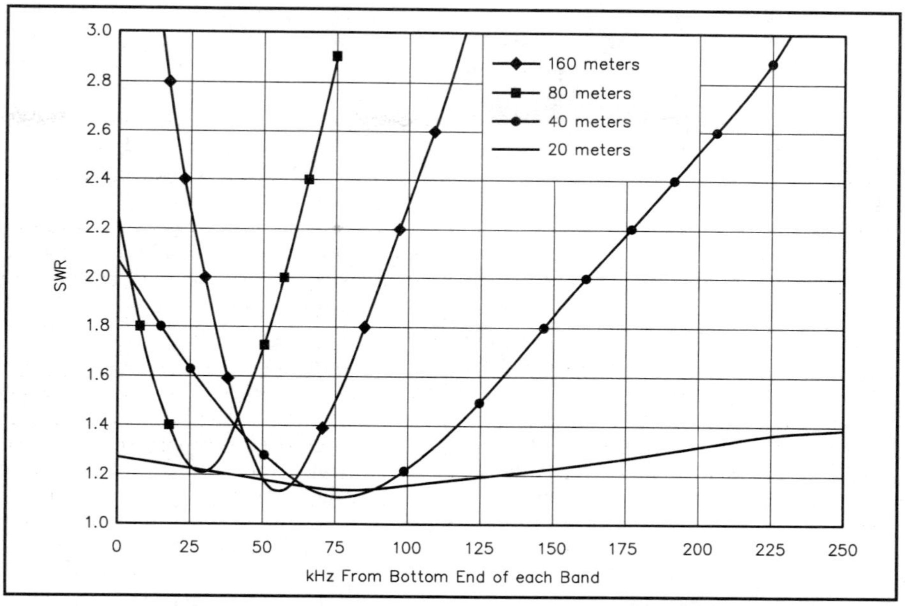

Figure 5—An SWR versus frequency chart for the multiband vertical antenna.

Trap Construction

The trap construction is shown in the drawing and photograph in Figure 4. Black 1⅞-inch (OD) ABS serves as the coil form. The coil windings are secured at the ends using studs made from #10-24×1½ screws, spaced 1½ inches apart. I used brass hardware, but stainless-steel hardware would be better, where corrosion is a problem. Each capacitor is housed inside its coil form and connects under the screw heads with ring terminals.

Before winding the coil, prepare the capacitor for installation inside the form. A 500 pF, 2500 V mica capacitor is required. Traps must withstand high voltages and a 2000 V to 2500 V rating should be sufficient for power levels in the 100 to 150 W range. If you have problems coming up with high-voltage capacitors, consider connecting several lower-voltage capacitors in series, perhaps four 2000 pF, 500 V silver-mica capacitors. When series connected, they are equivalent to a single 500 pF, 2000 V capacitor. Silver-mica capacitors have low dielectric loss, tight tolerances and are available from a number of electronic suppliers who carry surplus or new components.

The coils are made from nine turns of #15 AWG enameled wire (#14 or 16 AWG should also work), and should have a finished turn spacing of about eight turns per inch. Strip the insulation off one end of the wire and anchor the wire to a stud. Wind the coil tightly and closely space the turns. Strip the free wire end and secure it to the other stud.

The coil turns will be spread during the tuning operation. As the turns are spread, the windings may loosen. If this happens, simply grip a turn with some pliers and twist it to form a wire kink on the form surface. This will tighten the coil winding; repeat the action for more turns until the winding is tight.

ABS end caps help keep water out of the trap. Connect the top-hat wires to the coil-end studs with heavy-duty ring terminals. (See Figure 4B.) Secure each top-hat wire to the trap end caps with a heavy-duty, UV-resistant nylon cable tie to reduce the load on the ring terminals.

Trap Tuning

Before venturing outside, the traps must first be tuned. There are several ways to do this using a dip meter or one of the various antenna-measurement devices offered by MFJ or Autek. I use a Heathkit dip meter in conjunction with my station transceiver to get an accurate frequency indication.

You must decide where in the 80-meter band you want to tune the antenna and traps. Since I operate mostly at the low end of 80 meters, I tuned the traps to 3.530 MHz. When the coil turns are close up against each other, the trap resonates below the 80-meter band. As the turns are spread apart, however, the resonant frequency rises. Use a pocketknife and work toward an even spacing of about ⅛ inch between turns. Spread and compress the windings until you achieve the desired resonant frequency. Once the tuning is complete, apply a few lines of epoxy lengthwise, across the windings, to hold them in place. After the epoxy cures completely, the traps may be installed in the top-hat wires.

Antenna Tuning

Antenna tuning can be accomplished using a low-power transmitter and an SWR bridge, or one of the mentioned antenna-measurement instruments. For 80 through 20 meters, no impedance matching is required: Simply adjust the appropriate wire's length or change the capacitor value. For 160 meters, both the top hat length and feed-point matching must be set.

Begin tuning on 80 meters. Adjust the wire length (between the top of the mast and the traps) until resonance is achieved. A low SWR should accompany the resonant point. Depending on the desired frequency and the environment, the wire lengths should be near those shown in Figure 1. Start with a wire that is a little too long to avoid splicing wires. For my installation, the mast plus 80-meter top hat lengths are:

$$L\ (ft) = 230\ /\ f\ (MHz) \quad \text{(Eq 1)}$$

The required length will be shorter if the top hat is installed horizontally.

Once satisfied with the 80-meter tuning, adjust the overall top hat length for resonance on 160 meters. This is accomplished by shortening or lengthening the long, folded wire beyond the traps. As mentioned earlier, the R_A on 160 meters will be far less than 50 Ω, which may make it difficult to find the initial resonant frequency with an SWR meter. I suggest connecting the matching transformer of Figure 2, with the 12.5-Ω tap selected. This should provide a reasonable match for most installations and allow the minimum-SWR point to be determined. The optimum tap selection can be made after the antenna is resonant at the target frequency. For 160 meters, the total length of the mast plus the full top hat, followed the equation:

$$L\ (ft) = 255\ /\ f\ (MHz) \quad \text{(Eq 2)}$$

If you must add long lengths to the 160-meter top hat, the folded ends are too close to the ground.

A space is intentionally provided between the top-hat wire ends and the mast by the three-foot ropes shown in Figure 1. This helped reduce a peculiar end effect that made tuning less predictable. Therefore when adjusting the top hat length, change the end of the wire, but do not reduce the wire-to-mast spacing.

The 40 and 20-meter bands are each tuned in similar fashion and do not require raising and lowering the mast. Figure 1 shows approximate values for the series tuning capacitors: 200 pF for the 40-meter element and 150 pF for the 20-meter element. These values provide resonance in the CW subbands, with good bandwidth. You can easily determine exact values by substituting a variable capacitor—such as a 365-pF receiving capacitor—for the fixed value. Using an insulating knob on the capacitor, adjust it until the SWR minimum is at the chosen frequency. Remove the capacitor from the circuit, measure its value and replace it with the closest available fixed-value capacitor. A silver-mica capacitor works nicely in this application, and since it is used at a low-voltage point, it need not have a high working voltage like those of the traps.

Once you are satisfied with the tuning on each band, secure the guys and end ropes.

Results

The 2:1 SWR bandwidth of an antenna is always of interest. It doesn't indicate radiation performance, but rather the frequency range that can be covered without using a Transmatch. Wide SWR bandwidth can also indicate excessive loss, so keep this in mind. Figure 5 shows SWR plots for each of the bands. The curves are based on measurements made at the antenna feed point.

Just for the sake of curiosity, I checked the SWR on each of the other HF bands, as well. A couple of surprises did show up: The SWR on 30 meters is a little less than 2:1, and it's slightly greater than 2:1 in the FM portion of 10 meters. Although it was not intended for 30 meters, the antenna performs well on the band.

The proof of any antenna design comes down to how well the antenna performs at your particular station, with its particular constraints. I use 16 buried radials at this location—ranging in length from 66 to 130 feet. This is a minimal ground system, but the antenna is located in an area covered with dense desert vegetation and it precludes the installation of a more extensive return system. Despite this limitation, the antenna does a credible job. To date, I've worked all states and continents, except Europe, on 160 meters, using 80 to 90 W on CW. The DX performance has been even better on 80 meters, where stations on all continents have been contacted. Using this latest version and various previous generations of verticals, I have managed to work DXCC on 40 meters with an output power less than 100 W. That's not a bad accomplishment for an antenna "that radiates equally poorly in all directions."

Notes

[1]Thomas Kuehl, AC7A, "Build Efficient, Short Vertical Antennas," *QST*, March 1998.

[2]Antennas West is no longer in business. Antennas and More sells all Antennas West products. Contact them at 1038 S 350 E, Provo, UT, 84606; tel 888-277-5718, 801-373-8426 (both lines voice and fax); e-mail **support@antennasmore.com**; URL **www.antennasmore.com**.

[3]R. Dean Straw, N6BV, Editor, *The ARRL Antenna Book* (Newington: ARRL, 1997), ARRL Order No. 6133. See Chapter 22. ARRL publications are available from your local ARRL dealer or directly from the ARRL. Mail orders to Pub Sales Dept, ARRL, 225 Main St, Newington, CT 06111-1494. You can call us toll free at tel 888-277-5289; fax your order to 860-594-0303; or send e-mail to **pubsales@arrl.org**. Check out the full ARRL publications line on the World Wide Web at **http://www.arrl.org/catalog**.

[4]R. Dean Straw, N6BV, Editor, *The ARRL Antenna Book*, see Chapter 3.

Thomas, AC7A, was first licensed in 1969 at age 15, with the call WN6BQI. Early on, he developed an appetite for the technical side of the hobby that remains today. A graduate of Cal-Poly San Luis Obispo, he has spent the better part of the last 19 years employed by Burr-Brown Corporation as a Product Engineer. When not assisting his wife with raising their three children, you may find him chasing DX on the low bands or pruning antenna wires in the backyard. You can reach Thomas at 620 S Avenida Princesa, Tucson, AZ 85748; e-mail **Kuehl_Tom@BBrown.com**.

By Thomas Kuehl, AC7A

From *QST*, March 1998

Build Efficient, Short Vertical Antennas

Want to fit a 160 meter antenna on your postage-stamp lot? AC7A shows how to maximize efficiency and bandwidth.

Many of us lovingly admire the large antenna arrays that adorn the covers of Amateur Radio magazines. Nonetheless, today's hams must often rely on reduced-size antennas as their only practical means to get on the air. Mobile antennas, rubber-duckies and trap antennas are good examples.

To reduce antenna size, we must compromise. Generally, very short antennas provide narrower bandwidths, present tougher matching problems and have greater losses than their full-sized counterparts. Greater losses reduce radiation efficiency. This affects the amount of signal lost while transmitting *and receiving*.

You may have little control over the losses in commercially produced antennas, but home-built antennas may offer you more control than you think! This article evaluates the problems associated with reduced-size antennas and the methods you can use to optimize antenna performance.

Antenna efficiency is related to the radiation resistance (R_R)[1] and the total loss resistance (R_L) by:

$$\text{Efficiency}\left(\eta\%\right) = 100\%\left[\frac{R_R}{\left(R_R + R_L\right)}\right] \qquad \text{(Eq 1)}$$

An antenna that is 50% efficient yields only half as much signal (received or radiated) as would be by an ideal, lossless version of the same antenna. The rest of the signal is lost as heat.

In a full-size resonant antenna (such as a dipole), wire resistance is the principal loss. If the antenna is made from wire of appreciable size and good conductivity, the losses are very low and the radiation efficiency high—approaching 100%. A shortened antenna, however, needs some sort of loading and/or matching network to work with a transmitter. Loading and matching can be a significant source of loss, whether incorporated in the antenna itself or provided by a tuner and transmission line.

[1]Notes appear on page 32.

Hams often use a monopole antenna, aka "vertical," for reduced-size applications. Although I am a true fan of the vertical, it is an excellent candidate for losses: losses contributed by loading components *and* losses associated with conductivity of the ground that surrounds it. For now, let's say that the ground loss is the earth resistance in proximity to the antenna, through which the return currents must flow. We will look at this in more detail later.

Figure 1 shows a simple model (a series-tuned LCR circuit) for the vertical antenna, which can serve as an equivalent electrical circuit. Although the model shows the circuit as lumped elements, the elements are actually distributed along the antenna. As with any uniform straight wire, the inductance is a function of the conductor length and diameter. Capacitance (C) forms between the antenna element and the somewhat conductive ground surrounding it, with the air between them serving as the dielectric. We can view the total capacitance as a sum of incremental capacitances, formed by small antenna segments and the ground. The capacitance of each segment varies with its height. For equal-sized segments, those closer to the ground exhibit more capacitance than those at the top of the antenna. The capacitance measured at the bottom of the antenna appears as the sum of the entire capacitance. This is also true for the distributed inductance.

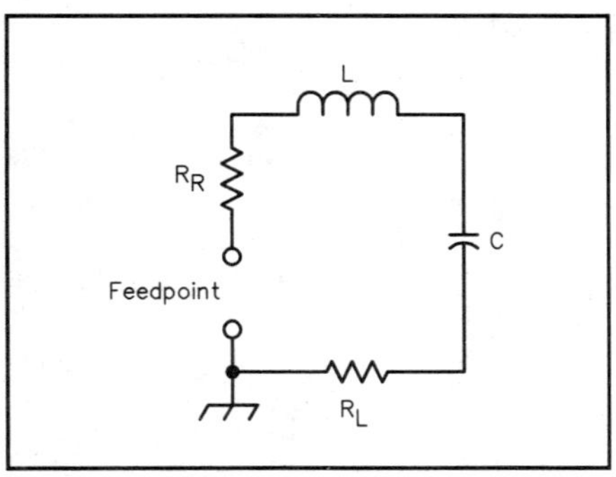

Figure 1—A series RLC circuit models an antenna simply.

Resistances in the model define the antenna efficiency. To maximize efficiency, we must minimize R_L, which consists of the total ohmic losses in L, C, the antenna conductors and any ground-return loss.

Despite the simplicity of the model, it is a good tool for analyzing the electrical characteristics of a full, or reduced-size antenna.

As in any series-resonant circuit, antenna resonance occurs when the capacitive reactance (X_C) and inductive reactance (X_L) are equal, leaving only the circuit resistances. The resonant frequency is:

$$f = \frac{1}{2\pi\sqrt{LC}} \qquad \text{(Eq 2)}$$

where f is in Hertz, L in Henries and C in Farads.

When we reduce the antenna conductor length, there is less self-inductance and less distributed capacitance. Because L and C are reduced, the antenna is no longer resonant at the original frequency, but at some higher frequency. To re-establish resonance at the original frequency, we must add inductance, capacitance or both to the antenna.[2] Additionally, the reduced length reduces R_R, which falls off rapidly as the element becomes shorter.

A full-size, λ/4-vertical has an electrical length of 90° Reduced-size antennas are often in the range from 10° to 45°. Figures 2 and 3 are plots of the theoretical R_R and X_C for a reduced-size (short) vertical antenna. The curves are based on the equations presented many years ago by Belrose.[3] The curves show how quickly R_R decreases and X_C increases (or C decreases) as an antenna becomes shorter. From this, you can see that very short antennas can present considerable matching and loading challenges.

The two reactance curves plotted in Figure 3 show the effect of conductor diameter on the reactance developed by the antenna. A larger diameter element exhibits more capacitance (less capacitive reactance) and requires less loading to achieve resonance. Element diameter has virtually no affect on radiation resistance.

Bandwidth

When we shorten an antenna it becomes highly reactive, and the radiation resistance drops to very low values, as shown in Figures 2 and 3. In effect, the loaded short antenna is a high-Q circuit, and a change in frequency results in a dramatic change in the feed-point reactance (X). The rate of impedance change dictates how quickly the SWR changes as the frequency moves from the resonant point. Any loss resistance tempers this behavior, but lowers efficiency. An efficient reduced-size antenna exhibits a narrower bandwidth than its full-size counterpart.

A common measure of antenna bandwidth is the frequency range where the SWR is 2:1, or less. Figure 4 is a bandwidth comparison between 44-foot verticals using inductive loading at the center, flattop capacitive loading and a full-size vertical of the same diameter. All are resonant at 1.825 MHz, matched for a 1:1 SWR at resonance. The ground resistance R_G was set to 5 Ω to simulate a very efficient radial system.[4]

The full-size vertical has a bandwidth of about 165 kHz. Its bandwidth is primarily a function of its length-to-diameter ratio, where low ratios provide wider bandwidths. A large-diameter antenna exhibits greater distributed capacitance and less distributed inductance than one with a smaller diameter. This results in an antenna that exhibits less reactance change with frequency, ie, a low-Q antenna.

Because of the heavy loading and high-Q nature of the inductively loaded antenna, it has a very narrow bandwidth: about 18 kHz. Broad bandwidth for an antenna of this type is evidence of loss resistance in the loading coil, ground or both. Remember that a 50 Ω dummy load has an almost infinite bandwidth, but an RF radiation efficiency of zero! We'll discuss the cap-loaded curve of Figure 4 later.

Matching

Even moderately efficient reduced-size antennas have a total feed-point resistances (R_R + R_L) that are less than the common 50 Ω coax characteristic impedance. To minimize SWR, we need some kind of matching network or transformer. Regardless of the loading type, the same matching methods work with all the antennas I'll describe. Take care, however, to assure that the matching method does not add substantial loss or further reduce system bandwidth.

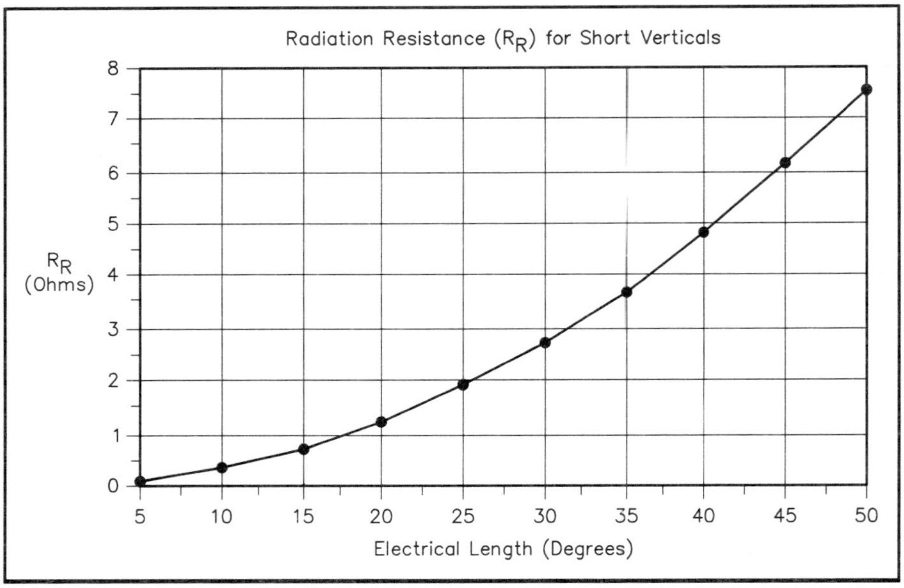

Figure 2—A plot of radiation resistance (R_R) versus antenna electrical length.

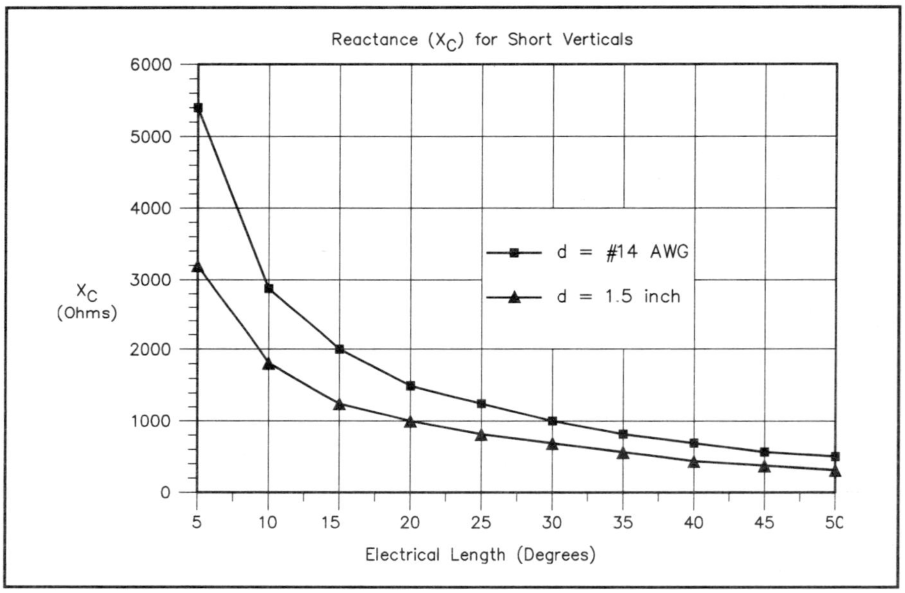

Figure 3—A plot of reactance (X_C) for short antennas. (Short antennas with no loading are always capacitive.)

Generally, a simple L-network can provide the necessary impedance transformation and tune out reactance. (Information about these networks appears in *The ARRL Handbook* and *The ARRL Antenna Book.*[5]) These networks consist of a single capacitor and inductor and provide matching that exhibits bandwidths much wider than that of most loaded antennas. With any matching network, seek out low-loss components, especially the coil.

I prefer to use a broadband matching transformer made with a low-loss ferrite toroid core such as an Amidon FT-140-61.[6] Figure 5 shows a parallel-wound transformer that offers a variety of matching impedances. I find that such a transformer is completely satisfactory for the 100 to 150 W range. My reduced-size vertical for 160 through 20 meters uses this type of transformer for matching on 160. After I have determined that an antenna is resonant, I simply move a clip-lead from tap to tap, until I find the lowest SWR. If you need to know the total feed-point resistance, simply use the transformer turns ratio to determine it. That can be helpful when evaluating loading or ground-system changes.

We can match a capacitively loaded antenna very well by feeding it at a point other than the base. Several years ago, Carl Eichenauer's *QST* article described a top-feed method.[7] The base of the vertical is grounded, and the feed line connects between one of the top-hat wires and the top of the vertical element. For that particular antenna—which has dimensions similar to example two, which I'll describe later—I found that the feed-point impedance is a nearly perfect match to 50 Ω. The coax runs down through the center of the vertical element.

When dealing with reduced-size, λ/2 antennas, R_R is usually much lower than 50 Ω and R_L is small. If $R_R + R_L$ (Z_A) is close to 12 or 13 Ω, a broadband 4:1 ferrite matching transformer is suitable. If Z_A is much greater or much less, use the transformer shown for the vertical as a conventional step-down transformer. If you know Z_A, use the same number of primary turns (12) and follow the transformer equation:

$$N_{sec} = N_{Pri}\sqrt{\frac{R_{Sec}}{R_{Pri}}} \qquad \text{(Eq 3)}$$

Center the secondary over the primary winding. If Z_A is unknown, determine the turns ratio empirically.

You could avoid both antenna loading and matching transformers by using a convenient, short, center-fed dipole antenna with open-wire line and a tuner. Other than the inconvenience of adjusting a tuner and the possible interaction of open-wire line with nearby conductors, this approach offers much flexibility.

Ground

If you choose a dipole antenna, ground loss is of minimal consequence. Vertical monopole antennas rely on ground as a current return path, and R_G is usually their greatest loss. This resistance results from the limited conductivity of the earth surrounding the antenna. You can visualize this as a field of series-connected resistors, emanating outward from the base of the antenna. The antenna return currents must flow through these resistors, which convert some power to heat (loss).

The key to minimizing ground loss is to "short out" as many of these resistors as possible. We do that by installing radial wires at the base of the antenna, arranged like the spokes of a wheel. More radials provide more low-resistance paths for return currents. For practical reasons, however, their number is usually few.

A ground system of radial wires is a completely satisfactory way to reduce ground-return resistance. The points of diminishing return for number of radials, radial lengths and radiator length all interact. For example, *The ARRL Antenna Book* tells us that for 15 radials, radial and radiator lengths greater that λ/8 are ineffective. Set radial length and orientation to fit the available space. More short radials are better than few long ones. As a minimum, 20 radials is a good objective, while more than

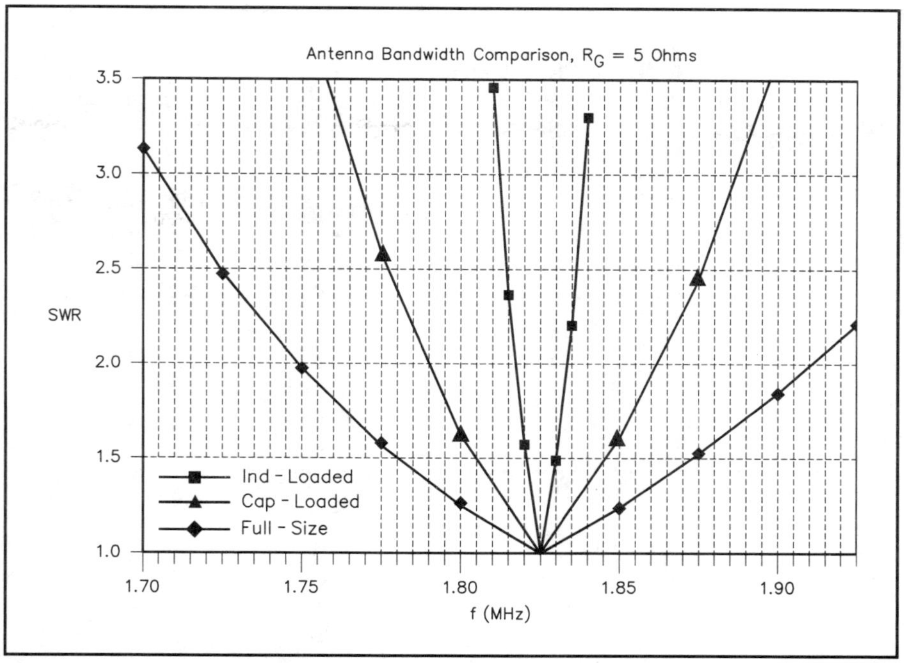

Figure 4—A plot of SWR versus frequency for full-size, inductively loaded and capacitively loaded antennas. Ground losses are assumed to be very low (5 Ω).

80 is likely of little benefit.[8] Don't depend on ground rods; they are an ineffective return in many locations. Jerry Sevick (W2FMI) and Al Christman (K3LC), have both written excellent articles that give detailed information about ground systems.[9, 10]

If ground radials are not practical at your location, there's another option that may even further reduce losses. For a vertical mounted tens of feet above the ground, three or four λ/4 radials at the antenna mounting point (on a roof, or strung through trees) provide an effective low-loss return path. This configuration is known as a *ground plane* or counterpoise of *elevated radials*. Raising the antenna above ground reduces the capacitance to ground, and the return currents flow through the radials instead of lossy ground. Research shows that four elevated radials are as effective as 120 radials at ground level.[11] So, if you can accommodate elevated radials, they can save you lots of wire and effort. Exactly how far must we elevate the radials? Al Christman cites heights of 10, 16 and 25 feet for very good, average and very poor soil at 1 MHz. Those heights decrease as frequency increases. Higher is better, but a few feet is sufficient on higher frequencies (provided that they are high enough so people and animals can walk under them safely).

If you cannot elevate the antenna, you can still use elevated radials. Al Christman describes radials that sweep upward from the base of a vertical to a convenient height, where they become horizontal.[12] This is the best of all worlds: A convenient ground-mounted antenna with no wire to bury and the performance of elevated radials!

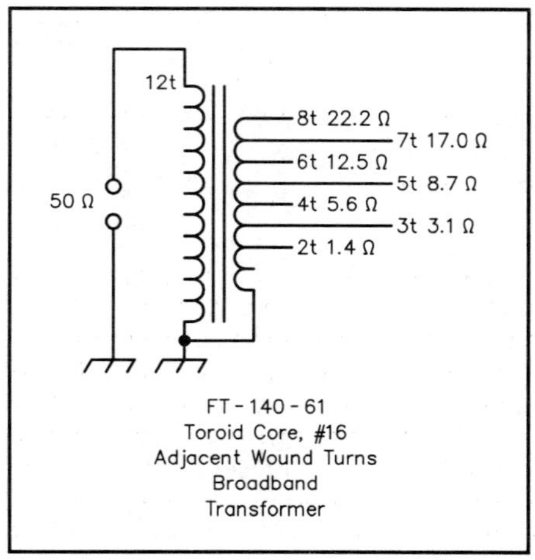

Figure 5—A schematic of AC7A's broadband transformer.

Inductive Loading

A shortened antenna resonates at a higher frequency than its full-size counterpart because it has less inductance and capacitance. We can add inductance, capacitance or both to compensate for the reduced self inductance and capacitance. We can insert an inductor at any desired point along the antenna length, but we should choose the location wisely. The inductance required to achieve resonance varies with the inductor's position along the antenna. We need less inductive reactance to achieve resonance when we place the inductor in a high-current area than when it's placed in a low-current area. The inductive load is usually a coil, but it could also be a transmission-line stub.

Table 1
Inductively Loaded Short Vertical Performance

f = 1.825 MHz, length = 44 feet, diameter = 1.5 inches

Load Location	X_L (Ω)	L (μH)	Q	R_S (Ω)	R_R (Ω)	$R_G=10\Omega$: R_S+R_G (Ω)	$R_G=10\Omega$: Efficiency (%)	$R_G=5\Omega$: R_S+R_G (Ω)	$R_G=5\Omega$: Efficiency (%)
Bottom	610.5	53.2	∞	0	2.9	10	22.5	5	36.7
			200	3.1	2.9	13.1	18.2	8.1	26.5
One-Quarter	782.2	68.2	∞	0	4.4	10	30.6	5	46.8
			200	3.9	4.4	13.9	24.0	8.9	33.1
Center	1124.4	98.1	∞	0	6.1	10	37.9	5	55.0
			200	5.6	6.1	15.6	28.1	10.6	36.5
Three-Quarters	2029.7	177.0	∞	0	8.1	10	44.8	5	61.8
			200	10.2	8.1	20.2	28.6	15.2	34.8

The main disadvantage of inductive loading is loss in the coil. The quality factor, Q, is a measure of coil performance. Q is a unit-less number based on the ratio of coil reactance to resistance.

$$Q = \frac{X_L}{R_S} \qquad \text{(Eq 4)}$$

X_L is the inductive reactance of the loading coil. R_S the series resistance of the coil, is the total of the wire resistance, loss resistance of the coil form and the loss of any other lossy material coupled to the coil. A high Q is desirable from a loss standpoint, but less loss implies a narrower antenna bandwidth because greater losses increase the SWR bandwidth.

Let's consider a very short 160 meter vertical to illustrate the effect of loading on antenna efficiency. The length is 44 feet, with an element diameter of 1½ inches. This represents a practical configuration with an antenna length of about 0.082 λ.

In the inductive loading example, we'll consider a loading coil at four common positions: base (minimum reactance), low (25%, 11 feet), middle (50%, 22 feet) and high (75%, 33 feet). The example doesn't consider end loading (at the top) because the loading inductance for that case is impractically large.

I used *EZNEC*[13] to determine the proper load inductance and the resulting feed-point radiation resistance (R_R) by an iterative process. I verified the R_R and coil requirements using the equations presented by Belrose; the results agree closely.

Table 1 gives the loading details and antenna efficiencies for each of the loading points, with Qs of infinity (loss-less coil) and 200 (a moderate value). I did this to show how coil Q affects antenna efficiency. For comparison purposes, the example shows losses and efficiency for 5 and 10 Ω ground resistances.

The table reveals a number of interesting characteristics of short, inductive-loaded antennas:

- The inductance required to resonate the antenna nearly doubles as the loading point moves from the base to the antenna's center. It doubles again as the loading point moves from the center to the 75% point. Above the 75% point, the inductance requirements increase even more rapidly; that is why end-loading is impractical.
- The R_R is very low; therefore, *any* loss resistance significantly limits antenna efficiency.
- R_R increases as the coil position moves up the antenna, but so does the coil R_S, eventually limiting any further possible gains. In fact, as we move the load from 50% to 75%, R_S begins to dominate and can decrease efficiency. This is especially evident when ground resistance is low (5 Ω).

Center loading is just about optimum for inductive loading. Mobile antenna designers have recognized this and settled on it as a good trade-off between antenna efficiency and mechanical stability.

In the case of a dipole antenna, the overall losses are lower because ground is not as a significant return-current path. Therefore, we need consider only the coil losses. A dipole that is twice the length of the vertical and loaded at 50% on each side has twice the R_R and R_S of the center-loaded vertical antenna in Table 1. Using the efficiency equation, the loaded-dipole efficiency is:

$$\eta = 100 \times \left(\frac{2 \times 6.1}{(2 \times 6.1) + (2 \times 5.6)} \right) = 52.1\% \qquad \text{(Eq 5)}$$

Compared to the 36.5% efficiency of the center-loaded vertical with a 5 Ω ground resistance; this is a significant gain. There are other tradeoffs, however. Horizontal dipoles at low heights have a lot of radiation at high take-off angles. As dipole height increases, so does radiation at lower angles (maximum lobe at about 30° for λ/2 high, about 15° for 1λ high).[14] Therefore, we must mount a dipole at a much greater height to achieve the low radiation angle of a vertical antenna. That height may be difficult to achieve if you are already pressed for space. Installing a loaded dipole vertically would help retain the low-angle radiation characteristic.[15]

Capacitive Loading

In the early days of radio when most of the operations were limited to frequencies below 1.5 MHz (200 meters), operators needed very tall antennas or needed to load them to reduce the height requirements. If you peruse early *QST*s, you'll likely see a common antenna of the time. It consists of a vertical wire connected to a number of parallel horizontal wires at the top, as shown in Figure 6. The horizontal wires are connected at the center and at their ends, which are supported by tall poles. This is *not* a horizontally polarized antenna with a single-wire feed; it is a capacitively loaded vertical. The horizontal wires provide capacitance to resonate the short vertical wire. The result is a vertical antenna of practical proportions.

Today, capacitive loading of fixed vertical antennas often takes the form of a "hat" structure at, or near, the antenna end. Mobile antennas often use a metal disk, or a spoked wheel, placed some distance above the loading coil. The capacitance added by the hat reduces the inductance needed to achieve resonance. When compared to simple inductive loading, this combination requires less inductance (less R_S), which increases efficiency.

For home installations, capacitive loading usually consists of simple wire structures similar to the horizontal wires used in the early antennas. Any wire configuration exhibits capacitance, but—as you will see—some are more effective than others.

You may recall seeing an 80 meter dipole pressed into service as a vertical on 160 meters. (An example appears in an old *ARRL Antenna Book.*[16]) This feat is achieved by shorting the feed-line conductors together at the transmitter and working the vertical against a ground system. The feed line—preferably vertically oriented—becomes the radiator and the horizontal dipole wires become the capacitive top hat.

Because the dipole wires are physically 180° apart, their currents are phase canceling and do little radiating. The antenna is truly a vertical polarized radiator on 160 meters. If the feed line is long enough, the overall system may even be resonant. If not, a tuner performs any necessary matching.

A comparison with the common inverted **L** antenna can help us get an intuitive handle on the behavior of top hats. If the antenna is simply an inverted **L**, the total length of the two sections is approximately λ/4, 128 feet. In the case of a flattop, however, the second horizontal wire exhibits capacitance to ground. The extra capacitance requires a small reduction in the horizontal wire lengths to maintain resonance at the original frequency. More wires present additional capacitance, which further reduces the wire lengths. Resonance is all a matter of radiator length and top-hat area (capacitance).

In Table 2, I've evaluated a number of different top-hat wire schemes using *EZNEC*. I have found that the modeling works very well. My former and current antennas use top hats that I first modeled using *EZNEC*. I recommend this approach; it will save you much work and wire.

We will evaluate different top-hat structures and discuss their performance individually. Their various configurations show flexibility in top-hat design. The radiator here is the same 44 foot, 1½ inch diameter vertical section used for the inductive loaded examples. All top-hat wires are #14 AWG wire.

The first example is a top hat made from two equal-length, opposing wires, a "flat top." It could be the 80 meter dipole from the earlier example, with the wire lengths set for resonance. With top-hat wire lengths of 78.5 feet, the antenna is resonant at 1.825 MHz. The vertical radiating section is still 44 feet, as in the earlier examples, but now the R_R has risen from 2.8 Ω (the unloaded value) to 9.9 Ω, a very significant increase that increases efficiency. Using a ground resistance of 5 Ω for comparison, the efficiency of the antenna is now 66.4%! This is almost twice the efficiency of the best-case, inductively loaded antenna!

R_R is much greater than that of the inductively loaded examples because the length from the base to the tip of each top-hat wire is much closer to λ/4. The top loading increases the current at the base of the capacitive-loaded antenna, and the vertical section exhibits a more uniform current distribution along its length.[17,18]

Although the flattop is only one-third of the height of a full-size vertical, it produces a bandwidth of 75 kHz. (See Figure 4.) This is due to the higher R_R and greater distributed capacitance, which lowers the antenna Q. Clearly, this is another advantage of capacitive loading.

It occurred to me that I could place capacitive loading at some point other than the top. This is equivalent to sliding the top hat down the vertical element. At first, this sounded like it might hold some promise for simplifying the top-hat installation, but beware! As an example, let's slide the flattop from example one down to the 75% point (33 feet) on the vertical element. Modeling shows that the horizontal wires must be 92 feet, and R_R falls from about 10 Ω, to 6 Ω. Why did the R_R drop if the vertical element is still 44 feet tall? A study of this arrangement reveals that the top hat essentially chokes off the current above the top hat. The current in the top 11 feet, above the top hat, drops to about one-tenth of its former value; rendering the length above the top hat nearly ineffective.

Table 2

Capacitively Loaded Short Vertical Performance

f = 1.825 MHz, len = 44 feet, diam = 1.5 inches

Example	*Top-Hat Wires* Count	Slope	Length (ft)	R_R Ω	***Efficiency (%)*** R_G = 10 Ω	R_G = 5 Ω
1	2	0°	78.5	9.9	49.7	66.4
2	2	−25°*	82	3.8	27.5	43.2
3	4	0°	55	9.8	49.5	66.2
4	4	−25°	59.5	5.0	33.3	50.0
5	4†	0°	29.5	9.9	49.7	66.4
6	4†	−25°	33	5.9	37.1	54.1
7	2††	0°	52.5	9.7	49.0	65.8
8	2††	−25°	57.5	5.5	35.5	52.5

*Negative angles indicate that wires slope down from antenna top
†Wire ends are connected together, see text
††Wires are folded, see text

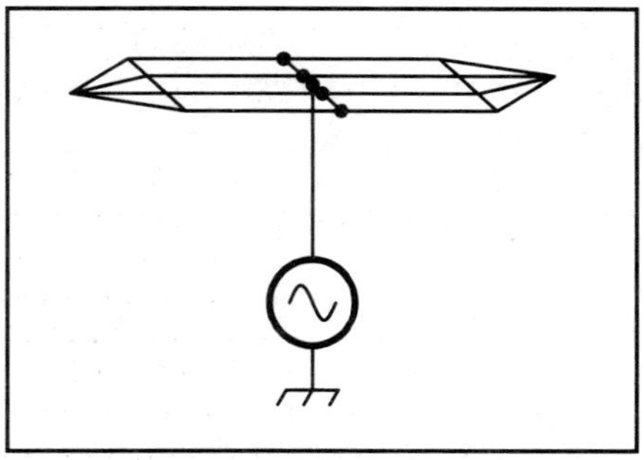

Figure 6—An early capacitively loaded vertical antenna.

In the second example, the top-hat wires are sloped down 25° below horizontal. This might occur when the vertical element supports the top hat. The convenience comes at a price: R_R drops to 3.8 Ω; efficiency drops to 43.2%, and the top-hat wires grow to 82 feet. Thus, a drooping top hat has two undesirable side effects: It decreases R_R and increases top-hat wire length. This is similar to what occurred with the top hat below the top of the antenna.

I have seen top-loaded designs with the top-hat wires steeply angled to the ground. The top-hat wire length requirement gets increasingly worse, and R_R drops to *very* low values; usually less than 2 Ω. Therefore, if possible, keep the wire ends away from the ground. Higher is better, even if the antenna takes on a **Y** shape.

Why does R_R drop when the top-hat wire ends approach the ground? Here's one way to look at it: The average height of a top-hat wire is lower when the wire slopes downward. For example, if a top hat attaches to the vertical at 45 feet and the ends are at 15 feet, the average height is 30 feet. Just as when the top hat was lower on the vertical, there is a corresponding decrease in R_R. I have found a reasonably close correlation of R_R values for horizontally positioned top hats and sloping top hats when the average heights are equal.

A sloping top-hat wire does not choke off the top portion of the antenna, but affects it differently. As the wire ends move downward, the top-hat currents are no longer horizontal, so they have both horizontal *and vertical* components. The vertical components flow in the opposite direction of the current in the main vertical element, so these currents produce a field that is counter to, and cancels some of, the main electromagnetic field. This is equivalent to removing part of the element's length, and that brings the undesirable effects.

Both of these top-loading examples require a wide top-hat span that you may not be able to accommodate. We can reduce their lengths substantially by adding more wires to the top hat.

Our third example has four equal-length wires situated in a horizontal plane. The wire length for resonance is now only 55 feet, and there is nearly no change in R_R from the first flattop example. I have found that the wires need not be exactly 90° apart. Moving each set of two closer together requires only minor lengthening of the wires. As a matter of fact, they can all be rotated toward one side, but their currents no longer cancel, and the antenna shows both vertical and horizontal radiation, much like an inverted **L**.

Let's continue working with a single center support. In example four, the four top-hat wires slope down 25° from the horizontal. This requires a small lengthening

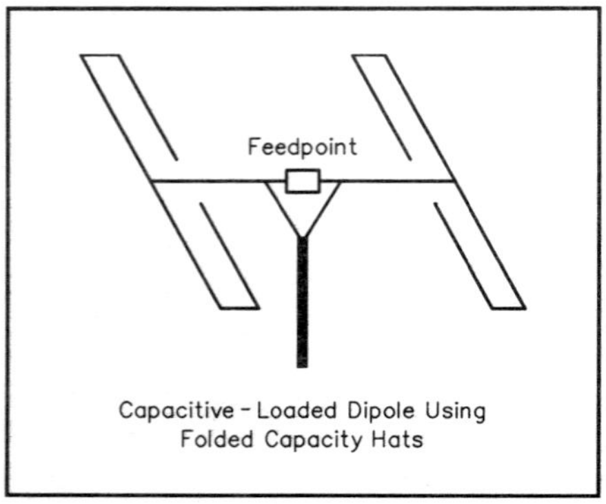

Figure 7—A capacitively loaded dipole using folded capacity hats.

(to 59.5 feet) to regain resonance, and R_R drops to 5 Ω. Since the average height of four sloping wires is higher than that of the two wire example, R_R decreases less, and the efficiency is better.

In example five, a loop of wire connects the four top-hat wires at their ends. The top hat is now taking on the form of a traditional skeletal top hat. The four wires emanating from the vertical element are much reduced in length (29.5 feet), and R_R is 9.9 Ω, as for flattop. This demonstrates an important point: Connecting the ends of the top-hat wires yields a very substantial gain in capacitance. Tying the ends together and adding more "spoke" wires produces a structure of ever-increasing capacitance, which can continue to form a solid plate. Such increased top-hat area allows us to significantly reduce the top-hat dimensions.

Example six shows the effect of allowing the whole skeletal top hat to droop by 25°. Once again, there is a small increase in top-hat wire lengths, but the radiation resistance drops to 5.9 Ω. So we're approaching good efficiency while using a single center support.

Because long top-hat spans won't fit on small home lots, I searched for a way to maximize capacitance with a simple flattop structure. The result is the folded flat top of example seven. The top-hat wires are slightly longer than half the full length, and fold back in their original direction. Wire spacing does not appear to be particularly critical; it can be from inches to feet. The main effect of the spacing is some variation in overall length requirement. One small peculiarity that I have observed is an end effect. If the ends of the top-hat come too close to the vertical radiator, antenna tuning becomes somewhat unpredictable. Leave some space between the end of the top-hat wire and the vertical element (3 or 4 feet for this example) to keep the tuning predictable. Table 2 shows that the folded flattop R_R is nearly identical to that of the full-size flattop, so the efficiencies are comparable.

I do not have two supports available at my location. That is why I'm interested in single-support scenarios. So, just as in the former examples, I brought the ends of the folded flattop down about 25° from horizontal. As seen in example eight, this reduced R_R to about one-half of the folded flattop value, to 5.5 Ω. This is comparable with the R_R of the four-wire top hat in example six. Considering my constraints, this configuration provides an acceptable compromise between installation and efficiency.

It should be apparent by now that capacitive loading can provide a reduced size antenna with good efficiency. Our radio forefathers knew what they were doing.

Center-fed antennas, such as dipoles, can be capacitively loaded at their ends, and are good candidates for locations where full-sized antennas won't fit. One possible configuration is the end-loaded dipole shown in Figure 7. It's like two verticals with folded top hats mounted back to back. Some signal cancellation will occur, but the efficiency should be high.

Summary

By no means restrict your antennas to one type of loading; both inductive and capacitive loading can be combined to suit your particular needs. The loading may be conventional coils and top hats, or even sections of transmission lines and folded wires that exhibit the desired loading characteristics.

The possibilities for new, smaller antennas are limited only by our imaginations. The fundamentals I've presented should clearly show that a little forethought can make a reduced-size antenna perform well.

Notes

[1]The radiation resistance is the equivalent resistance that would dissipate the power the antenna radiates, with a current flowing in it equal to the antenna current at a current loop (maximum). The loss resistance is the total of all ohmic losses in the antenna system, including any loading networks and any ground return path. The loss resistance of a half-wave antenna is ordinarily small, compared with the radiation resistance, and can usually be neglected for practical purposes.

[2]Although the author chooses to restore resonance, that step is not required if transmitter can tolerate the reactive load or if a reactive network performs the impedance matching. (We can usually design reactive networks to accommodate antenna reactance.)—*Ed.*

[3]John S. Belrose, "Short Antennas for Mobile Operation," *QST*, Sep 1953, pp 30-35.

[4]*The ARRL Antenna Book* (18th Ed., ARRL Order #6133). Table 1 (p 3-3) indicates that 60 0.2 λ radials yield 5 Ω ground loss and 1 dB power loss at low elevation angles with a λ/4 radiating element; 24 λ/8 radials yield 10 Ω ground loss and 2 dB power loss. See Note 8 for references, Note 5 for ordering information.

[5]ARRL publications are available from your local ARRL dealer or directly from ARRL. Mail orders to Pub Sales Dept, ARRL, 225 Main St, Newington, CT 06111-1494. You can call us toll-free at tel 888-277-5289; fax your order to 860-594-0303; or send e-mail to **pubsales@arrl.org**. Check out the full ARRL publications line on the World Wide Web at **http://www.arrl.org/catalog**.

[6]Amidon Inc, 3122 Alpine Ave, Santa Ana, CA 92799; tel 714-850-4660, fax 714-850-1163.

[7]Carl Eichenauer, "A Top-Fed Vertical for 1.8 MHz—Plus 3," *QST*, Sep 1983, pg 25-27.

[8]*The ARRL Antenna Book* (18th Ed.) p 3-2, discusses the interaction of radial quantity, length and radiator length. Table 1 (p 3-3) indicates that 90 λ/4 radials provide 2 Ω of ground loss for 0.5 dB power loss at low elevation angles with a λ/4 radiating element. Twenty-four λ/8 radials yield 10 Ω ground loss and 2 dB power loss. Table 1 is after J. Stanley's "Optimum Ground Systems for Vertical Antennas" (*QST*, Dec 1976, pp 13-15), which draws heavily from *Radio Broadcast Ground Systems*, available from Smith Electronics Inc, 8200 Snowville Rd, Cleveland OH 44141.

[9]Jerry Sevick, "Short Ground-Radial Systems for Short Verticals," *QST*, Apr 1978, pp 30-33.

[10]Al Christman, " Ground Systems for Vertical Antennas," *Ham Radio*, Aug 1979, pp 31-33.

[11]Ibid.

[12]Al Christman, "More on Elevated Radials," Technical Correspondence, *QST*, Mar 1993, p 72.

[13]*EZNEC* is available directly from Roy Lewallen, W7EL, PO Box 6658, Beaverton, OR 97007; tel 503-646-2885, fax 503-671-9046; e-mail **w7el@teleport.com**; URL **ftp://ftp.teleport.com/vendors/w7el/.**

[14]*The ARRL Antenna Book* (18th Ed.) p 3-10, Fig 12. For ordering information, see Note 5.

[15]Attempts to mount the dipole vertically would open a new complex issue. The dipole is 88 feet long. Its ends are high-voltage points that must be well clear of the ground for safety. Hence, we need a support that is nearly 100 feet tall. To maintain the antenna pattern, low-impedance feed lines must leave the antenna center perpendicular to the antenna or coaxially with the antenna. The first option requires feed-line support high in the air. The second option is difficult to construct. End feeding solves these problems but requires feed-line isolation and a matching network, which adds losses.—*Ed.*

[16]Figure 38B, *The ARRL Antenna Book*, 17th Ed.,1994, pg 4-23. *This illustration does not carry over to the 18th edition.*

[17]George Kennedy, *Electronic Communications Systems*, (New York: McGraw-Hill, 1970), p 369.

[18]You can download the *EZNEC* antenna definitions package from the ARRL "Hiram BBS" (tel 860-594-0306), or the ARRL Internet ftp site: **oak.oakland.edu** (in the **pub/hamradio/arrl/qst-binaries** directory). In either case, look for the file KUEHL.ZIP.

Thomas, AC7A, was first licensed in 1969 at age 15, with the call WN6BQI. Early on, he developed an appetite for the technical side of the hobby that remains today. A graduate of Cal-Poly San Luis Obispo, he has spent the better part of the last 19 years employed by Burr-Brown Corporation as a Product Engineer. When not assisting his wife with raising their three children, you may find him chasing DX on the low bands or pruning antenna wires in the backyard. You can reach Thomas at 620 S Avenida Princesa, Tucson, AZ 85748; e-mail ***Kuehl_Tom@BBrown.com.***

By Brian V. Cake, KF2YN

From *QST*, April 2004

The "C Pole"—A Ground Independent Vertical Antenna

KF2YN takes the vertical to new heights with this folded design that doesn't require a counterpoise.

When I moved to my new home on the coast of northeast Florida, it was into a deed-restricted community, where "unsightly antennas" were forbidden. I enjoy occasional operation on the HF bands (principally 14 MHz and above) and the location was just begging for the use of vertical antennas, where the proximity to the water would help with good low radiation angles. The verticals could be hidden in the upper deck support structure and everybody would be happy, including my wife.

Unfortunately, the old saw about vertical antennas radiating equally poorly in all directions has a lot of truth to it and losses in the ground system can eat up much of your power. I made the mistake of attempting to measure the ground conductivity in my backyard. That was after I compared the on-air performance of a vertical half-wave dipole for 10 meters with a simple quarter-wave vertical with no radials. I was shocked at the quarter-wave vertical performance. I was even more shocked when I measured 30 kΩ between deep rods spaced 2 feet apart in my back yard.

Conventional solutions to this problem involve the use of radials or counterpoises, but I didn't want to sprinkle the lawn with wires. A full-size vertical dipole, at 30 plus feet for 20 meters, is too high for this location.

With all of these considerations in mind I went looking for another solution, and found an interesting configuration. It is ground independent, has a ground-level 50 Ω feed point, is less than half the height of a full-size half wave dipole, is very efficient, and has a 2:1 SWR bandwidth of about 3 percent. It can be suspended from any convenient support, rolls up into a tiny space and makes a good Field Day antenna.

Basics

The antenna consists of a vertical half-wave dipole that has been folded virtually in half, as shown in Figure 1. By erecting this just above ground level the ground currents are reduced dramatically over those of a quarter-wave ground-mounted monopole. The H-plane radiation pattern for this antenna is virtually omnidirectional.

As shown in Figure 1, the antenna is symmetrical about the feed point and is known as an open folded dipole. The feed-point impedance can be altered by changing the ratio of the diameters of the vertical wires. My intention, however, was to use suspended wire as the elements. The antenna can be analyzed in much the same way as a conventional folded dipole, and it turns out that it can be treated as a short dipole loaded by means of a short-circuited length of transmission line. I decided to take the easy way out and model it using *EZNEC*, however.

There are two practical problems with the antenna in Figure 1: The feed-point impedance is too low and the feed point is in the wrong place. The feed-point impedance depends on the geometry, but for spacing between the vertical wires of about 20 inches on 15 meters the feed-

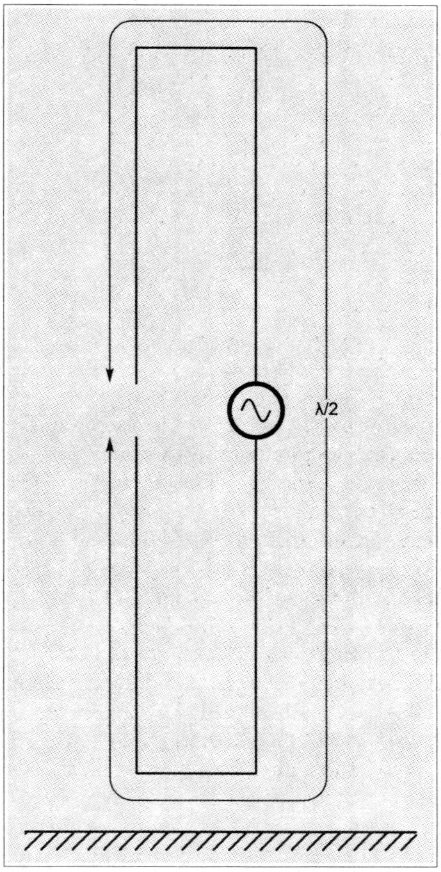

Figure 1—A vertical half wave dipole bent virtually in half.

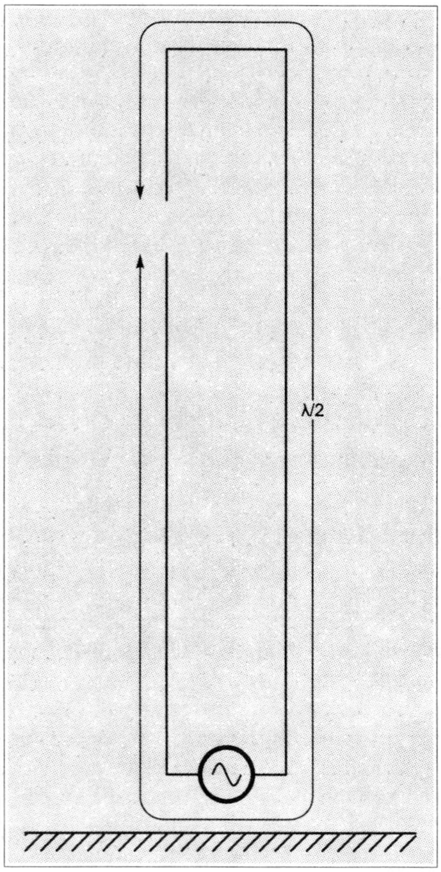

Figure 2—The bent dipole with a shifted feed point.

Table 1
Dimensions (in Inches) of the Modeled Antennas
Wire diameter is 1/16-inch.
Height of lower horizontal wire is 12 to 24 inches (non-critical).
See Figure 3 for dimensional details.

Band (Meters)	*2:1 SWR Bandwidth (kHz)*	*Dimension* A	*Dimension* B	*Dimension* C	*Dimension* D	*Dimension* E
20	400	177	85	84	8	40
17	540	137	66	67	4	31
15	600	124	60	60	4	20
12	800	100	53	43	4	23
10	800	87	46	37	4	20

point impedance is about 25 Ω. This has to be transformed up to 50 Ω. Also, it is highly desirable to have the feed point at ground level, since otherwise the feed cable has to be dressed away from the antenna such that currents are not induced into the feeder. These currents can lead to undesirable effects of RF in the shack and a modification of the radiation pattern. A ground-level feed point is nicer too, because the cable can be buried a short distance under the lawn.

Both of these problems can be fixed by rearranging the antenna as shown in Figure 2. Moving the feedpoint away from the voltage node at the antenna center increases the feed point impedance and an exact match to 50 Ω can be obtained by shifting the position of the gap at the dipole ends. Unfortunately, doing this places the feed point at a position where there is a substantial common-mode potential. That is to say, the two antenna feed-point terminals have the same potential on them relative to ground (in addition to the normal differential potential across the feed point), and this potential can be several hundred volts for an input power level of 100 W. If the coax is connected directly to the feed point, the natural resonance of the antenna is destroyed and it becomes useless. There are several ways to solve this problem, including the use of an inductively coupled loop, but I chose to use a balun.

The Balun

The only problem with the balun is that it has to work with a high common-mode potential at the feed point and this can lead to trouble. In particular, some baluns that use ferrite cores can cause power loss and intermodulation distortion under conditions of high common-mode potential. This fact is not emphasized in the balun literature, but is important for all antennas with a feed point that is not at a voltage node, such as unbalanced dipoles, off center fed dipoles and multiband long wires. I have designed two different baluns for these antennas:

1) A simple air-core balun consisting of 60 turns of RG-58/U close-wound on a 2 inch diameter length of PVC pipe (about 33 feet of RG-58/U total) provides excellent choking action and reduces the line current to about 1/10 of the feed point current. This will work fine from 14 MHz to 30 MHz, but soaks up a fair bit of power, mostly in cable losses. The total losses are about 14% (about 0.6 dB) on 20 meters and rise to about 18% (about 0.8 dB) on 10 meters. This balun does have the advantage that a quick trip to RadioShack and your local hardware store can provide the materials you need.

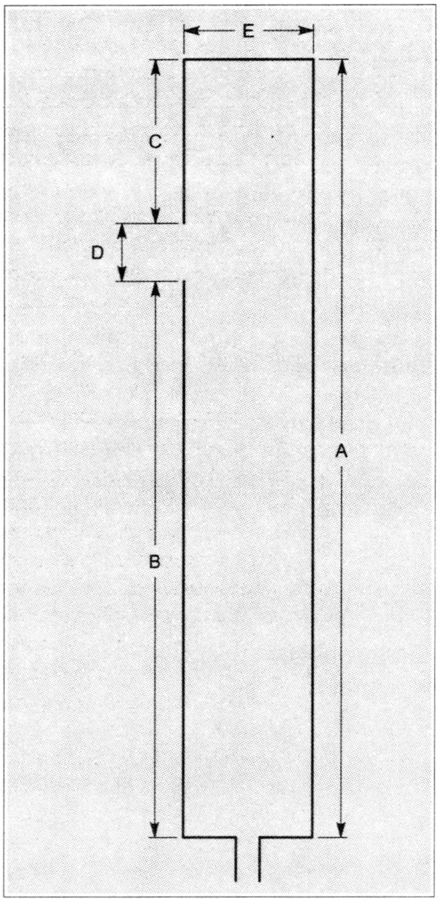

Figure 3—Dimensional details of the antennas. See Table 1.

2) An alternative design using ferrite toroids reduces the power loss by over a factor of two to <7% (0.3 dB) on all bands. Two different designs, one for 20 meters and the other serving 17 meters and above, are needed in order to keep the core power loss low. The 20 meter balun consists of 19 turns of RG-174/U coax on an FT-240-61 core. For higher frequencies use 15 turns of RG-174/U on an FT-240-67 core. It is possible that a close-spaced winding of the same number of turns of 14 gauge or similar wire will give lower loss than the RG-174/U and will also handle higher power, but I have not tried it.

Antenna Construction

The dimensions for the 20, 17, 15, 12 and 10 meter bands are shown in Table 1. Refer to Figure 3 for the dimensional key. You will note that the spacing between the vertical wires is 20 inches for 15 and 10 meters, but 40 inches for 20 meters. This is because I wanted to squeeze the antenna into available space. Twenty inches would also work fine for 20 meters as long as you adjust the vertical length of the wires.

The dimensions given are those used in the models. Note that the actual dimensions will vary from these. This is because of the effects of support structure, the proximity to objects nearby and the effects of ground on the feed point impedance. My antennas for both 15 and 20 meters resonated too low in frequency and had to be shortened quite a bit. This was partly because I used PVC covered wire and partly because they were hung close to the upper deck. The dielectric constant of the wood reduced the antenna's resonant frequency.

The antenna element can be made of anything from thin, PVC covered copper wire up through aluminum tubing, with suitable small changes in the element lengths. Thin wire will have higher losses than fat wire and will also have a higher common mode potential at the feed point. On 15 meters the loss attributable to the element resistance is 0.57 dB (14% power loss) when 32 mil (20 gauge) copper wire is used. This drops to 0.3 dB (7% loss) when using 1/16 inch copper, and to 0.15 dB (3.5% loss) for 1/8 inch copper. I used PVC covered wire because bare copper wire looks ugly after exposure to a beach atmosphere for even a short while. Also note that the balun choke reactance will change the resonant frequency of the antenna somewhat. An inductance of 25 μH on 20 meters will shift the resonant frequency upward by about 1.7%, or 250 kHz.

Assembly

The assembly method I used is shown in Figure 4. Here, suitable lengths of 3/4 inch schedule-40 PVC pipe are used as the top and bottom spreaders, with the element wire simply pushed through the tubes. The spacer in the gap is a 6 inch piece of the same tubing, with holes drilled right through at 4 inch spacing (10 inches and 8 inches respectively for the 20 meter version). Start construction by cutting each piece of the element wire to the dimensions shown, plus 2 feet or so. This additional length allows for securing the wires to the spacer and for adjustment of the resonant frequency and of the SWR. Make the spreaders and the spacer as shown. Drill suitable holes in the top spreader for a suspension cord, if that is the way you are supporting it. Drill two holes near the center of the lower spacer to allow the wires to exit the spreader at the feed point. It is also advisable to drill a few holes in the spreaders to allow water to exit.

Lay the spreaders out in their approximate positions on the ground and thread the wires through them. Temporarily secure the wires at the feed point with tape. Make the balun and solder the feed point wires to the balun cable. Once you have made any adjustments necessary you should seal these joints against the ingress of water. Hoist the antenna into position, and pull on either vertical wire in order to get the lower spreader horizontal. The base of the antenna can be anything from 1 foot or so off the ground to as high as desired. The ground rod shown is not essential—it is useful for anchoring the balun so the base of the antenna doesn't flap in the breeze.

You can now test for resonant frequency and SWR. It is unlikely that you will get it exactly right the first time but, if you are using bare copper wire and the antenna is in the clear, then with the dimensions given it should be pretty close. If the resonant frequency is too low, lower the antenna (or use a stepladder), untwist the wires above and below the spacer, and shorten the element by an equal amount on either side of the spacer, then retwist the wires.

Once the resonant frequency is right, check the SWR. If your SWR meter indicates that the feed-point resistance is too high, then it is necessary to raise the position of the spacer. This is easily done by untwisting the wires, and moving the spacer farther up the top wire. Be sure to keep the total wire length unchanged, or the resonant frequency will shift. Then retwist. A low feed-point resistance will require the spacer to be lowered. If your SWR meter does not indicate whether the resistance is high or low, you will have to guess which way to move the spacer.

This sounds like an arduous setup procedure but it is actually very quick and easy to do, especially where the spacer is easily accessible from the ground. With the antenna vertical it should be possible to get an SWR of very close to 1:1 on your favorite frequency. If you can't get the SWR down below 1.5:1 then suspect that something is wrong. Also, once you have the antenna in position, check the 2:1 SWR bandwidth. It should be roughly that shown in the dimension table. If it is substantially wider than this, you should suspect that losses are higher than they should be. Finally, when all appears well, find an unused frequency and apply power for a minute or so, then check that the balun is not getting hot. This is a big advantage of a ground-mounted feed point!

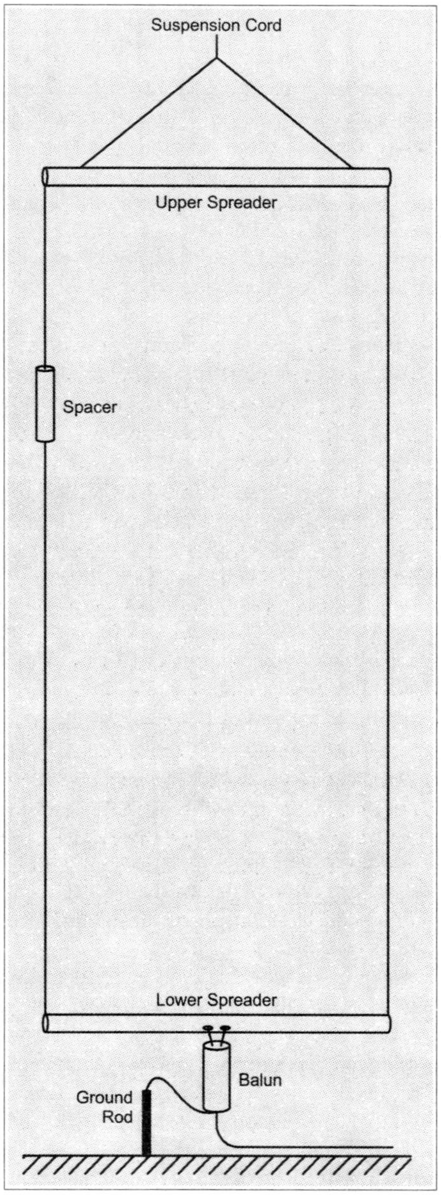

Figure 4—The antenna constructional details.

Additional Notes

It is likely that a "sloper" version will work well and it is obviously easy to support, but I have not tried it.

If you are running more than 100 W out I suggest you use air-core baluns. As pointed out earlier, ferrite-core baluns can be problematic in situations where there is considerable common-mode potential. The hysteresis losses are strongly dependent on frequency and on the flux density. For a given frequency, the hysteresis losses are proportional to somewhere between the square and the cube of the flux density. This can cause distortion and heating problems. The ferrite-cored baluns described here will loaf along at 100 W but I wouldn't recommend that you go too much above that.

The name I came up with for this antenna is the "C pole," because of its shape and because the popular "J pole" is so-named for its resemblance to the shape of that letter.

Does it Work?

The evidence I have that this is an effective antenna is part scientific and part observational. The scientific part refers to the observation that the antenna Q, as measured by the 2:1 SWR bandwidth, is as expected, and the expected Q includes the effects of identifiable loss mechanisms. The observational part is simple—it is very easy to work DX stations, even though I never run more than 100 W out. Part of that is no doubt because of my excellent location, but the antenna does play a significant role. I'm sure you can tell when you are using a good antenna system—with an effective antenna, operating is a pleasure, not a struggle. That is the way it has been with this antenna design, and if you put it together carefully I'm sure you will get the same enjoyment out of it that I have.

Brian Cake, KF2YN, was born in England and first licensed as G8AFH. He received an EE degree from City University, London. Brian is now retired after 15 years as Chief Technical Officer at LeCroy Corporation, where he was manager of the Advanced Development Group. He now lives on the Matanzas Inlet, near St Augustine, Florida where he designs and tests new antenna ideas and builds miniature live-steam locomotives. Brian can be reached at 248 Barrataria Dr, St Augustine, FL 32080; **bcake@bellsouth.net**.

By Al Christman, K3LC

From *QST*, October 2000

The Parasol: A 160-Meter Vertical Antenna with a Difference

This shorter, grounded Top-Band radiator can be scaled for use on other frequencies.

This article describes an unusual vertical-antenna design having several features that should be attractive to hams. First, its height is about half that of a conventional $^1/_4$-λ monopole. Second, the vertical portion of the antenna is grounded at its lower end, which eliminates the need for a base insulator. Third, the feedpoint is at the *top* of the structure, which means that virtually the entire length of the tower is available for other purposes. Two of these antennas have already been built and operated with satisfactory results on the AM broadcast band in northwestern Pennsylvania. Here I'll describe 160-meter band applications of the antenna, but you can easily adapt the design to other frequencies by scaling the dimensions accordingly. To construct the computer models shown in this article, I used *EZNEC Pro*,[1] with a *NEC-4* calculating engine.[2]

Basic Concept

I christened this antenna the "Parasol," because of its shape. It consists of three basic parts: the vertical portion, the ground system and the parasol section. At the low-frequency end of the spectrum, the vertical portion would probably consist of tower sections, but at higher frequencies, aluminum tubing, solid rods, or even wire can be used. The physical height of the vertical portion is typically on the order of $^1/_8$ λ, although somewhat longer or shorter dimensions can be used. In general, the gain increases with increasing height. The vertical section is grounded at the bottom, so no base insulator is needed, but one or more driven ground rods may be added for lightning protection. If metallic guys are used, break them into nonresonant lengths with insulators and insulate them at their tower-attachment points.

To minimize losses, a good ground system is needed, and the classical AM-broadcast configuration of 120 $^1/_4$-λ buried radials works well. Radials may be laid on the ground surface if burying them is not an option. Although the exact dimensions of the ground screen aren't critical, performance improves as the number and/or length of the radials is increased. As an alternative, gull-wing elevated radials[3,4,5] may be used with the Parasol antenna. Connect these bent radials to the grounded tower base, sloping them outward and upward from that point until they reach their final elevation (perhaps 15% of the tower height), where they are bent into a horizontal position for the remainder of their length. A slope angle of 45° is fine, but steeper or shallower angles will also work. Typically, eight of these gull-wing radials provide satisfactory performance if they are equally spaced around the tower and are long enough—a span of $^1/_4$ λ for each radial is good. Keeping all radial lengths equal helps ensure a circular azimuthal radiation pattern. Elevated radials that are horizontal throughout their entire length can also be used, but this appears to reduce the gain slightly; such radials would not be bent at their inner ends, but would be electrically connected to the tower at some distance above the ground.

At the apex of the tower, several downward sloping parasol wires are added, which serve as the ungrounded part of the antenna. Three such wires are generally sufficient, although more can be used. If only two parasol wires are installed, the azimuthal-plane radiation pattern becomes less circular. All of the parasol wires are anchored to, but insulated from, the tower top, and should be spaced symmetrically in azimuth around the tower. The wire lengths can be chosen arbitrarily, or they may be pruned in the field to achieve a feedpoint impedance that is purely resistive (ie, zero reactance). Computer simulation indicates that resonance can be achieved when these three parasol wires are shortened to less than $^1/_8$ λ, although this has not been verified experimentally. The outer (lower) ends of the parasol wires should be connected to their guy cables with insulators. The guying material may be nonmetallic, or conductive (steel) suitably segmented with additional insulators.

Generally, the antenna gain increases as the parasol conductors are made more horizontal. Therefore, make the downward slope angle of these elements as small as possible, within the limitations of the property. Instead of using sloping wires for the parasol conductors, a horizontal structure composed of metal tubing could be built. Such a top hat would resemble a wheel with metal spokes, whose outer ends can also be joined together by a perimeter conductor. As the number of spokes is increased, their resonant length decreases, thus shrinking the required dimensions of the parasol structure.

No matter which parasol design is used, the center of this symmetrical assembly must be located at the tower apex, but insulated from it. If sloping parasol wires are used, then equal-length, insulated pigtail leads should extend from the upper ends of these wires to a common point, where they are bonded to the center conductor of the transmission line. If the tower has a top plate, a suitable chassis connector for the transmission line may be installed in the center of the plate, facing downward, to terminate the

[1]Notes appear on page 42.

feeder. The pigtail leads would extend upward from the inner ends of the parasol wires and be soldered to the connector's center pin. The outer shield of the feed line, of course, is bonded (via the connector) to the metal tower itself, thus completing the electrical circuit.

Secure the transmission line to the tower throughout its length. It may be helpful to electrically bond the line's outer shield to the tower at ground level, midway up and at the apex. Ideally, the feed line should be run on the *inside* of the tower, but this may not be feasible. If a horizontal parasol top hat is employed, its hub must be placed on an insulator located at the tower apex, forming the feedpoint of the antenna. Typically, a section of the desired transmission line would be installed which is long enough to reach from the tower-top feedpoint to the ground. At the lower end of this line, a simple L-network can then be used to match the resulting input impedance to that of the station coax. As an alternative, an impedance-matching network could be placed at the feedpoint at the tower apex.

No input impedances have yet been measured on existing antennas, but depending on the system parameters, resistance values in the range of 15 to 60 Ω have been obtained by computer analysis. Resistances of 50 to 60 Ω are indicated when the tower height is around 0.15 λ. The input reactances predicted by *EZNEC* vary widely, depending not only upon the length of the parasol wires, but also upon the assumed diameters for the various conductors. Overall, increasing the tower height and/or making the parasol conductors more horizontal will yield increases in antenna gain and input resistance along with a decrease in the resonant length of the parasol conductors.

Reference Antennas

To serve as a standard against which the performance of several Parasol models could be compared, I computer-simulated an AM broadcast-style, ground-mounted, ¼-λ vertical monopole antenna for use on 160 meters. This reference antenna was placed over real earth with a soil conductivity of 0.005 Siemens/meter and a dielectric constant of 13. The vertical element is constructed from two-inch diameter tubing, while the radials are made of #12 AWG wires. The ground system consists of 120 radials buried six inches; a five-foot-long, ½-inch diameter ground rod is also included. Except for the ground rod, each conductor was specified to be 134.368 feet long, which is a ¼ λ at 1830 kHz. *EZNEC-4* predicts a peak gain of 1.28 dBi at a takeoff angle of 23° for this antenna and an input impedance of 40.83 + *j*22.89 Ω.

Because of software limitations, all of the metallic elements in any particular computer model must be composed of the same material. Since many Top-Band ham antennas are made from galvanized steel tower sections and copper wires, I decided that each *EZNEC-4* antenna simulation should be constructed entirely of aluminum, which has a conductivity between that of copper and steel.

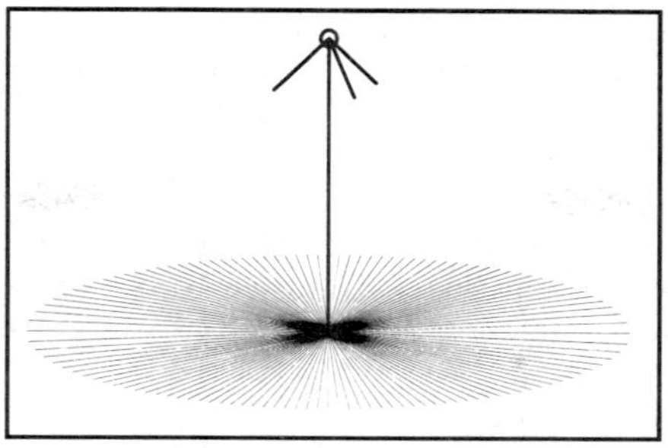

Figure 1—A Parasol antenna composed of a 70-foot tower with 120 buried ¼-λ radials. The three parasol wires are each 46.533 feet long, and slope downward at an angle of 30° below the horizon. The small circle at the apex of the tower denotes the feedpoint.

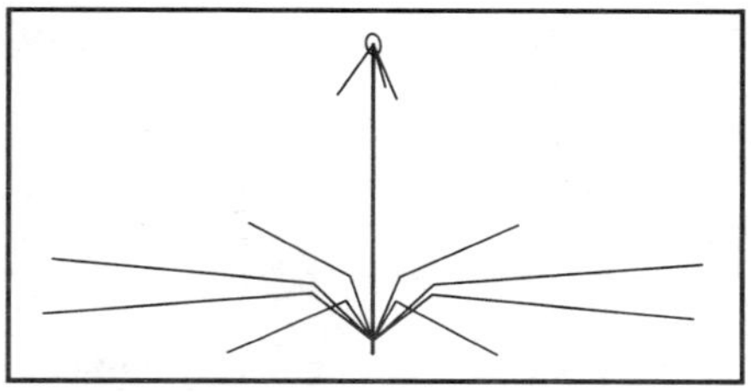

Figure 2—A Parasol antenna which again utilizes a 70-foot tower, along with three 51.894-foot parasol wires sloping downward at 30° below the horizon. The ground system is composed of eight gull-wing radials whose wing span is 134.4 feet (¼ λ at 1830 kHz). These bent radials slope upward from ground level at an angle of 45° until reaching a height of 12 feet, at which point they become horizontal for the remainder of their length. The small circle at the apex of the tower denotes the feedpoint.

I also modeled a typical elevated vertical antenna. I combined a two-inch-diameter monopole with four horizontal #12 AWG radials, with the base of the antenna 30 feet above ground. All wire lengths were set to 134.368 feet, or ¼ λ at the operating frequency of 1830 kHz. Now the gain slips to 1.12 dBi at an elevation angle of 19°, while the feedpoint impedance becomes 34.18 + *j*1.99 Ω.

Computer Models of Parasol Antennas with Quarter-Wave Radials

I computer-simulated several versions of the Parasol to determine the effects of changes in the ground system and to see how they would compare with the reference antennas mentioned earlier. In each case, the apex of the structure was fixed at 70 feet, which is just over ⅛ λ at 1830 kHz. The tower is made from a two-inch-diameter aluminum conductor, and there is also a five-foot-long, ½-inch-diameter aluminum ground rod.

The inner ends of three aluminum #12 AWG parasol wires were placed at the top of the tower and spaced uniformly around its axis at 120° intervals. From the tower apex, each of these wires slopes toward the ground at an angle of 30° below the horizontal. The feedpoint is between the junction of the three parasol wires (coax center conductor) and the top of the tower (coax outer conductor).

For the first Parasol model, an extensive ground system of 120 buried ¼ λ (134.368 feet) radials was used, just as for the initial reference antenna described earlier (see Figure 1)The three parasol wires were trimmed to achieve resonance, obtained when each one measured 46.533 feet. The resulting gain was 1.26 dBi at a takeoff angle of 24°, and the input resistance was 28.62 Ω. The azimuthal-plane radiation pattern is highly circular, with a maximum gain variation of less than 0.01 dB at any compass heading.

The second Parasol model is similar to the first, with the exception of the ground system. If an extensive number of buried or on-the-ground radials cannot be installed, the Parasol may be equipped with gull-wing elevated radials. In this example, four #12 AWG radials were attached to the tower-base at ground level, and spaced 90° apart. These wires slope upward at an angle of 45° until reaching a height of 12 feet. The radials then extend outward horizontally at a height of 12 feet until they reach a maximum distance of 134.368 feet (¼ λ) from the tower axis. Because they are bent, the physical length of these wires is a bit more than ¼ λ. Resonance occurred when the length of the parasol wires was 51.595 feet, at which point the maximum antenna gain was 0.51 dBi at an elevation angle of 24°. With only four radials, the azimuthal-plane gain varies from 0.48 to 0.51 dBi, for a noncircularity of 0.03 dBThe input resistance is 16.59 Ω.

The third Parasol model uses eight gull-wing radials (see Figure 2) instead of the four that were included in version two mentioned earlier. Now the elevated radials are spaced 45° apart, but their bent configuration and length are the same as before. The vertical tower and ground rod are also unchanged, but the resonant length of the three sloping parasol wires increases slightly to 51.894 feet. Now the peak gain rises to 0.87 dBi at 23° take-off angle, while the input resistance is 15.16 Ω. Doubling the number of elevated gull-wing radials from four to eight increases the maximum gain by 0.36 dB, and also reduces the noncircularity in the azimuthal-plane radiation pattern to 0.01 dB.

The final Parasol model in this section again includes eight ¼-λ radials, but now they are completely horizontal throughout their length. The inner ends of these eight wires are connected to the vertical portion of the parasol 12 feet above the ground and, of course, they're uniformly spaced at 45° intervals around the tower. To achieve resonance here, the length of the parasol wires had to be stretched to 60.012 feet, at which point the gain was found to be 0.76 dBi at an elevation angle of 23°. The input resistance is just 12.44 Ω, and the noncircularity of the radiation pattern in the azimuthal-plane is 0.03 dB.

Table 1

Comparisons of power-gain and input-impedance values for two reference antennas and four versions of the Parasol antenna, all of which employ full-size quarter-wave radials. Each Parasol configuration includes three wires that slope downward from the tower apex at an angle of 30° below the horizon. The length of the parasol wires was adjusted to achieve resonance in each case. At the operating frequency of 1830 kHz, a quarter-wavelength is equal to 134.368 feet.

	Power Gain (dBi)	Take-Off Angle (Degrees)	Feedpoint Impedance (Ohms)
Ground-mounted, 1/4-λ, base-fed vertical monopole with 120 1/4λ radials	1.28	23	40.83 + *j*22.89
Elevated (H=30 feet), 1/4-λ, base-fed vertical monopole with four 1/4λ radials	1.12	19	34.18 + *j*1.99
70-foot Parasol, 120 buried 1/4-λ radials; 46.533-foot parasol wires	1.26	24	28.62 + *j*0.0006
70-foot Parasol, eight 1/4-λ elevated gull-wing radials; 51.595-foot parasol wires	0.51	24	16.59 + *j*0.0034
70-foot Parasol, eight 1/4-λ elevated gull-wing radials; 51.894-foot parasol wires	0.87	23	15.16 + *j*0.0019
70-foot Parasol, eight 1/4-λ horizontal elevated radials; 60.012-foot parasol wires	0.76	23	12.44 – *j*0.0044

Table 2

Comparisons of power-gain and input-impedance values for one reference antenna and four versions of the Parasol antenna, all of which employ radials that are only 70 feet long (just over 1/8 λ). Each Parasol configuration (with one exception noted by the asterisk [*]) uses three wires that slope downward from the tower apex at an angle of 30° below the horizon. The length of the parasol wires was adjusted to achieve resonance in each case. At the operating frequency of 1830 kHz, a quarter-wavelength is 134.368 feet.

	Power Gain (dBi)	Take-Off Angle (Degrees)	Feedpoint Impedance (Ohms)
Ground-mounted 1/4-λ, base-fed vertical monopole with 120 buried 70-foot radials	0.79	22	45.05 + *j*20.35
70-foot Parasol with 120 buried 70-foot radials; 46.758-foot parasol wires	0.59	24	33.15 + *j*0.0036
***70-foot Parasol with 120 buried 70-foot radials; parasol wires are 50.478 feet long, with a down-sloping angle of 45°**	0.24	24	27.93 – *j*0.0037
70-foot Parasol with eight 70-foot gull-wing, inductively loaded elevated radials; 51.397-foot parasol wires	–0.43	25	22.90 – *j*0.0049
70-foot Parasol with eight 70-foot, totally horizontal, inductively loaded elevated radials (parasol wires are 61.69 feet long)	–1.15	25	20.24 + *j*0.0069

Table 1 summarizes the results thus far, and a number of observations can be made. First, all four versions of the Parasol antenna generate less gain than the classic 1/4-λ base-fed vertical monopole with a traditional ground system of 120 buried 1/4-λ radials. However, the performance of a Parasol with the same extensive ground system falls short of the reference antenna by only a small amount, and it actually exceeds the gain of an elevated vertical with full-size 1/4-λ radials and monopole. Second, a Parasol with eight 1/4-λ gull-wing elevated radials is only about 0.4 dB behind the AM-broadcast-style antenna. Surprisingly, radials that are completely horizontal don't do quite as well as those configured as gull wings. Lastly, the Parasol with many buried radials has a feedpoint resistance approaching 30 Ω, while those using gull-wing radials have values for R_{in} that are only about half as great.

Computer Models of Parasol Antennas with Shortened Radials

Next, let's take a look at what happens to the performance of a Parasol antenna when its radials are truncated to lengths less than 1/4 λ. Because the Parasol in these Top-Band examples is only 70 feet high (just over 1/8 λ), we can use computer analysis to learn how well this antenna works when the radials are also 70 feet long. For our reference antenna, a full-size 1/4-λ base-fed vertical element will be combined with a ground system of 120 buried 70-foot radials. The gain of this antenna is found to be 0.79 dBi at 22° take-off angle, while the feedpoint impedance is 45.05 + *j* 20.35 Ω.

If a 70-foot-tall Parasol is teamed with the same type of ground system described earlier (120 buried 70-foot radials), then the maximum gain is 0.59 dBi at a 24° take-off angle. To resonate the antenna, the parasol wires must have a length of 46.758 feet, which leads to an input resistance of 33.15 Ω. In this case, the down-slope angle of the three parasol wires was held constant at 30° below the horizontal. This means that either their earth-anchor points must be *outside* the limits of the smaller ground screen, or else they must be tied to elevated supports such as posts or trees. If we lower the outer ends of the parasol wires until their down-slope angle is 45°, then ideally their earth-anchor points would coincide with the perimeter of the 70-foot-radius ground screen (assuming no sag in the parasol wires). By doing so, the antenna gain falls to 0.24 dBi at an elevation angle of 24°, a drop of roughly 1/3 dB from that which was attainable when the down-slope angle of the parasol wires was only 30°. In this scenario, the parasol wires must be lengthened to 50.478 feet to achieve resonance, with a resulting input resistance of 27.93 Ω.

Next, eight gull-wing radials were used for the ground screen, but their outer tips were extended to a horizontal distance of only 70 feet from the tower centerline. Recall that, because these radials are bent, their actual length is somewhat greater than 70 feet. Because the major portions of these gull-wing radials are raised to a height of 12 feet above ground, supports are needed to maintain this elevation at their outer ends. Thus, the down-slope angle of the parasol wires was kept at only 30°, with the expectation that similar elevated supports would also be used to anchor them. After trimming the parasol wires to resonance, the resulting antenna had a peak gain of only –3.32 dBi, which is quite low.

At this point, I decided to tune the shortened radials in an attempt to improve the antenna performance. With this in mind, the eight radials were divided into two adjoining groups of four. Next, each set of four adjacent gull-wing radials was converted into one leg of a very low dipole antenna by shorting their inner ends together, just above ground level. Then, equal loading inductances were added to both dipole legs to obtain resonance at 1830 kHz, which required a 7.06 μH inductor at the inner end of each leg. With the two loading inductors now in place, the parasol wires had to be retuned (to a length of 51.397 feet) to resonate the entire antenna at 1830 kHz. Now the gain rose to –0.43 dBi at 25° take-off angle, with an input resistance of 22.9 Ω.

Last of all, eight shortened radials that were completely horizontal were used to form the ground system of the Parasol antenna. Each radial was 70 feet in length, and their height above ground was fixed at 12 feet, as usual. The peak gain for this configuration was very disappointing, at just –6.76 dBi. So, once again, a low dipole was created by joining the inner ends of four adjacent radials together to form one leg of a 12-foot-high dipole. This time, two 9.13 μH loading inductances were needed to resonate the radials at 1830 kHz. When the parasol wires were trimmed to resonance, the antenna yielded a maximum gain of –1.15 dBi at 25° take-off angle, and an input resistance of 20.24 Ω.

Table 2 lists the outcome for all of the "short-radial" antennas that have been examined. First, we can see that the classic AM broadcast-style antenna loses almost exactly half a decibel of gain when shortened radials are used instead of full-size quarter-wave wires (gain = 0.79 dBi for 70-foot radials, versus 1.28 dBi for radials that are 134.368 feet long). The 70-foot-tall Parasol antenna generates a gain of 0.59 dBi, which is only 0.2 dB down from the full-size $^{1}/_{4}$-λ tower over the same (truncated) ground screen. However, when we increase the down-slope angle of the three parasol wires from 30 to 45°, the gain of the Parasol drops by another third of a decibel, to 0.24 dBi. Note that this version of the Parasol is roughly half as tall as the traditional $^{1}/_{4}$-λ base-fed vertical monopole, and its ground screen consumes about one-fourth of the acreage of the typical quarter-wave-radial counterpoise, yet its gain is just 1.04 dB less than that of the normal antenna (at 1.28 dBi). If we substitute elevated radials that are inductively loaded, the gain drops still further, although gull-wing radials are superior to those that are horizontal throughout their entire length.

Parasol Arrays

Computer simulation indicates that multiple Parasol elements can be used effectively together. A 4-square array composed of conventional elevated quarter-wave monopoles is shown in Figure 3, where 12 quarter-wave horizontal radials are arranged in a tic-tac-toe fashion at 30 feet above ground. As usual, the vertical elements are two inches in diameter, while the radials are #12 AWG. This antenna yields excellent performance, with a forward gain of 6.37 dBi at 19° take-off angle, and an azimuthal-plane F/B of 31.46 dB at the same elevation angle.

If we lower this entire antenna by 30 feet, so the vertical elements have their insulated bases at ground level, then we can still use elevated radials if we convert them into gull-wing types (see Figure 4). With the modified radials sloping upward from the ground at a 45° angle until reaching their final height of 30 feet, we must extend their length in order to keep the same overall wing span as before (134.368 feet, or $^{1}/_{4}$ λ at 1830 kHz). In

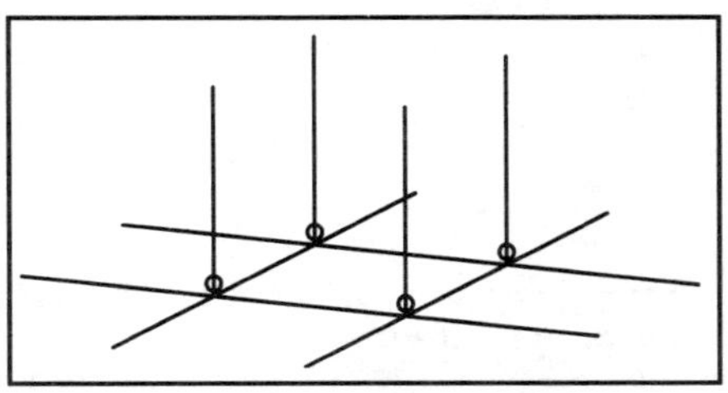

Figure 3—An elevated, base-fed, four-square phased array, using 12 horizontal radials arranged in a tic-tac-toe fashion. The length of each radial and each vertical element is $^{1}/_{4}$ λ at 1830 kHz, while the elevated bases are 30 feet above the ground. The small circle at the base of each tower denotes the feedpoint.

Table 3

Comparisons of forward gain and azimuthal-plane F/B for several versions of the four-square phased-vertical array. All of the Parasol configurations use 70-foot-tall vertical elements in combination with three parasol wires that slope downward at an angle of 30° below the horizon.

Power Gain (dBi)	Take-Off Angle (Degrees)	Azimuthal-Plane F/B (dB)
Elevated (H=30 feet) $^{1}/_{4}$-λ base-fed elements with 12 elevated, horizontal $^{1}/_{4}$-λ radials		
6.37	19	31.46
Ground-mounted, $^{1}/_{4}$-λ base-fed elements with 12 elevated, gull-wing $^{1}/_{4}$-λ radials		
5.16	19	22.11
70-foot-tall parasol elements with 12 elevated, gull-wing $^{1}/_{4}$-λ radials (parasol wires are 51.595 feet long)		
4.83	22	23.86
70-foot-tall parasol elements with 12 elevated, horizontal $^{1}/_{4}$-λ radials (parasol wires are 60.012 feet long)		
4.74	23	25.26
70-foot-tall parasol elements with 196 buried, $^{1}/_{4}$-λ radials (parasol wires are 46.533 feet long)		
5.16	23	23.43

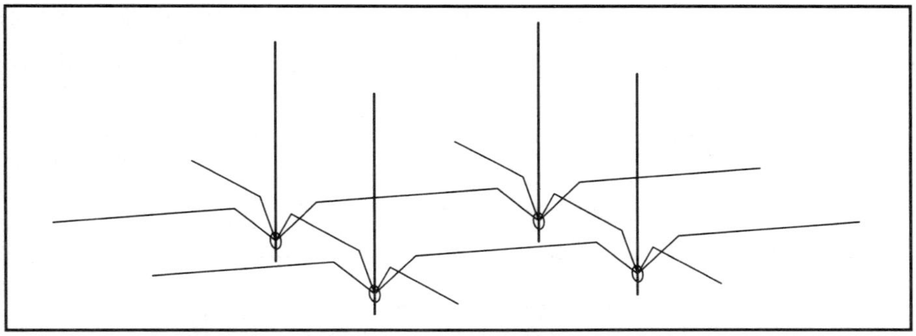

Figure 4—A ground-mounted, base-fed, four-square phased array, using 12 gull-wing radials arranged in a tic-tac-toe fashion. The height of each vertical element, and the wing-span of each radial, is $^{1}/_{4}$ λ at 1830 kHz, while the horizontal portion of each radial is elevated 30 feet above the ground. The small circle at the base of each tower denotes the feedpoint.

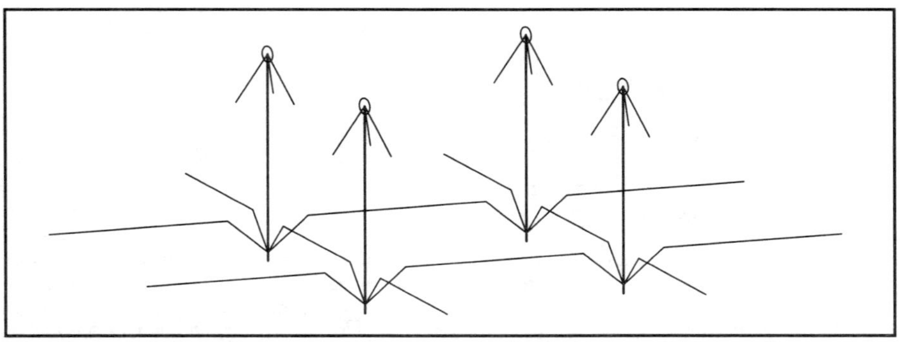

Figure 5—A four-square parasol array using 70-foot towers and 12 gull-wing radials. Each radial has a wing span of 134.4 feet ($^{1}/_{4}$ λ at 1830 kHz); its horizontal portion is elevated 12 feet above ground. Each element has three 51.595-foot parasol wires that slope downward at an angle of 30° below the horizon. The small circle at the apex of each tower denotes the feedpoint.

this configuration, the gain falls to 5.16 dBi at 19° elevation angle, and the corresponding F/B in the azimuthal plane is now 22.11 dB.

A Parasol four-square for the Top Band, using 70-foot towers is, of course, considerably shorter. Figure 5 is a drawing of such an array, with 12 gull-wing radials whose horizontal sections are only 12 feet off the ground. Again, the radials have been extended so their span remains at 134.368 feet. As usual, each element has three parasol wires (length = 51.595 feet) sloping downward at 30° below the horizon. The gain here is 4.83 dBi at 22° take-off angle, and the F/B is 23.86 dB in the azimuthal plane. Adding four more gull-wing radials (one oriented along the bore-sight in each of the four directions of fire) bumps the gain up only slightly, to 4.89 dBi. On the other hand, if the four "shared" radials are omitted, so that each vertical element has just two gull-wing radials all to itself, then the forward gain drops to 4.65 dBi.

Returning to the design shown in Figure 5, the 12 gull-wing radials may be replaced by elevated $^1/_4$-λ radials that are completely horizontal throughout their lengths. When this is done, the forward gain falls slightly to 4.74 dBi at an elevation angle of 23°, but the azimuthal-plane F/B improves to 25.26 dB.

If better performance is desired, we can resort to the classical tradition of "lots of radials in/on the ground." Figure 6 shows a parasol 4-square that is similar to that of Figure 5, but the elevated radials have been removed, and the length of the parasol wires is adjusted to 46.533 feet. Now each element has 49 buried radials that are spaced 3° apart, spread out in an arc directly in front of each tower. In this manner, none of the radials overlap each other, and all can be a full-size quarter-wave long. This array yields a gain of 5.16 dBi at 23° elevation angle, with an accompanying F/B of 23.43 dB in the azimuthal plane.

Table 3 provides the important data for the various four-square arrays outlined here. Notice that none of the Parasol four-squares is as good as the full-size elevated array, which is roughly 94 feet taller. However, the parasol array which includes four sectors of buried radials has just as much forward gain as the ground-mounted base-fed four-square with quarter-wave vertical elements and full-size gull-wing elevated radials. Further, even the worst of the parasol arrays has roughly 3.5 dB of forward gain compared to a single (omnidirectional) AM broadcast-style antenna, with the added benefit of 20+ dB of F/B. The elevation- and azimuthal-plane radiation patterns for most of these arrays are shown in Figures 7 and 8 respectively.

Computer Models Using Other Metals

If a Parasol antenna is constructed for operation on the Top Band, it is likely that a galvanized (zinc-coated) steel tower, with a height of 60 to 80 feet, would serve as the vertical element. *EZNEC4* models of all-zinc versions of the antennas described in this article show gain values that are within 0.1 dB of those given for the equivalent aluminum designs.

On-the-Air Tests

Two different versions of the Parasol antenna were built at commercial AM broadcast stations in northwestern Pennsylvania. The first was constructed in the fall of 1998, and the second during the latter part of 1999. Subjectively, both of these antennas performed satisfactorily, although no detailed measurement data was collected in either case. A Top-Band four-square parasol array

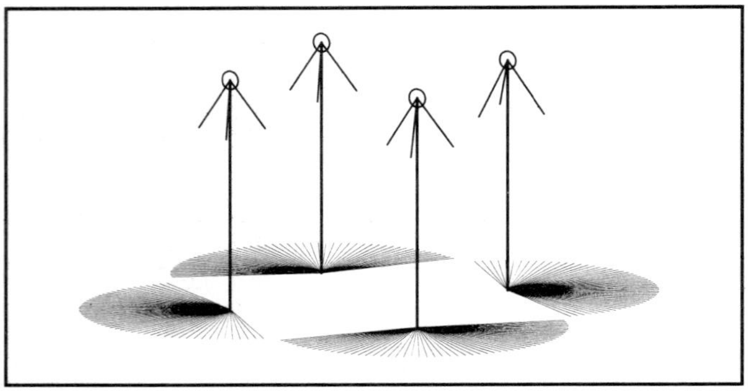

Figure 6—A four-square parasol array using 70-foot towers and buried radials. The ground screen for each element consists of 49 buried $^1/_4$-λ radials, that are spaced 3° apart in azimuth and cover a 144° sector directly in front of that element. Each antenna also has three 46.533-foot parasol radiator-wires that slope downward at an angle of 30° below the horizon. The small circle at the apex of each tower denotes the feedpoint.

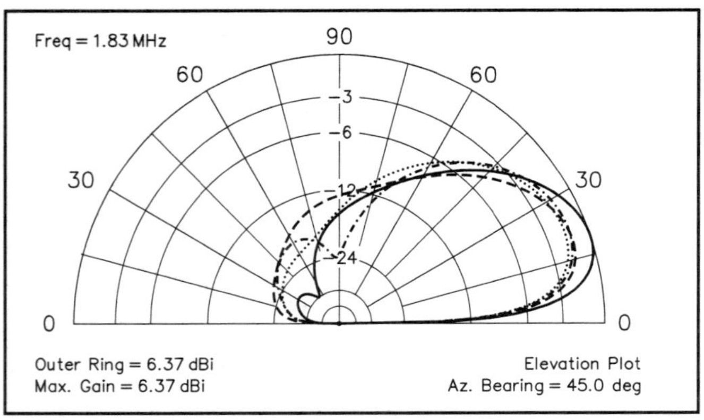

Figure 7—Elevation-plane radiation patterns for several of the four-square phased arrays discussed in the text are represented by different line types: The solid line represents the elevated base-fed array (H = 30 feet) using $^1/_4\lambda$ vertical monopoles and 12 $^1/_4$-λ radials. The dashed line is the pattern of the ground-mounted base-fed array utilizing quarter-wave vertical monopoles and 12 gull-wing radials with a wing span of $^1/_4$-λ. The dotted line is the pattern of the Parasol array using 70-foot towers and 12 gull-wing radials with a wing span of $^1/_4\ \lambda$. The combination dashed/dotted line is the pattern for the Parasol array using 70-foot towers and 196 buried $^1/_4\lambda$ radials (49 radials per element).

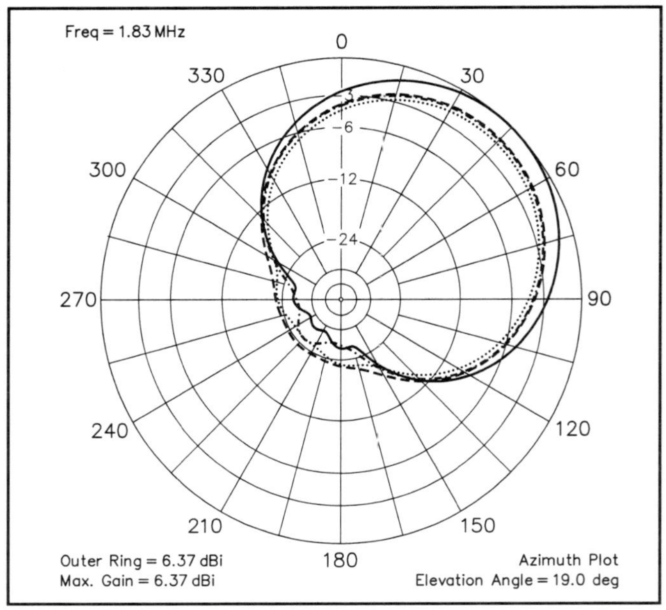

Figure 8—Azimuthal-plane radiation patterns for several of the four-square phased arrays discussed in the text are represented by different line types: The solid line represents the elevated (H = 30 feet) base-fed array using quarter-wave vertical monopoles and 12 $^1/_4$-λ radials. The dashed line is the pattern of the ground-mounted base-fed array using $^1/_4$-λ vertical monopoles and 12 gull-wing radials with a wing span of $^1/_4\ \lambda$. The dotted line is the pattern of the Parasol array using 70-foot towers and 12 gull-wing radials with a wing span of $^1/_4\lambda$. The combination dashed/dotted line is the pattern for the Parasol array using 70-foot towers and 196 buried $^1/_4$-λ radials.

How Does the Parasol Work?

Imagine a horizontal wire located some distance above the ground. If this wire is a half-wavelength long at some particular frequency, then the distribution of current and voltage on it will be as shown in Figure A. Notice that the voltage is greatest at the ends of the wire, while the current peak is in the middle. If we place the feed point at the center of the wire, then the input impedance will be low, because the voltage is low and the current is high at this point (Z = (V/I)). If we move the feed point so that it is off center, then the input impedance will be higher because we will be driving the antenna at a location where the voltage is larger than before, while the current has diminished. The shape of the radiation patterns of the two antennas should be the same because the voltage and current distributions are unchanged. This isn't strictly true, in general, due to the presence of the feed line, loss in the wires, and so on.

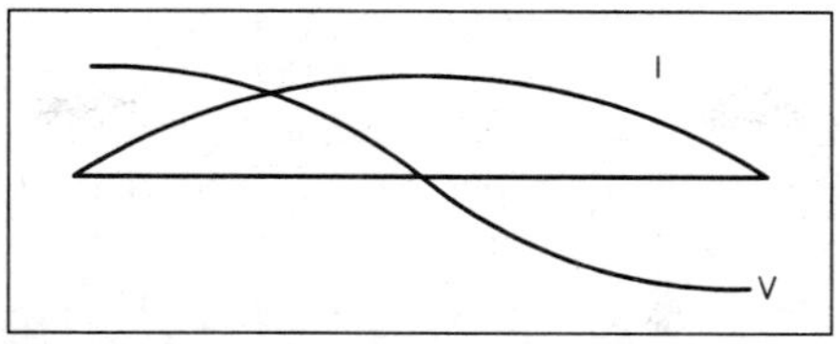

Figure A—The current and voltage distribution on a half-wavelength horizontal dipole system.

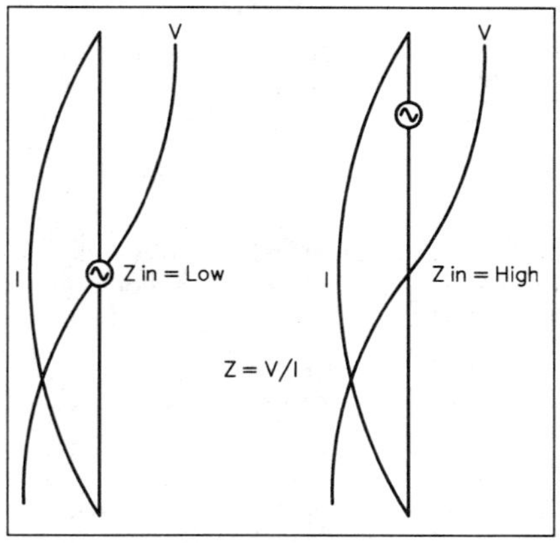

Figure B—Moving the feed point on our now-vertical half-wavelength dipole changes the input impedance from a low value (left) to a higher value (right), but doesn't alter the current and voltage distribution on the wire.

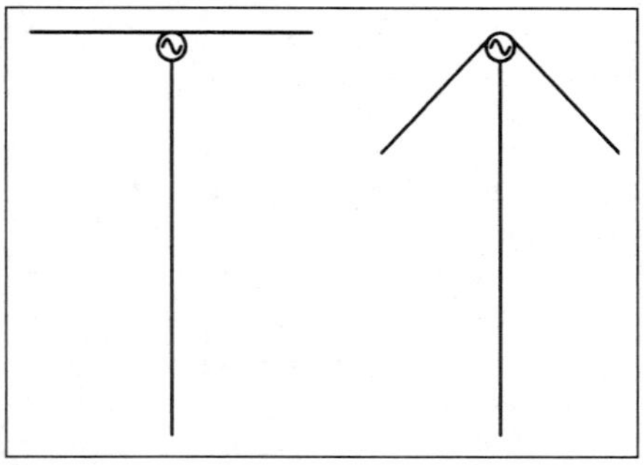

Figure C—The upper 1/8-wavelength of the vertical dipole is replaced by several symmetrically disposed horizontal top hat wires, each of which is 1/8-wavelength long (left). These top-hat wires can also be allowed to droop at bit, as shown on the right.

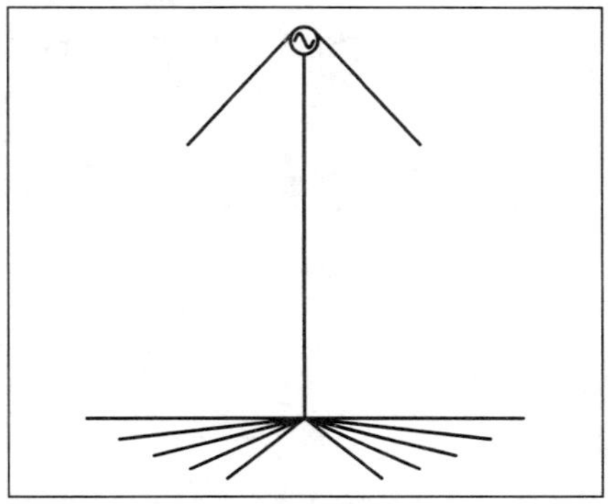

Figure D—The entire lower half (1/4-wavelength) of the original vertical dipole can be removed and the remaining portion placed directly over a ground plane consisting of in-the-ground, on-the-ground or elevated radials.

Let's take our half-wavelength horizontal dipole and move the feed point back to the center. Next, rotate it 90° (stand it on one end) so that the lower tip is a few feet above the ground. We can do this by hanging the wire from a high branch in a tall tree. Ideally, the distributions of voltage and current on the wire will be the same as when the wire was horizontal, although the ground and the tree may have some impact (which we will ignore). See Figure B. As a practical example, let's examine a half-wavelength of No. 12 aluminum wire at 14.175 MHz, suspended over average soil so that its lower end is 3 feet above the ground. When center fed, the gain is 0.39 dBi at 17° take-off angle and the impedance is 92.37 +j40.14Ω.

As was true for the horizontal antenna, we can easily move the feed point so that it is off center. This shouldn't change the radiation pattern shape or size since the distribution of current and voltage will remain as before. However, the input impedance will rise because of the increase in the ratio of voltage to current at the new feed point. Referring again to our half-wavelength center-fed 20-meter vertical dipole, let's move the feed point upward by 1/8-wavelength as shown on the right side of Figure B. Now the gain falls slightly to 0.23 dBi at 17° take-off angle, while the input impedance rises to 214.6 +j64.25Ω. When we compare the radiation patterns for these two vertical dipoles, we find that they are very similar in both size and shape.

Now suppose that our tree support isn't quite tall enough for our needs, and as a result, we aren't able to make the extreme top portion of our wire antenna completely vertical. To solve this problem, we replace the upper tip with a couple of horizontal wires, forming a "top hat." From experience, we expect that both the gain and the input impedance of this modified antenna will decrease somewhat because we have had to shorten it. Returning to our 20-meter off-center-fed antenna example, we will remove the upper 1/8 wavelength (the entire portion above the elevated feed point) and replace this single conductor with a top hat composed of three 1/8-wavelength horizontal wires that are spaced uniformly, 120° apart in azimuth, around the top of the remaining vertical wire. See the left-hand side of Figure C.

Notice that if you turn the drawing upside down, it looks just like a 3/8-wavelength tall "ground plane" with three 1/8-wavelength radials. Our 20-meter example antenna, when configured as shown, has a gain of 0.09 dBi at 19° take-off angle and an input impedance of 123.9 +j267.1Ω. The gain and the input resistance have fallen, as we expected. The input reactance has become quite large, however, since we used three 1/8-wavelength top hat wires to replace a single 1/8-wavelength vertical section. This makes the antenna look quite a bit "longer" and the input impedance becomes highly inductive as a result.

If we can't make the top hat completely horizontal, we can allow it to droop a bit, as shown in the right-hand side of Figure C. However, this will cause the gain to fall, since the current flowing in the top hat wires will now partially oppose the current flowing in the vertical wire. For our 20-meter antenna example, allowing the top hat wires to droop at an angle of 25° below the horizontal yields a gain of –0.01 dBi at a 19° take-off angle, and the input impedance falls to 105.6 +j238.1Ω.

Finally, we are ready to perform the very last step in the transformation of our half-wavelength vertical antenna into a "parasol." As is often seen in AM broadcast installations (as well as ham applications), the lower half of the half-wavelength vertical can be removed and replaced by a system of radials. The voltage is zero at the electrical center of the half-wavelength element, so the antenna may be grounded at that point. The radials may be buried in the ground, laid on top of the ground or elevated (see Figure D). To modify our 20-meter vertical with the drooping top hat, we chop off the bottom 1/4-wavelength portion of the vertical section, leaving only the upper 1/8-wavelength portion behind, along with the feed point and the three 1/8-wavelength parasol wires. If a classical AM-broadcast-style ground system of 120 1/4-wavelength buried radial wires is used, then the gain of this parasol antenna is 0.10 dBi at 29° take-off angle and the input impedance is 22.06 +j224.9Ω. As mentioned in the text, resonance is usually achieved when the parasol wires are shorter than 1/8-wavelength, and this is borne out here. Also, making the parasol wires more horizontal (and/or increasing the height of the vertical section) will generate more gain and a higher value for the input resistance.—*Al Christman, K3LC*

similar to the one shown in *Figure 6* is planned for installation at a station in northwestern Pennsylvania in the near future.

Summary

This article has provided information on the design of a low-profile vertically polarized antenna that is appropriate for a variety of applications. The structure is only about half as tall as a conventional $^1/_4$-λ monopole, and computer analysis indicates that its performance is very good, considering its size. Although this material has focused on Top-Band antennas, the parasol can be used on any frequency at which its particular design attributes are needed. In locations where antenna- or tower-height restrictions are strictly enforced, the parasol can provide relief. For example, a four-square parasol array for the 40-meter band would require elements that are less than 18 feet high. Although several of the antennas that are analyzed in this article make use of *buried* radials, it is likely that similar results would be obtained if *on-the-ground* radials were utilized instead.

To obtain maximum performance from these short antennas, it is important to pay close attention to system losses. Keep the contact resistance at all connections as low as possible. If galvanized tower sections are used, it is very important to maintain the integrity of the zinc coating. Further, when adjacent tower sections are joined together, a low-resistance joint is necessary. The use of conductive grease is helpful, along with tight mechanical bonding. The "parasol" wires make up the ungrounded portion of the antenna, so their electrical conductivity should be as high as possible. Use large-diameter conductors made of copper or Copperweld for these wires (or aluminum tubing if the frequency is high enough).

Finally, realize that computer models are imperfect representations of the real world. They cannot possibly include all of the features that are actually present, such as buildings, vegetation, other conductive objects, variations in wire diameter and composition, irregular terrain, changes in ground conductivity and permittivity, etc.

Notes

[1]*EZNEC* antenna software is available from Roy Lewallen, W7EL, PO Box 6658, Beaverton, OR 97007; tel 503-646-2885, fax 503-671-9046; **w7el@teleport.com**. Price: $89 postpaid; add $3 outside the US and Canada. Visa and MasterCard charge cards accepted.

[2]G. J. Burke and A. J. Poggio, *Numerical Electromagnetics Code (NEC) - Method of Moments*, Naval Ocean Systems Center, San Diego, California, January 1981.

[3]Al Christman and Roger Radcliff, "Using Elevated Radials with Ground-Mounted Towers," *IEEE Transactions on Broadcasting*, Volume 37, Number 3, September 1991.

[4]Al Christman, R. Paul Zeineddin, Roger Radcliff, and Jim Breakall, "Using Elevated Radials In Conjunction with Deteriorated Buried-Radial Ground Systems," *IEEE Transactions on Broadcasting*, Volume 39, Number 2, June 1993.

[5]R. Dean Straw, N6BV, "Antennas Here are Some Verticals on the Beach...", *The ARRL Antenna Compendium*, (Newington: ARRL, Vol 6, 1999), pp 216 to 225.

Al Christman, K3LC, has a PhD in electrical and computer engineering from Ohio University and is currently serving as a professor in the EE department at Grove City College in western Pennsylvania. Al is a Vietnam-era veteran, having served on active duty with the US Army. First licensed as WA3WZD in 1974, Al is an active DXer with more than 320 countries worked on 20-meter SSB. He and his wife, Lynn, recently celebrated the birth of their first grandchild. You can contact Al Christman, K3LC, at Grove City College, 100 Campus Dr, Grove City, PA 16127-2104.

The Ultimate Portable HF Vertical Antenna

Phil Salas, AD5X

From *QST*, July 2005

I've had a tremendous re-response to the portable antenna project published in the July 2002 *QST*. Many folks had problems locating some of the parts, however. It seems that sprinkler-system parts are not as common in other sections of the country as they are in Texas! Also, the original article required that you drill and tap screw holes in brass couplings and the sprinkler-system risers. I wanted to eliminate the screw-hole tapping to simplify the construction.

Over the years, the antenna evolved from the original design through changes that used fiberglass, aluminum tubing, and then brass tubing. The most recent design is longer, lighter and more compact (when disassembled) than the original antenna. It is also easier to build, and easier to the find parts. Band coverage is also increased to include 60 meters, as well as 40 through 10 meters!

The Ultimate Portable Antenna is designed for easy transport. It breaks down into multiple mast sections, a whip section, an air-wound center loading coil section, and a small base support. No piece is longer than about 20 inches, so it will easily fit into most suitcases. The lead photo shows the disassembled antenna, including guys and radials. The fully assembled antenna has a length of almost 16 feet! See Figure 1. That's me, standing next to the antenna set up on my front lawn, in Figure 2.

The key to efficiency for short antennas (such as a loaded vertical less than a quarter wavelength) is the length. The longer the antenna, the greater the radiation resistance and therefore the less impact you have on efficiency due to ground and loading coil losses. This antenna is almost $^{1}/_{4}$ λ on 20 meters. For operation on 17 through 10 meters, you will shorten the antenna to $^{1}/_{4}$ λ. The length of the complete antenna minimizes the required loading coil for 60, 40 and 30 meters. So let's build it!

AD5X

The complete unassembled antenna.

AD5X

Figure 1—The complete antenna setup in the author's front yard.

AD5X

Figure 2—Standing next to the antenna to illustrate the height of the tubing sections and coil.

Gathering the Parts

You can find most of the parts for this antenna at your local hardware store. The loading coil, coil taps, 10 foot telescoping whip, and SO-239 are available from MFJ (**www.mfjenterprises.com**). Be sure to mention this *QST* article when ordering the kit of parts (MFJ-1964-K) for a special price discount. Table 1 is the complete parts list.

A few notes about pipe sizes may be in order for anyone not familiar with plumbing fixtures. When you go to the hardware store to purchase the $^{1}/_{8}$ inch NPT nipples and couplers, don't expect to find anything measuring $^{1}/_{8}$ inch! The outside diameter of a $^{1}/_{8}$ inch pipe is 0.405 inches, or about $^{13}/_{32}$ inch. The standard threads are 27 turns per inch. (The $^{1}/_{8}$ inch designation comes from the approximate inside dimension of the pipe, although today you may find pipes with different wall thicknesses and the same outside dimensions. For this project you don't have to worry about the wall thickness of your fittings.) The NPT specification

Table 1
Parts List

Qty	Item
1	5 inch long × 2.5 inch diameter × 10 TPI air wound coil (MFJ-404-008)*
1	10 foot telescoping whip (MFJ-1954)*
1	SO-239 chassis mount connector (MFJ-610-2005)*
5	Coil clips (MFJ-605-4001)*
2	3 foot pieces of 3/8 inch diameter brass tubing (ACE Hardware) (McMaster-Carr 8950K581 — 6 foot length)**
1	3/8 inch diameter wood dowel. (36 inch length at Home Depot—only 3 1/2 inch needed.)***
1	3/4 inch PVC T — All of the PVC fittings are white schedule 40 pipe (Home Depot)
1	3/4 inch slip × 1/2 inch female pipe thread PVC adapter (Home Depot)
1	3/4 inch slip × 1/2 inch slip PVC adapter (Home Depot)
1	3/4 inch slip PVC plug (ACE Hardware)
1	1/2 inch NPT (male thread) to 1/8 inch NPT (female thread) brass adapter bushing (ACE Hardware)(McMaster-Carr 50785K64)
8	1/8 inch NPT brass couplings (Home Depot)(McMaster-Carr 50785K91)
4	0.7 inch long 1/8 inch NPT all-thread nipples (these are also called "close nipples") (Home Depot) (McMaster-Carr 50785K151)
2	#8 brass wing-nuts (ACE Hardware)
2	#8-32 × 3/4 inch brass machine screws (ACE Hardware)
2	#8-32 brass nuts (pack of 6) (Home Depot)
2	#8 copper-plated steel (or brass) split lock washers (ACE Hardware)
1	36 inch length of 1/8 inch diameter brass rod (Home Depot)
1	3/8-16 × 1 1/4 inch hex head bolt, zinc plated. Choose the longest you can that is threaded all the way to the head. (Home Depot)
1	3/8-16 × 12 inch hex head or carriage bolt, zinc plated. Choose the longest bolt you can find in this size if you can't find a 12 inch bolt (Home Depot)
1	3/8-16 coupler, zinc plated (ACE Hardware)
2	3/8-16 nuts, zinc plated (pack of 6) (Home Depot)
1	3/8 inch lock washer, zinc plated (pack of 10) (Home Depot)
4	#6 stainless steel 3/8 inch sheet metal screws (ACE Hardware)
3	#8 solder lugs (Home Depot)
6	#8 × 1 1/2 inch brass wood screws (pack of 6) (Home Depot)
90 feet	Wire (any gauge, insulated or not, for six 15-foot ground radials.)
1	Alligator clip and miscellaneous short pieces of connecting wire.

*The total retail price for all the MFJ parts is $53.80, plus shipping/handling. Mention this article when you order and MFJ will sell a kit containing the telescopic whip, coil, coil clips, and SO-239 connector for $39.95 plus $6 s/h. The kit is MFJ-1964K. You can order additional coil clips for $2.95 each.

You can use 3/8 inch aluminum tubing if you prefer. Aluminum tubing is about half the cost of brass, but you will either need to drill the brass couplings and aluminum tubing so they can be connected with stainless steel sheet-metal screws or solder them together using aluminum solder and a torch (see **www.solder-it.com).

***You can use 3/8 inch fiberglass rod instead of the wood dowel if you prefer, but fiberglass rod is more difficult to find. Check out bicycle flags and driveway marker stakes as potential sources of fiberglass rod.

The brass plumbing items are also available from McMaster-Carr, (**www.mcmaster.com**) a mail-order supplier with no minimum order requirement. They stock 6 foot lengths of 3/8 inch brass tubing as well as the close nipples, couplings and brass bushings. The Table lists part numbers to help you look up those items on the Web site.

means National Pipe Thread. There are several other thread specifications, but NPT is the most common.

I used 1/8 inch NPT close nipples for the antenna connectors. These are about 0.7 inches long, and are usually fully threaded over over the entire length. The PVC pipe fittings must be the white, schedule 40 PVC pipe. Do not use the thinner-walled pipe sometimes known as CPVC, which has a cream color or yellow tint. When fittings are designated as "slip," it means pieces are intended to be glued—they just slip together—rather than being threaded.

Brass Rod Preparation and Assembly

See Figure 3 for the assembly details. First cut three 18 inch sections of the 3/8 inch brass tubes with a hacksaw or tubing cutter and de-burr the tubing. The ends of the couplings that fit over the brass tubes must be reamed out with a 3/8 inch drill bit. Otherwise the couplings won't fit over the tubing. To do this, first screw a 1/8 inch NPT coupling on each end of a 1/8 inch NPT close nipple. Use wrenches to screw these together as tightly as possible. Next, clamp one of the couplings securely and ream out the opposite coupling with a 3/8 inch drill bit. (Use a drill press for this operation if at all possible.) Reverse, and ream out the other coupling. [See Figure 4 for one way to use a woodworker's clamp and a drill press.—*Ed.*] Now unscrew the couplings. One end of the nipples will break loose from one coupling, and the other end will stay tight in the remaining coupling. You'll now have a female and male end that will fit over each end of a section of brass tube, as shown in Figure 5. You will need four pair of these male/female brass connectors: three pair for the brass tubes and one pair

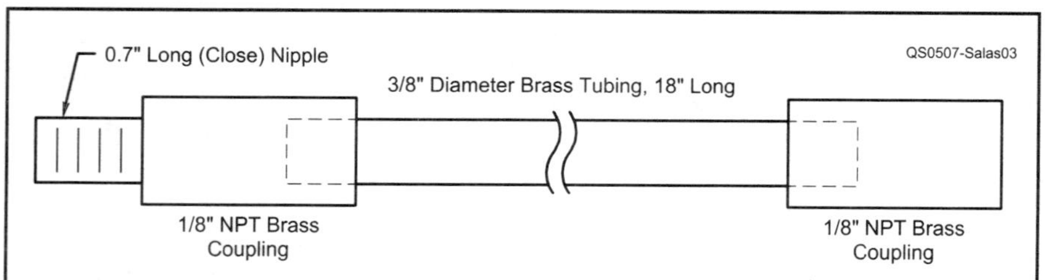

Figure 3—Assembly of the brass tubing sections.

Figure 4—A pipe coupling and nipple secured in a woodworker's clamp ready to drill out the end on a drill press.

AD5X

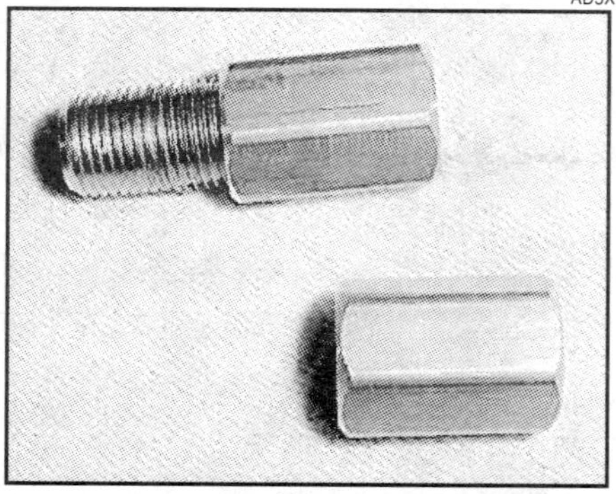

Figure 5—After the pipe couplings have been drilled out and one coupling removed from the nipple, the pair is ready to be installed on the ends of an 18 inch length of brass tubing.

for the loading coil assembly. If you'd like, you can solder the nipple/coupling assemblies together. The assembly tends to be very tightly secured even without soldering, however.

Now insert the male/female brass pairs just constructed over all three of the 18 inch brass tubes and solder the couplings directly to the tubes. This is easily done with a large soldering iron, or even better, with a torch and silver solder. Solder-It has a nice small butane torch that works well. See **www.solder-it.com**.

Loading Coil Assembly

Slide 1/8 inch NPT male/female coupling pairs over both ends of a 3/8 inch diameter, 3 1/2 inch long wood dowel. You will need to drill a 1/8 inch diameter hole completely through each of the 1/8 inch NPT brass couplings and dowel as shown in Figure 6. Next cut two 3 inch lengths of the 1/8 inch diameter brass rod. Insert one of these 3 inch sections through the holes on one brass coupling. Center the rod so that equal lengths are available on both sides of the coupling, and solder the rod to the coupling with a large soldering iron or torch. (Be careful not to burn the wood dowel with the torch!)

Now position a 3 inch length of the MFJ-404-008 coil such that the 3 inch brass rod just installed pokes through the last two turns on the coil. See Figure 7. Solder the coil turns to the rod. On the opposite end of the coil assembly, insert the remaining 3 inch brass rod through two adjacent turns on this end of the coil, through the brass coupling, and through the coil turns. Solder the coil turns to the brass rod, and then solder the brass rod to the coupling.

Next, indent every other turn on the coil with a small flat-head screwdriver. You should do this on opposite sides of the coil to give you plenty of adjustment capability.

Finally, on the end of the coil with the brass nipple (male end), solder a 6 inch piece of insulated wire terminated with an alligator clip. For extended outdoor use, you may wish to treat the wood dowel with varnish.

Top Whip

The MFJ-1954 10 foot telescopic whip comes with a standard 3/8-24 mounting thread. While the whip mounting stud is not a correct match for the 1/8 inch NPT coupling (27 turns per inch), I just thread the screw directly into the 1/8 inch NPT coupling. Because the 3/8 bolt is slightly smaller than the 1/8 inch pipe outside diameter, and the pipe coupling is tapered inside, they will go together. [See the "Alternative Construction Ideas" sidebar for a way to match the threads, if you are uncomfortable with the mismatch.—*Ed.*]

Base Assembly

For the base spike, I've used a 3/8-16 × 12 inch zinc-plated hex-head bolt. Only 1 1/2 inches of the bolt is threaded, and so I used the long smooth end of the bolt to go into the ground after cutting off the hex head. A damp cloth easily cleans the bolt after use. I also used a 3/8-16 × 1 1/4 inch zinc-plated hex-head bolt at the base of the PVC assembly, and a 3/8-16 zinc-plated coupler to attach the 1 1/4 inch bolt to the 12 inch bolt, as shown in Figure 8. This way you can leave the long bolt off if you want to bolt the base assembly directly to a metal plate or trailer mount, or screw on the long bolt for ground mounting.

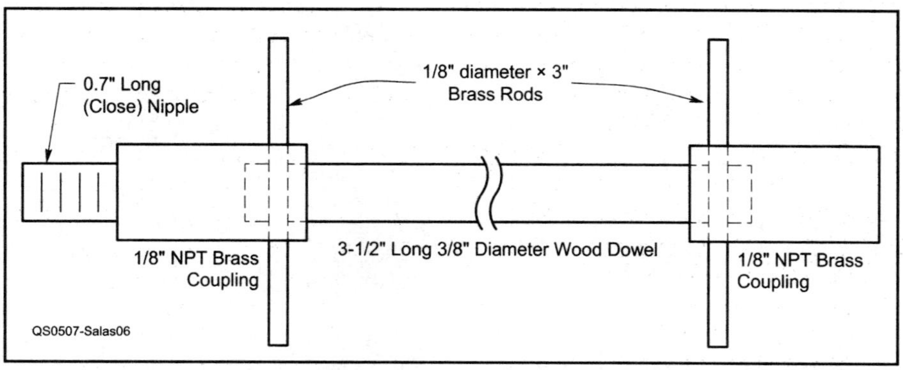

Figure 6—How brass couplings are installed on a length of 3/8 inch wood dowel and 1/8 inch brass rod in order to mount the coil.

AD5X

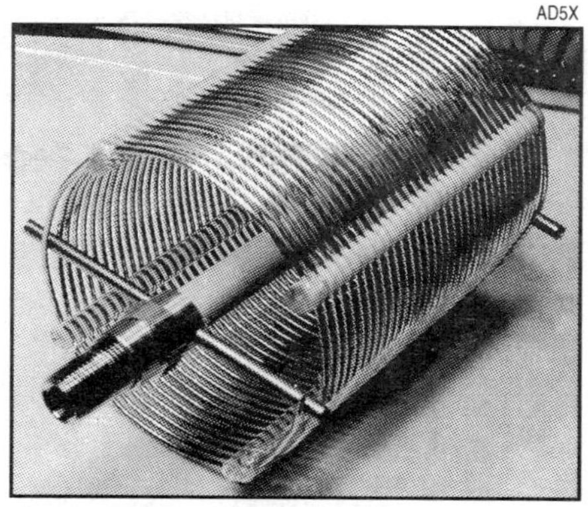

Figure 7—The loading coil assembly ready to be soldered.

Referring to Figure 8, drill a 3/8 inch diameter hole into the 3/4 inch PVC plug used for the base support 1 1/4 inch bolt. Cut off about half of the length of the 3/4 inch PVC plug to leave plenty of room inside the T for wiring. Solder a ground wire to the head of the 3/8-16 × 1 1/4 inch bolt as shown, or use a 3/8 inch solder lug. Insert the threaded end of bolt into the plug, and secure with a 3/8-16 nut and a lock washer. If you wish, you can glue the plug in place with PVC pipe cement instead of using the #6 stainless steel sheet metal screws shown. The screws make changing the support assembly easy in case you should ever want to, though. You might even consider using screws to attach the other two adapters into the T.

To prepare the 12 inch bolt, cut off the hex head and round this end with a file. Screw a 3/8-16 nut onto the threaded end, then add a lock washer and screw the 3/8-16 coupler onto the threads so the bolt is about halfway through the coupler. Tighten the nut against the coupler (with a lock washer between the nut and coupler). This 12 inch bolt assembly can now be easily screwed onto the 1 1/4 inch bolt on the base assembly for ground mounting.

Now place the SO-239 over the 1/2 inch hole in the 3/4 to 1/2 inch slip adapter and mark the location for the two #6-32 × 3/8 inch long stainless steel sheet metal screws that will hold it in place. (The adapter I used had a hex head on the outside lip, and by turning the SO-239 so diagonal mounting holes are over opposite points in the hex head, there should be enough material to hold the screws easily.) You'll see that these holes will be right in

Alternative Construction Ideas

As I began editing this article for publication in *QST*, I decided to build the antenna. After visiting three local Home Depot stores and most of the True Value and ACE Hardware stores in the area, I realized that builders may have difficulty finding some of the materials. For example, none of the stores I visited had any stock of brass or aluminum tubing in the size I wanted.

I contacted Phil Salas with a few questions about this article and the materials he specified. Phil informed me that since writing the article he had continued to try new ideas for building the antenna, and he found several alternative materials. His first suggestion was to use 1/2 inch aluminum tubing, with a wall thickness of 0.058 inch. He tapped the inside of that tubing for 1/8 inch pipe threads, and just turned the brass nipples into the pipe. This saves the step of drilling out the couplings to slip the 3/8 inch brass tubing inside. I was able to find some 1/2 inch aluminum tubing with a 0.050 inch wall thickness but not the thicker wall. Unfortunately, a 1/8 inch NPT nipple slips right inside that tubing, so I could not cut threads into the tubing. Perhaps 7/16 inch tubing with a 0.050 inch wall thickness would work, but I did not find any tubing in that size.

Phil also suggested McMaster-Carr (**www.mcmaster.com**) or similar mail-order/Internet stores as a good way to obtain the tubing and other plumbing parts. Their prices compare favorably with what I found in Connecticut, and with no minimum order they may save you some shopping time in a variety of hardware stores. Of course many of us probably enjoy browsing the aisles of our local hardware stores. You never know what new tool you may need or what other antenna projects may come to mind as you peruse the available materials. Phil has included McMaster-Carr part numbers for several of the pieces in Table 1.

I found a K & S Engineering display of hobby tubing at my local Do It Best hardware store. They ordered the 3 foot lengths I needed for a reasonable price. After building an antenna with that tubing, however, I realized that the 0.014 inch wall thickness was too thin to support the antenna without guy lines. Phil suggested that I could use that tubing by inserting a 1/4 inch dowel or 1/4 inch fiberglass rod, such as from a bicycle flag to stiffen the brass tubing. I ordered a 6 foot length of the brass tubing with a 0.032 inch wall from McMaster-Carr, as listed in Table 1. That proved to be a near ideal tradeoff between weight and strength.

I also discovered a much wider array of PVC pipe fittings than I had imagined. Many of the pieces appear to have similar specifications at first glance. For example, I found 3/4 inch × 1/2 inch adapters with male pipe threads on the 3/4 inch end and either male or female threads on the 1/2 inch end as well as one with both ends designed for slip fitting of the PVC pipe into the adapter.

There are many different ways to arrive at the end result you need. For example, if you can't find a 1/2 inch × 1/8 inch brass bushing for the top of the T, a 1/4 inch × 1/8 inch bushing will turn into a 1/2 inch × 1/4 inch bushing to achieve the same results.

All of my searching for parts shows that there may be a variety of materials that you could use to build a similar antenna. What you find in your area may not match what the author or I found. With a bit of creative shopping you will be able to find the parts you need to build the antenna.

The author suggests that you use #6 × 3/8 inch sheet metal screws to hold the pipe plug with the bolt into the bottom of the T. This will make it easier to disassemble that section if you have to fix a wiring problem inside the T. I decided that was a good approach for the fittings on the other two parts of the T as well. Photo A shows the T clamped in my woodworkers clamp, ready to drill the screw body holes.

WR1B

Photo A — The 3/4 inch PVC T is clamped in a woodworker's clamp, ready to drill the screw-body holes with a 9/64 inch drill bit. I had first drilled 5/64 inch pilot holes with the adapters and bottom plug installed in the T. Try to center the hole on the side of the T, and drill all the way through both sides. Then remove the adapters and plug, and drill the screw-body holes in the T.

Phil has found that threading the 3/8-24 stud on the bottom of the telescoping whip into the 1/8 inch NPT coupling is adequate. The perfectionist in me doesn't like the idea of mismatched threads. I decided to make a slightly different adapter than Phil suggests in his article. I used a 1/8 inch pipe coupling and nipple on the bottom of a 1 inch length of the brass tubing. (I drilled out the brass coupling with a 3/8 inch drill, the same as is done for the antenna tubing sections.) For the top section of my adapter I used a 3/8-24 coupling nut, which is about 7/8 inch long. I drilled out the threads on about half of that coupling, too, and pushed it onto the brass tubing. (I soldered these, and all other joints on my antenna.) Now all the threads mate properly. The 1/8 inch NPT to 3/8-24 adapter allows me to attach the whip to any antenna section, as desired. *—Larry Wolfgang, WR1B*

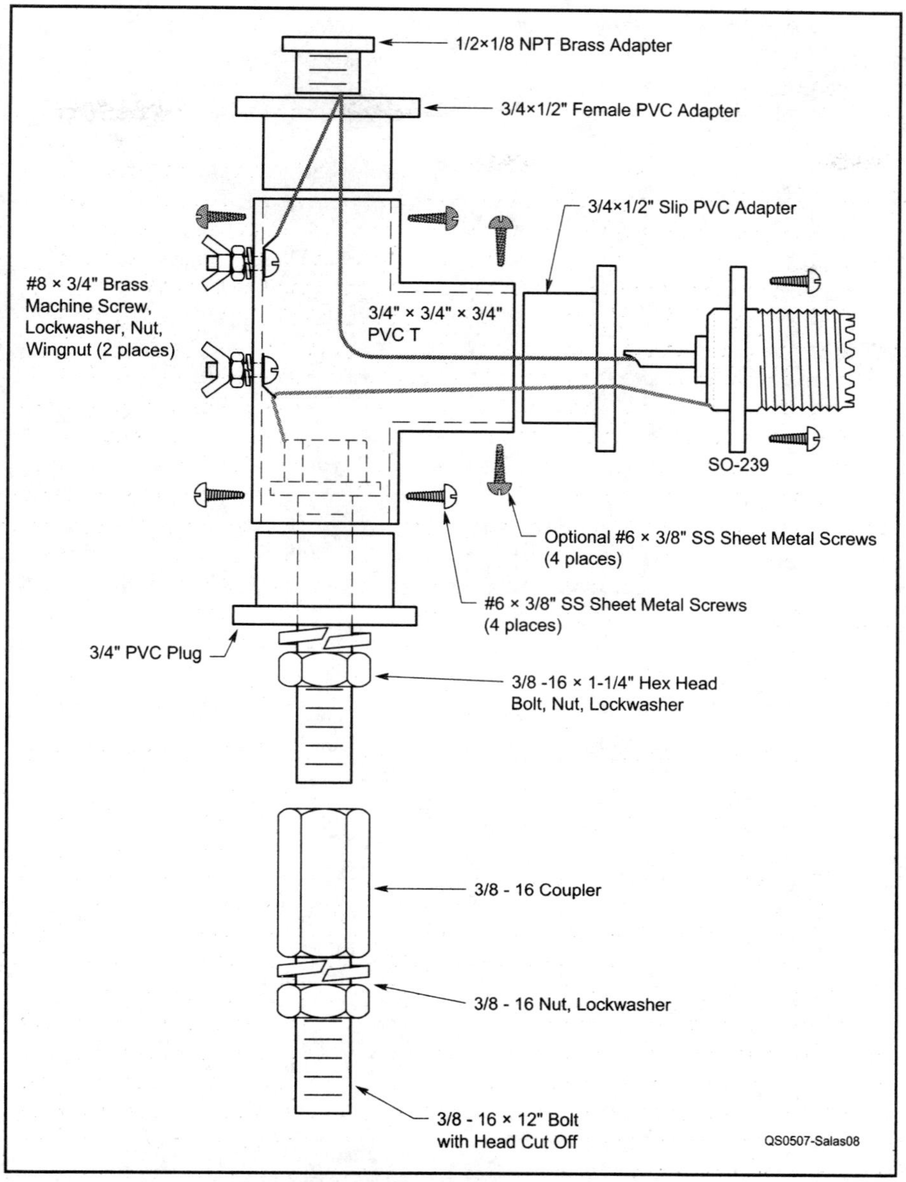

Figure 8—The antenna base and coaxial cable connector details.

the center of the PVC lip. Carefully drill two $^5/_{64}$ inch holes at these points. We will mount the SO-239 on the adapter later.

Place the $^3/_4$ inch PVC plug/spike assembly in the T and drill two $^5/_{64}$ inch diameter holes through the T and plug. [See sidebar Photo A for how I clamped the T in a woodworker's clamp to drill the holes. Try to drill straight through both sides in one pass to make alignment easier.—*Ed.*] Remove this assembly from the T and drill out these $^5/_{64}$ inch holes in the T to $^9/_{64}$ inch. Also drill out two holes in the SO-239 connector to $^9/_{64}$ inch, since the holes are not large enough to pass the #6 × $^3/_8$ inch sheet metal screws.

Use an $^{11}/_{64}$ inch drill bit to drill mounting holes for the #8-32 brass machine screws. I drilled the mounting holes close to the "top" and "bottom" of the T. Alternatively, if you align the bit with the inside edges of the side of the T and drill through the back of the T, you will be able to fit a screwdriver into the T to tighten this hardware.

Next we'll prepare the antenna interface at the top of the base. First, cut off part of the $^3/_4$ inch slip × $^1/_2$ inch female pipe thread PVC adapter so as to leave additional room in the T for wiring. Solder a piece of 14 gauge copper house wire directly to the inside lip of the $^1/_2$ × $^1/_8$ inch NPT brass adapter. You'll need a large soldering iron or a torch, since the brass adapter mass is pretty large. Screw this adapter tightly into the $^3/_4$ × $^1/_2$ inch PVC adapter.

Now solder a wire to the center conductor of the SO-239 connector as shown. This wire should be soldered to the wire stub on the $^1/_2$ × $^1/_8$ inch NPT brass adapter at the antenna interface, and then to the upper wing-nut assembly as shown. (Alternatively, you could use #8 solder lugs for these connections, and put the top #8-32 brass machine screw through the hole in the solder lug.) The $^3/_4$ × ½ inch PVC adapter can now be glued into place using PVC pipe glue. Solder a short piece of copper braid (from a piece of RG-58 cable) from the SO-239 ground (solder directly to the SO-239 body) to the brass ground screw, and finally to the wire soldered to the head of the 1$^1/_4$ inch bolt. (#8 solder lugs are a good alternative.) You can now complete the assembly of the base by inserting the PVC plug and 1$^1/_4$ inch bolt assembly into the T and installing the #6 stainless steel sheet metal screws as shown in Figure 8. Incidentally, the upper wing-nut assembly is used in case you need to add capacitive or inductive base matching should you want to improve the SWR on the lower bands. See Figure 9.

Ground Radial Network

The radial network is made up of six 15 foot radials, using 22 gauge insulated wire, though any gauge wire, insulated or not, can be used. I've found it best to make up three pairs of two wires each attached to a #8 spade lug on one end of each pair. This minimizes the hassle of deploying, and later rolling up, the radials. The three #8 lugs will attach to the ground screw on the base assembly. When the wires are rolled up, you should hold them together with twist-wraps. Solder a 1$^1/_2$ inch brass wood screw on the outer end of each radial. You can simply push these screws into the ground to hold the radials in place. Put a blob of hot glue on each wire/screw soldered interface to give it a little strain relief.

Guying

This antenna is self-supporting in a low breeze. In many cases, however, it will be necessary to guy the antenna because of its 16 foot length. For effective guying, I attached 9 foot lengths of nylon cord (3 pieces) just above the base of the 10 foot MFJ telescoping whip. Use a tie-wrap and close it just enough so that it won't slide over the base of the MFJ whip. Cut the 9 foot sections of nylon cord and heat the ends with a match to fuse the nylon so it won't unravel. Tie one end of a 9 foot section of nylon cord around each tie-wrap and secure with hot glue or epoxy. For the ground stakes, you can use the extra piece of brass tubing. (You only used 4$^1/_2$ feet of the 6 foot length.) Cut the remaining 18 inch piece of tubing into three 6 inch sections.

AD5X

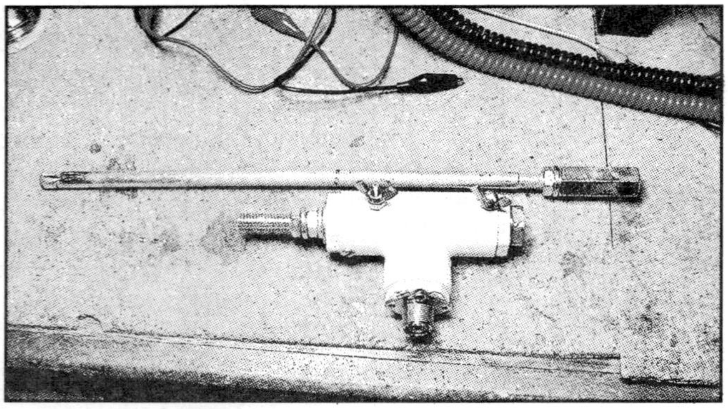

Figure 9—The completed antenna base and mounting spike.

AD5X

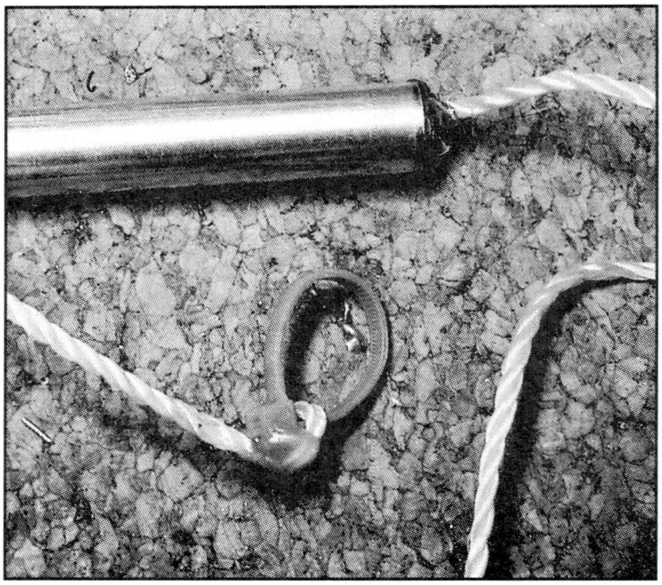

Figure 10—A set of 3 guys was made by using a wire tie (closed to slip over the top of the telescoping whip but not over the base of the whip). Hot glue was used to attach a nylon line to each wire tie. Hot glue was also used to attach the guy line to a left-over length of brass tubing.

Attach the end of the nylon cord without a tie-wrap to one end of each tube with hot glue. Also plug the open ends of the three tubes with hot glue. (You could also use some long spikes or bolts for stakes, if you are concerned that you won't be able to drive the brass tubing into hard ground.) For storage, wrap the nylon cord around each brass stake and hold it in place with masking tape. Figure 10 shows the details. See Figure 11 for a photo of the guys attached to the telescoping MFJ whip. When bolted to a trailer mount or plate, the antenna should really not need guying unless the wind is strong.

Antenna Assembly

To assemble the antenna, first screw the three brass-rod sections together, and then screw these into the top of the base assembly. Push the entire assembly firmly into the ground, keeping it as vertical as possible. Next, screw the loading coil and telescoping whip assemblies together. Slip the three guy tie-wrap/nylon cord assemblies over the whip and extend the telescoping whip. Screw this entire top assembly into the female end of the top brass tube. You only need to turn the brass fittings finger tight. Finally, push the guy rods in the ground and extend the six radials. Attach the common ends to the ground screw on the base assembly. Add coax, and you are ready to tune your antenna.

Antenna Tuning

Begin with 60 meters to determine the tap points on the coil. If your rig cannot tolerate a 2:1 SWR, you may need to add a 330 pF, 300 V, silver mica capacitor across the two wing-nut assemblies. See Figure 12 for a close-up view. This capacitor is fine for both 60 and 40 meters. The SWR on 30 meters will be closer to 1.7:1. If you need to improve this, use a 220 pF capacitor. You should not need any capacitors for 20 through 10 meters. Both my IC-706MKIIG and SG-2020 work fine into a 2:1 SWR.

To begin, use an antenna analyzer set to 5340 kHz to locate the coil tap point that gives the best SWR. Mark this tap point. Repeat the procedure at 5380 kHz. Move to 40 meters and repeat, again selecting two taps on 40 meters to give you the band coverage you desire. Repeat again for 30 meters—only a single tap is required for

AD5X

Figure 11—A close-up of the guy lines attached to the telescoping whip. The coil and coil clips detail can also be seen.

AD5X

Figure 12—The antenna base, with the ground radials and a matching capacitor attached to the brass screws on the PVC T.

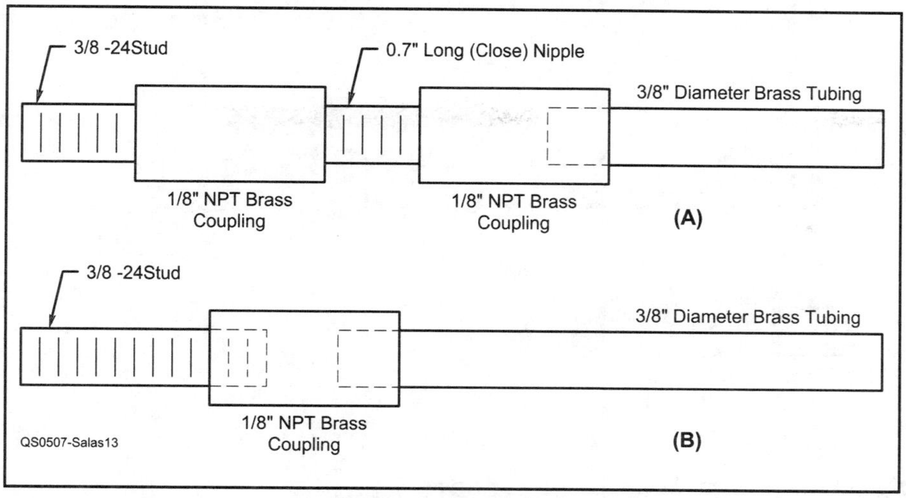

Figure 13—(A) shows how a 1/8 inch NPT brass coupling can be used to adapt the pipe threads to a 3/8-24 standard antenna mount. (B) shows an alternative attachment, if one only wants to attach the antenna to a standard mount, and not use the PVC base to mount the antenna on the ground.

this band. For 20 meters, you will find that only the top turn of the coil is necessary for resonance. The antenna is almost a quarter wavelength long on 20 meters.

For 17, 15, 12 and 10 meters, you will need to short out the loading coil and remove sections of the brass tube—then adjust the whip for resonance. On these bands, the antenna will be 1/4 λ long. Remove two brass sections for 17 and 15 meters, and all three sections for 12 and 10 meters. Use a permanent black marker pen to indicate the correct band positions on the telescoping whip.

Finally, attach the MFJ-605-4001 coil clips to the coil tap points determined above. You may wish to solder the clips in place to make the entire assembly a little more robust. From this point forward, you can just go back to these tap points, or re-adjust the top whip as necessary, and not have to worry about making SWR measurements.

Mounting Options

You can easily make a 3/8-24 threaded interface so that the antenna can be mounted on a standard 3/8-24 antenna mount. This would be useful for those with a standard ball mount on their car who want to use this extended-length antenna when stopped. As mentioned earlier, the 1/8 inch NPT thread is 27 turns per inch, and it is slightly larger than 3/8 inch diameter, with a slight taper. While the 3/8-24 standard stud will screw into an 1/8 inch NPT thread coupling, the 1/8 inch NPT nipple will not screw into a 3/8-24 threaded coupling. Therefore, an adapter is necessary if you want to mount this antenna to a standard 3/8-24 antenna mount.

One way to make an adapter is to purchase a 3/8-24 bolt and screw it tightly into a 1/8 inch NPT coupling. Cut off the head of the 3/8-24 bolt with a hacksaw and file carefully so that you don't damage the threads. Running a 3/8-24 die over the threads will clean up any damage you may have done. You can now either screw this assembly onto the 1/8 inch NPT nipple on the bottom brass tube section, as shown in Figure 13A, or screw the 3/8-24 bolt directly into the bottom 1/8 inch NPT coupling as shown in Figure 13B (if you'll never need the 1/8 inch NPT interface on the bottom antenna section).

Conclusion

Because of the interest in my original portable antenna, I've evolved that design into an antenna that is longer, lighter, more compact and easier to fabricate, and gives you more mounting options. You can also experiment with the antenna length. For example, you can remove a section or two, use more or fewer sections, decrease or increase section lengths, or place the loading coil in different positions. With the loading coil described in this article, you have quite a bit of latitude in the antenna length for a given band. For best efficiency though, try to keep the antenna as long as possible and the coil as high as possible.

Don't hesitate to make changes based on hardware availability. Try aluminum or copper tubing, or even wire wrapped 3/8 inch fiberglass or wood dowel. It's fun to design antennas "on the fly" while standing in the plumbing section of your hardware store. This makes for interesting discussions with the clerks, however!

Finally, I want to express my appreciation to Martin F. Jue, K5FLU, and Richard Stubbs, KC5NSZ, of MFJ, for working with me to provide a reasonably priced kit of parts to make this both an affordable and fun project to build.

Phil Salas, AD5X, is an ARRL Life Member. He's been licensed for 41 years, and enjoys HF operating (mostly CW). Phil's wife Debbie (N5UPT) and daughter Stephanie (AC5NF) are obviously very understanding of this hobby! Phil holds a BSEE from Virginia Tech, and an MSEE from Southern Methodist University, and is now fully retired after 33 years in the telecommunications industry. Phil can be reached at 1517 Creekside Dr, Richardson, TX 75081-2913 or at **ad5x@arrl.net** *if you have any questions or comments.*

By Rudy Severns, N6LF

From *QST*, July 2000

Verticals, Ground Systems and Some History

What makes a vertical antenna cook? Here you can gain some insight as to what this popular antenna likes and dislikes.

Over the past 100 years, beginning with Marconi and continuing to this day, vertical antennas and their associated ground systems have received considerable attention. Many fine articles and technical papers have explained the finer points of vertical antenna operation. Sometimes we forget the information's origins—and sometimes the wisdom gets a little distorted. Occasionally it's worthwhile to revisit the earlier work and recognize how the old relates to present-day applications.

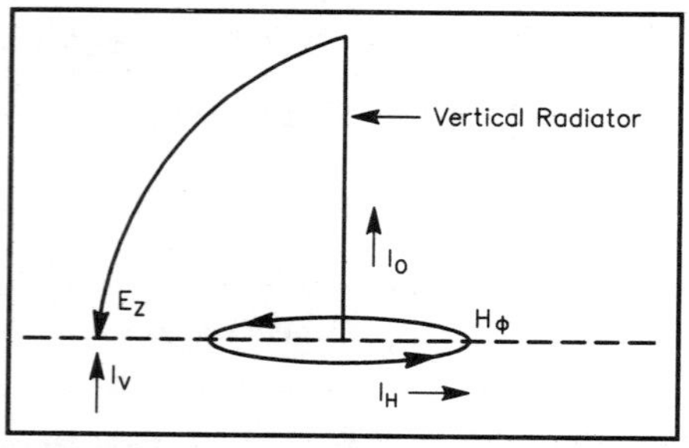

Figure 1—Fields and ground currents near the base of a vertical antenna.

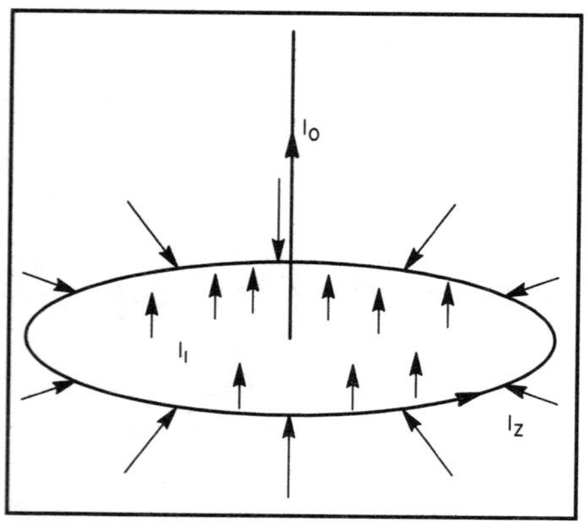

Figure 2—Definition of the current zone near the base of a vertical antenna. I_z represents the total current flowing through a zone at a given radius (r_1) by assuming the current is u niform to a depth of one skin depth (δ) as shown in Figure 13.

Research

A few years ago, I decided to get on 160 meters and wanted an effective antenna. I decided on a vertical of one form or another, but soon realized that I really didn't have a good understanding of how to get the best performance from a vertical. That led me to research the amateur and professional literature and discover a treasure trove of information.

Examining these early papers, I was struck by the depth of understanding and the quality of the work, both analytical and experimental. These papers represent a tremendous amount of effort—especially when you realize that up until a few years ago, all the computations were done manually with nothing more advanced than a pencil, a slide rule or a mechanical adding machine! Today, personal digital computers, equipped with a variety of software quickly manipulate the most complex expressions. With the software, it's easy for us to examine and manipulate mathematical expressions derived in earlier work and mine them for new understanding and insights. We now have antenna-modeling programs that are nothing short of magical, although their magic must be used with some caution. It's important to not only have a fundamentally solid understanding of antennas, but the modeling programs as well.[1]

[1]Notes appear on page 56.

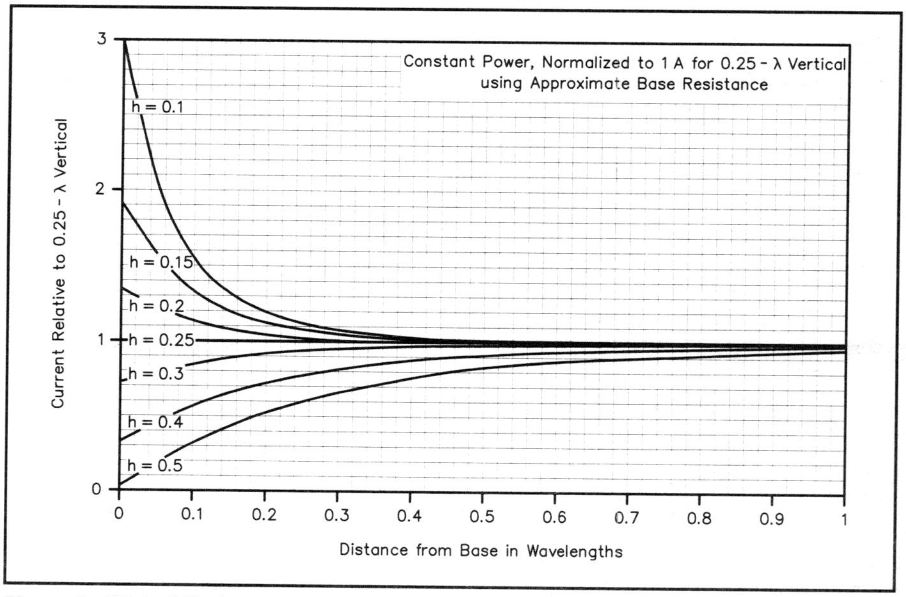

Figure 3—Plot of the current in amperes at the base of a vertical as a function of height and radius in wavelengths. The current in the base of the 0.25-λ antenna is assumed to be 1 A and the currents in the other antennas are adjusted to maintain the same input power.

What follows is a short tour of some of the earlier work that explains some of the lore of verticals and where it came from. I put the math in an Appendix and generated graphs for the discussion. All the graphs were done using a spreadsheet. After reading this article, I recommend you explore for yourself using the equations in the Appendix. The integration of power for Figure 6 was done with *Maple*; *MathCad* or *Mathmatica* would also do fine. You can also do integration with a spreadsheet.[2]

George Brown

In the mid-1930s, radio broadcasting was coming of age and the Institute of Radio Engineers (IRE) proceedings had many papers on vertical antennas and associated ground systems. One of the more influential writers of the time was George H. Brown. A series of papers written by Brown and his colleagues[3-10] at RCA have proved over time to be the most influential. The 1937 IRE paper (see Note 9) has been repeatedly referred to in Amateur Radio publications and is the basis for many later articles.[11-19] (References 16 and 19 have extensive bibliographies for further study.) At the time, these papers were so influential that they became the basis for the FCC standards for broadcast antenna installations! The way we think about verticals today has, in large part, been shaped by this work.

George Brown received his PhD from the University of Wisconsin-Madison in 1933. The core of his dissertation[20] is an analysis of the fields and ground currents associated with a vertical antenna with an extensive buried-radial ground system. This became the basis for much of the work that followed. Brown's work contains a great deal of analysis in addition to experimental results.

Papers on broadcast verticals were not Brown's only contributions to antenna art. He is credited with inventing the ground-plane antenna and wrote numerous other papers on antenna subjects. In later years, Brown was the director of the RCA Sarnoff laboratory. Although not a ham, George Brown contributed enormously to Amateur Radio.

A Closer Look at Verticals

A vertical antenna has two field components that induce currents in the ground around the antenna. Figure 1 shows (in a general way) the electric (E, V/m) and magnetic (H, A/m) field components in the region near the antenna. Because the soil near the antenna usually has a relatively high resistance, both of these field components can induce currents (I_V and I_H) in the ground surrounding the antenna resulting in losses. The worms may enjoy the heated ground, but the power dissipated there subtracts from the radiated power, weakening the signal. As indicated in Figure 1, the tangential component of the H field ($H\phi$) induces horizontal currents (I_H) flowing radially and the normal component of the E field (E_z) induces vertically flowing currents (I_v). Actually, things are a bit more complex than this, but we don't need to thrash that to understand conceptually what's going on. Introducing a system of ground wires, buried or elevated, modifies the current flowing in the ground and (hopefully) reduces loss.

Brown's work was primarily concerned with broadcast antennas in the 0.5 to 1.5 MHz range, although some of his experimental work was carried out at 3 MHz. To make the analysis tractable he made several assumptions:

- The ground system would consist of a large number of radials buried a short distance below the surface.
- The ground characteristics were predominately resistive, ie, dominated by conduction currents, so displacement currents could be ignored.
- Because of the extensive ground screen and its shallow depth, the E-field losses were assumed to be small.

For his work, these assumptions were good approximations, but they are not en-

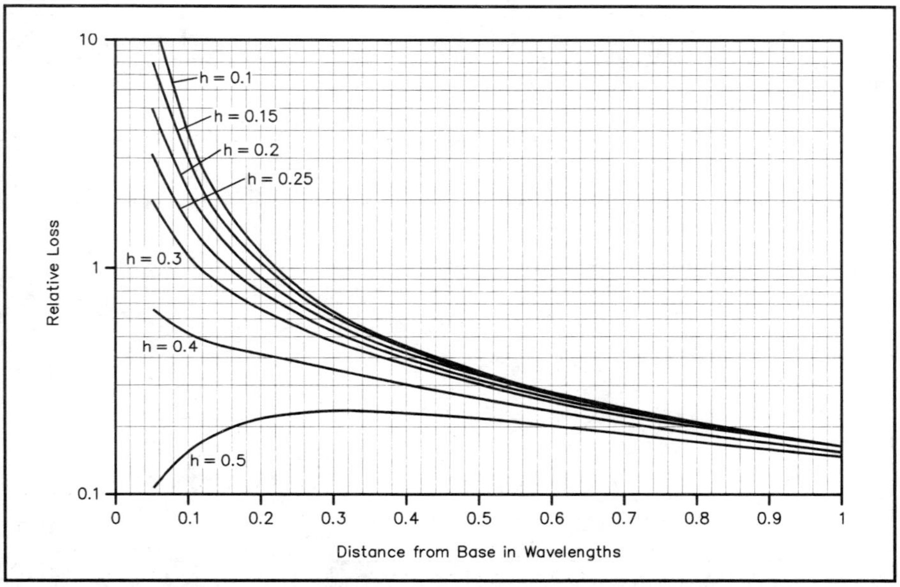

Figure 4—Relative ground loss for several different height verticals. The loss is normalized by allowing the expression which takes into account skin depth and ground conductivity to be equal to 1.

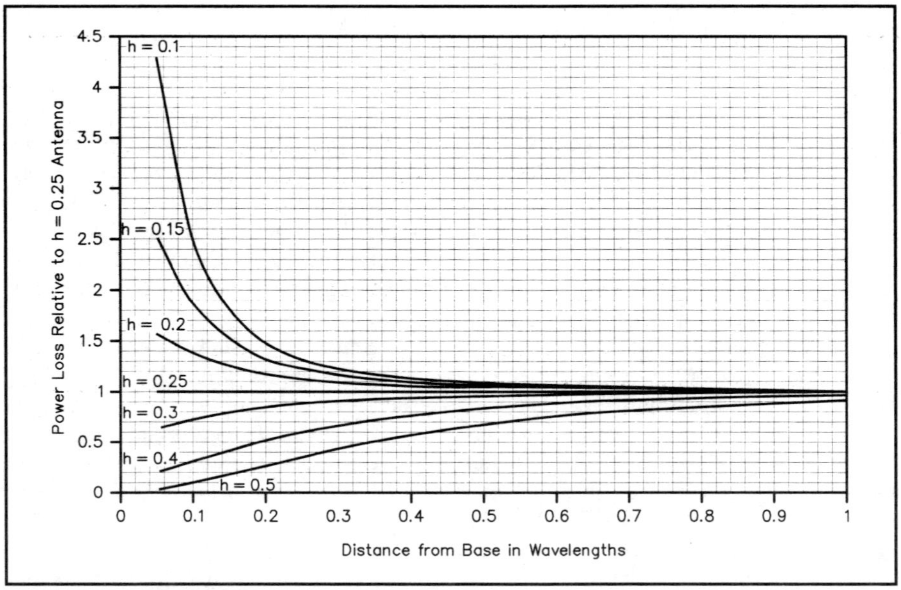

Figure 5—Ground loss at a given radius relative to a 0.25-λ vertical.

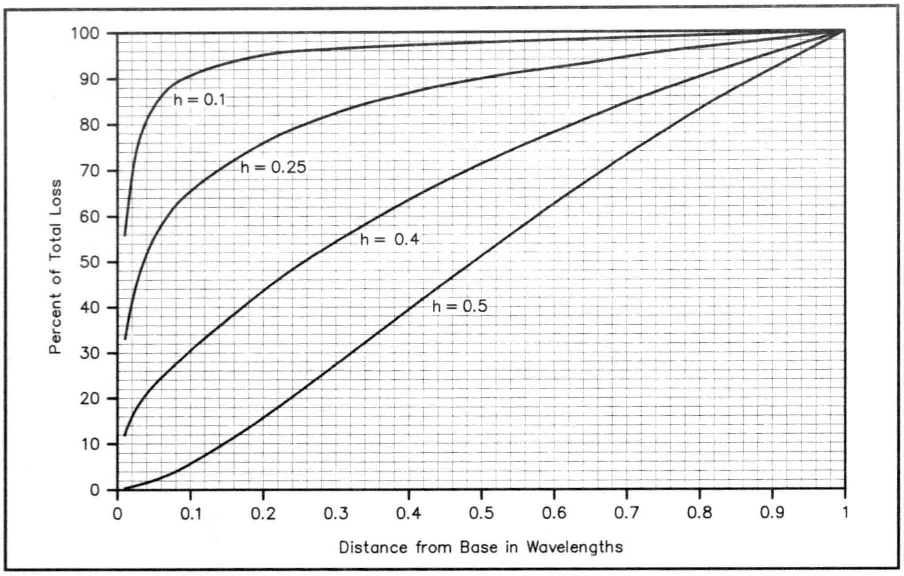

Figure 6æPercent of total ground loss within a given radius (in wavelengths) relative to the total loss at 1-l. This is a measure of the effectiveness of a ground system of a given radius.

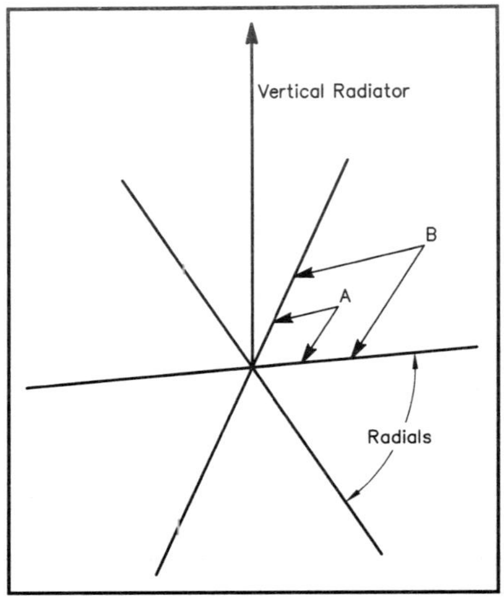

Figure 8—Current entering the ground between radial wires.

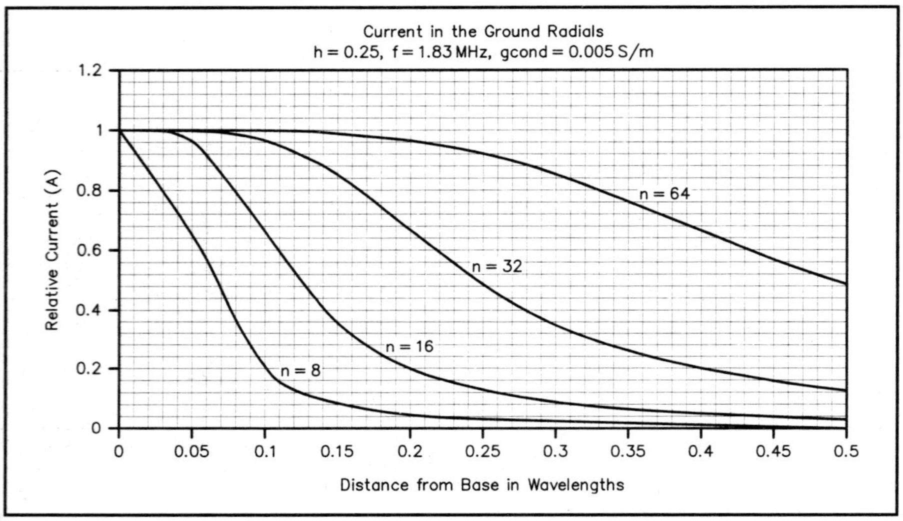

Figure 7—Total current in the radials (I_w) as a function of radius from the base of a 0.25-λ vertical operating at 1.83 MHz and with a ground conductivity of 0.005 S/m (average ground).

tirely valid for HF amateur verticals with small numbers of radials and certainly not valid for elevated radials. Nonetheless, his work is a very good place to start. At the end of the discussion we will look again at these assumptions.

Figure 2 is a sketch of current flow in the antenna and the surrounding ground. I_z represents the total current flowing through a cylindrical zone at a given radius. I_1 represents the current returning to the antenna in addition to the base current. I_0 is the current at the base of the antenna. Brown derived an equation (see the Appendix) that describes the ground current as a function of antenna height and distance from the base of the antenna. The heights I will be using in the following discussion are the effective *electrical* heights. For example, if you use some top loading on the vertical, the effective electrical height is greater than the physical height. For the following graphs, I have used simplified expressions that use the effective height. It is important to recognize that simply adding a top hat to a vertical of given physical height can reduce the ground losses. We will be able to see this from the effect of height on ground-current amplitudes. Simply moving a loading coil from the antenna's base further into the antenna reduces ground losses because it reduces ground-current amplitude.

Figure 3 is a graph of this current (I_2)for several effective heights. The currents have been adjusted for constant input power (about 37 W) at the base of the antenna, with 1 A into a 0.25-λ vertical as the reference. This graph clearly shows the high currents flowing in the ground near the base of a short antenna. Compared to a 0.25-λ vertical, the 0.1-λ vertical has three times the ground current; as you further shorten the antenna, the ground current increases rapidly. Keep in mind that the ground loss is proportional to the *square* of the current (I^2R), so the power loss in the immediate region of the base is *much* higher for the shorter antenna.

One way to visualize the relative losses is to calculate them. This is where a spreadsheet really helps. If you take the currents given in Figure 3, square them and divide by the circumference of a circle at a given distance from the base—taking into account the ground resistance and the current's depth of penetration—you know the power loss at a given radius. Figure 4 is a graph of the power loss as a function of the distance from the antenna base. This shows that the losses are high near the base, are greater for shorter antennas and taper off rapidly as distance from the base increases. Note also that for a 0.5-λ vertical, the maximum loss occurs about 0.3-λ away from the base! The ground system in this region may profit from some additional attention. You may ask "Who uses 0.5-λ verticals, especially on 80 or 160 meters?" What about 0.5-λ slopers hung from towers? Even though they are typically not connected directly to ground, they would benefit from a ground system under them. John Devoldere, ON4UN, makes this point in his book (see Note 19). For simplicity, in Figure 4, I have assumed that the depth of current penetration into the soil and the soil conductivity are normalized to 1. For the actual losses in real ground at amateur operating frequencies, the proper equations are in the Appendix if you would like to graph them for yourself. We can also generate a graph showing the loss relative to the 0.25-λ vertical as shown in Figure 5.

Now we can take the next step and integrate the total loss inside a given radius to get a feeling for how large we should make our ground systems. Figure 6 is a graph of the total loss within a given radius, relative to the total loss inside a 1-λ radius for *each* antenna height. I chose the 1-λ radius as the reference because it contains most of the near-field loss and also represents a practical maximum radial length for most installations (560 feet on 160 meters!)The absolute value of the total loss is, of course, higher for a short antenna when compared to a taller one. For the 0.1-λ-high antenna, if we have a good ground screen out to a distance of 0.1-λ, we'll eliminate over 90% of the ground loss! This is where the idea comes from that for short antennas we should concentrate our ground systems inside a short radius. A larger ground system will do no harm; in fact, it reduces the loss even more, but if we have a limited amount of wire, we are much better off to *use many short radials instead of a few long radials*. Note that this graph assumes a large number of radials (more than 100). If only a few radials are used, the effectiveness of the ground system is reduced, although for short antennas it is not necessary to use as large a number of radials.

We can see why this is so by using another of Brown's equations, the one for the current in the radials as a function of radial length and number of radials (see Appendix). Figure 7 is a graph of the current in the radials as you move away from the base of a 0.25-λ vertical with various numbers of radials. The vertical has a 1 A current in the base and (from Figure 3) the total current (I_z) is constant as you move farther out. What we see is the current in the radials (I_w) falling off. The fewer the radials, the more rapidly the current decreases with distance from the base. The total current is still 1 A, but the remainder (I_e) is flowing in the ground and inducing losses. If you use only a few radials it does no good to make them very long because the outer portions of the radials pick up very little current.

What's happening here? Figure 8 is a sketch of a radial system with current entering the ground at two points (A and B). Current reaching the ground at point B has to flow much farther in the soil than current at point A before reaching a radial. The farther from the radiator you go, the greater is the distance between each radial and its neighbor and the farther is the distance the current must flow in the soil. There comes

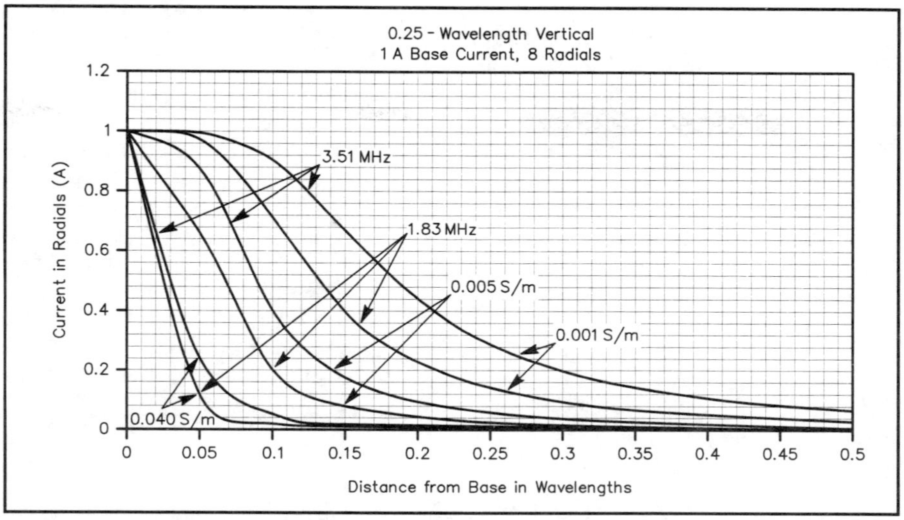

Figure 9—The effect of ground conductivity and frequency on the current in radial wires 1 A of base current and eight radials.

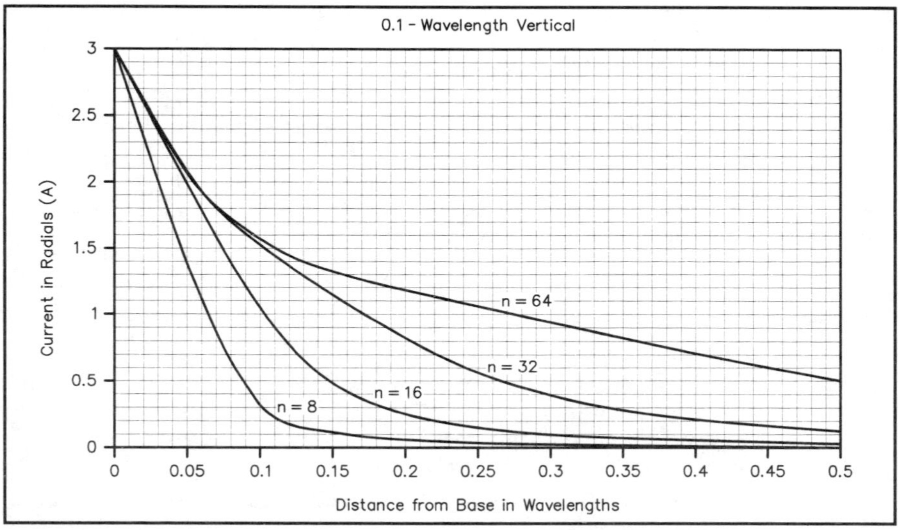

Figure 10—Radial-wire currents of a 0.1-λ vertical for several different numbers of radials (n).

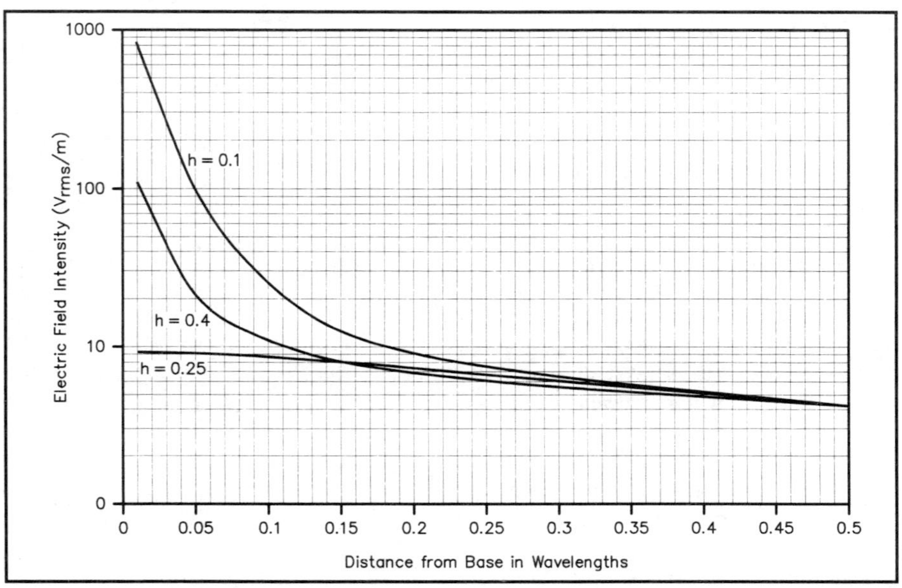

Figure 11—Electric-field intensity near the base of a vertical operating at 1.830 MHz with 1500 W input.

a point where the distance between the radials is so great that the radials are no longer effective. The more radials you use the closer together they will be (at a given radius) and the farther out will be the point at which the radial is no longer effective.

Now that we have Brown's equations in our spreadsheet we can explore further the effects of ground conductivity and frequency on radial number and length. In Brown's time this would have been very laborious, for us it is just a few mouse clicks! Figure 9 is a graph for a 0.25-λ vertical with eight radials, at 1.83 and 3.51 MHz for three different ground conductivities. Notice that as the ground improves (higher conductivity) the current in the radials falls more rapidly. This seems paradoxical: To get the full benefit of the radial system, you have to have more radials as the ground improves! Notice also that as frequency is increased, longer radials can be used effectively.

What about the change in radial current for shorter or longer antennas? That's easy. We just multiply the current values in Figure 3 times the values in Figure 7. Figure 10 is an example for a 0.1–λ vertical. Again we see the advantages of using lots of relatively short radials with a short vertical.

Electric Fields Near the Base

Another consideration is the intensity of the electric field (E) in the region around the base of the antenna. Figure 12 is a graph of E near the base of several verticals of different heights with an input power of 1500 W at 1.830 MHz. Notice how high the field is for the 0.1-λ antenna: about 100 times the value for the 0.25-λ vertical. This is an important consideration for any conductors or structures close to the base of the antenna. Large potentials can be induced into them. These fields can even ignite tall grass! Notice also that as the antenna height exceeds 0.25-λ, the field intensity again increases. The old-fashioned 0.25-λ vertical has many advantages.

A Word of Caution

George Brown's work has proven to be very useful and has been the basis for many articles in amateur publications. However, we have to keep in mind the assumptions Brown made (listed earlier) and remember that his concern was for *broadcast* applications. One assumption he made is that the ground characteristic is primarily resistive. This is a good approximation for most grounds at 160 and even 80 meters, but at higher frequencies, the ground behaves as though there is *capacitance in parallel with the resistance*: ie, there will be displacement as well as conduction currents.

For frequencies above 4 MHz, Brown's equations still give us a good qualitative feeling for what's going on and the overall guidance they offer is still valid. But Brown was careful to point out that you shouldn't rely on the absolute numbers. The need to consider displacement currents can be illustrated by looking at curves for skin depth in soil as a function of frequency and ground characteristics (the generating equations are in the Appendix). Figure 12 is representative of skin depths for typical soils. The graph is an extension of one given in *QST* by Charlie Michaels, W7XC (see Note 18). The dashed lines represent skin depth when conductivity only is considered. The solid lines represent skin depths using the com-

Appendix

Definitions

I_o = current in the base of the antenna or at the current loop in the case of the 1/2λ antenna
I_z = zone current at radius $r_1 = I_w + I_e$
I_e = total current in the earth at radius r_1
I_w = total current in radial wires at radius r_1
f = frequency in Hertz
f_{MHz} = frequency in MHz
E = electric field intensity
h = height of antenna in wavelengths
r_1 = distance from base in wavelengths
s = soil conductivity in Siemens/meter [S/m]
n = number of wires in the radial system
r_2 = radius of radial wires in cm

Zone Currents

$$\frac{|I_Z|}{|I_o|} \equiv I_n = \frac{1}{\sin 2\pi h}\sqrt{\left[\sin 2\pi p \pm \sin 2\pi r_1 \cos 2\pi h\right]^2 + \left[\cos 2\pi p \pm \cos 2\pi r_1 \cos 2\pi h\right]^2}$$

$$p \equiv \sqrt{r_1^2 + h^2}$$

(Equation 1)

Current Distribution in Radial Wires

$$\frac{I_e}{I_w} = j\left(\frac{3.6\sigma\pi^4 r_1^2}{f_{MHz}\, n^2}\right)\left[\log\left(\frac{3\times 10^4 \pi r_1}{f_{MHz}\, n r_2}\right) \pm 0.5\right] = j\left|\frac{I_e}{I_w}\right|$$

$$\left|\frac{I_w}{I_z}\right| = \frac{1}{\sqrt{1+\left|\frac{I_e}{I_w}\right|^2}}$$

$$\left|\frac{I_e}{I_z}\right| = \frac{1}{\sqrt{1+\left|\frac{I_w}{I_e}\right|^2}}$$

(Equation 2)

Electric Field Intensity

$$|E| = \left(\frac{2 f_{MHz} I_o}{\sin 2\pi h}\right)\sqrt{\left(\frac{\cos 2\pi \sin 2\pi r}{r_1} \pm \frac{\sin 2\pi p}{p}\right)^2 + \left(\frac{\cos 2\pi h \cos 2\pi r_1}{r_1} \pm \frac{\cos 2\pi p}{p}\right)^2}\left[\frac{V}{m}\right]$$

(Equation 3)

plete equation for skin depth in a general medium. What has been added is the *permittivity of the soil*, which is related to capacitance. For seawater, the conductivity dominates at any frequency below 2 meters. For very good soil, we see that conductivity still dominates over the HF range, but for average or poor soils, the expression for skin depth considering only conductivity gives a depth that is progressively much too large, especially for poor soils. This alters the ground-current distributions from those predicted by Brown; the actual losses may be higher.

If we look at most amateur literature concerning ground characteristics, we see that the emphasis is on measuring ground resistance and the effect of ground resistance on losses, with little said about the permittivity. This is a direct reflection of Brown's work and his concern with broadcast frequencies. We have been following his lead for the last 60 years. In reality, for most soils at HF, we need to take into account the permittivity of ground. Unfortunately, measuring the complex impedance of soil is considerably more difficult than measuring just soil conductivity. W7XC's article partially corrected this and was incorporated in later editions of the *ARRL Antenna Book*, but we still have some work to do.

Brown also assumed that the E-field losses were small. (In his 1935 paper and his thesis, he does compute the electric-field intensity, but then points out that these ground losses are small when a shallow, dense, buried radial system is used with a 0.25-λ vertical. For systems with many buried radials, this is a good approximation. However, when there are only a few radials, or when the radials are elevated above ground, the E-field loss may not be small at all. The importance of E-field losses to amateurs has been pointed out by Clay Whiffen, KF4IX, and Ben Zieg, K4OQK.[21] They showed the increased loss possible when the top of a vertical (where there is a very high electric field) is placed close to a tree. We also know that the outer ends of elevated radials have very high potentials and can induce E-field losses in the ground, grass, shrubs and sod beneath the radial system.

When we compare buried radials with elevated radials we find that the current distribution is very different between the two types of radial systems (see Note 14). Making buried radials longer may not help much if only a few radials are used, but it doesn't hurt. Buried radial systems with a radius greater than 0.5λ can be very effective if enough radials are used. However, as Burke and Miller[22] have shown, making *elevated* radials longer than 0.3λ can lead to greatly increased loss when only a few radials are used. Larger numbers of elevated radials do reduce this loss and allow larger elevated ground systems to be effective. It is important that we *do not directly equate* buried and elevated ground systems on the basis of Brown's work. They are different animals, both of which certainly have their place.

Skin-Depth Equations

The exact expression for penetration or skin depth in a general material is given by:

$$\delta = \left(\frac{\sqrt{2}}{\omega\sqrt{\mu\varepsilon}}\right)\left[\sqrt{1+\left(\frac{\sigma}{\omega\varepsilon}\right)^2}\pm 1\right]^{\pm 1/2} \quad \text{(Equation 4)}$$

where:

δ = skin depth in meters

$\omega = 2\pi f$

$\mu = \mu_o\mu_r$

$\mu_o = 4\pi 10^{-7}$ Henry/meter

μ_r = relative permeability

$\varepsilon = \varepsilon_o\varepsilon_r$

$\varepsilon_o = 8.85\times10^{-12}$ Farads/meter

ε_r = relative permittivity

For most soils, $\mu_r \approx 1$ (unless you set up shop in an open-pit iron mine!). For good conductors:

$$\frac{\sigma}{\omega\varepsilon} >> 1 \quad \text{(Equation 5)}$$

Which allows the equation for δ to be simplified to:

$$\delta = \frac{1}{\sqrt{\pi\sigma\mu f}}\ \text{m} \quad \text{(Equation 6)}$$

where:

f is in Hertz

Ground loss

Ground loss for a ring of soil (dr) at a given radius (r_1) from the base can be calculated with the aid of figure 13. If we assume that the average current is uniform to one skin depth (δ), the loss in the ring will be:

$$\frac{dP}{dr} = \frac{I_e^2}{2\pi\delta\sigma r} = \frac{f_{MHz}\, I_e^2}{600\pi\delta\sigma r_1} = \left(\frac{f_{MHz}}{300\delta\sigma}\right)\left(\frac{I_e^2}{2\pi r_1}\right)\left[\frac{W}{m}\right] \quad \text{(Equation 7)}$$

where:

δ and r are in meters and r_1 is in wavelengths (λ).

$$\lambda = \frac{300}{f_{MHz}}\,[\text{m}] \quad \text{(Equation 8)}$$

The graph in Figure 4 assumes that

$$\frac{f_{MHz}}{300\delta\sigma} = 1 \quad \text{(Equation 9)}$$

and that r_1 is in wavelengths.

A Final Word

I hope you will find this information useful. If you really want a thorough understanding of the topic, you should graph these equations yourself and read the listed references.[22] The *QST*, *ham radio* and *CQ* articles are quite easy to follow; even Brown's papers are no great chore to read. Some modeling with *NEC* or *MININEC* software will give you even more insight. Particularly on

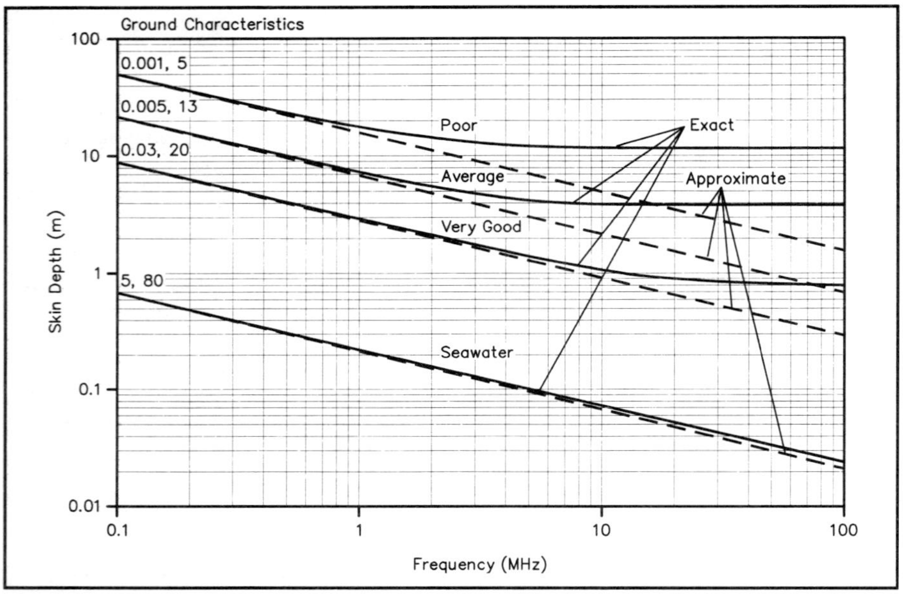

Figure 12—Skin depth in soil of various characteristics as a function of frequency.

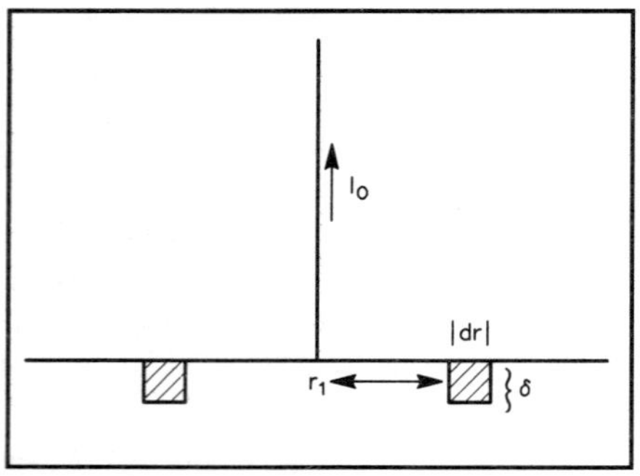

Figure 13—Calculation of ground loss in a small ring of soil at a given radius.

the lower bands, verticals can be very effective, but you have to understand what you are about to get good results.

Acknowledgement

The equations used here have been taken directly from Brown's 1935 (see Note 4) and 1937 (see Note 9) papers. All I have done is to restate them in a form handy for spreadsheet manipulations and to repeat some of his observations and conclusions.

Notes

[1]See also L. B. Cebik, W4RNL, "*NEC-4.1: Limitations of Importance to Hams*," *QEX*, May/Jun 1998, pp 3-16 and John Rockway and James Logan, N6BRF, "Wire Modeling Limitations of *NEC* and *MININEC* for *Windows*," *QEX*, May/Jun 1998, pp 17-21.

[2]Byron Gottfried, *Spreadsheet Tools For Engineers*, McGraw-Hill Basic Engineering Series Tools, 2000, see Chapter 9. A compressed *Excel* file (*SEVERNS.ZIP*) is available from **http://www.arrl.org/files/qst-binaries/**.

[3]G. H. Brown and Ronold King, "High-Frequency Models in Antenna Investigations," *IRE Proceedings*, Vol 22, No. 4, Apr 1934, pp 457-480.

[4]George H. Brown, "The Phase and Magnitude of Earth Currents Near Radio Transmitting Antennas," *IRE Proceedings*, Vol 23, No. 2, Feb 1935, pp 168-182.

[5]H. E. Gihring and G. H. Brown, "General Considerations of Tower Antennas for Broadcast Use," *IRE Proceedings*, Vol 23, No. 4, Apr 1935, pp 311-356.

[6]G. H. Brown, "A Critical Study of The Characteristics of Broadcast Antennas as Affected by Antenna Current Distribution," *IRE Proceedings*, Vol 24, No. 1, Jan 1936, pp 48-81.

[7]G. H. Brown, "Directional Antennas," *IRE Proceedings*, Vol 25, No. 1, Part 1, Jan 1937, pp 78-145.

[8]G. H. Brown and John G. Leitch, "The Fading Characteristics of the Top-Loaded WCAU Antenna," *IRE Proceedings*, Vol 25, No. 5, May 1937, pp 583-611.

[9]G. H. Brown, R. F. Lewis and J. Epstein, "Ground Systems as a Factor in Antenna Efficiency," *IRE Proceedings*, Vol 25, No. 6, Jun 1937, pp 753-787.

[10]G. H. Brown, "A Consideration of the Radio-Frequency Voltages Encountered by the Insulating Material of a Broadcast Tower Antenna," *IRE Proceedings*, Vol 27, No. 9, Sep 1939, pp 566-578.

[11]F. R. Abbott, "Design of Optimum Buried-Conductor RF Ground Systems," *IRE Proceedings*, Vol 30, No. 7, Jul 1952, pp 846-852.

[12]R. E. Leo, W7LR, "Vertical Antenna Ground Systems," *ham radio magazine*, May 1974, pp 30-35.

[13]John O. Stanley, K4ERO/HC1, "Optimum Ground Systems for Vertical Antennas, *QST*, Dec 1976, pp 13-15.

[14]Jerry Sevick, W2FMI, "Short Ground-Radial Systems for Short Verticals," *QST*, Apr 1978, pp 30-33.

[15]Archibald C. Doty, K8CFU (now W7ACD), John A. Frey, W3ESU, and Harry J. Mills, K4HU, "Efficient Ground Systems for Vertical Antennas" *QST*, Feb 1983, pp 20-25.

[16]Archibald C. Doty, K8CFU, "Improving Vertical Antenna Efficiency," *CQ*, Apr 1984, pp 24-31.

[17]Brian Edward, N2MF, "Radial Systems for Ground-Mounted Vertical Antennas," *QST*, Jun 1985, pp 28-30.

[18]C. J. Michaels, W7XC, "Some Reflections On Vertical Antennas," *QST*, Jul 1987, pp 15-19.

[19]John Devoldere, ON4UN, *Low-Band DXing*, (Newington: ARRL, 3rd ed), 1999, Chapter 9.

[20]G. H. Brown, "A Theoretical and Experimental Investigation of The Resistance of Radio Transmitting Antennas," PhD thesis, University of Wisconsin-Madison, June 29, 1933

[21]Clay Whiffen, KF4IX, and Ben Zieg, K4OQK, "Trees and Verticals," Technical Correspondence, *QST*, Nov 1991, p53

[22]G. J. Burke and E. K. Miller, "Numerical Modeling of Monopoles On Radial-Wire Ground Screens," *IEEE Antennas and Propagation Society International Symposium Proceedings*, Jun 1989, pp 244-247

Rudy Severns, N6LF, was first licensed as WN7WAG in 1954 and has held an Extra class license since 1959. He is a consultant in the design of power electronics, magnetic components and power-conversion equipment. Rudy holds a BSE degree from the University of California at Los Angeles. He is the author of two books and over 80 technical papers. Rudy is also an IEEE Fellow. You can contact Rudy at PO Box 589, Cottage Grove, OR 97424; **rudys@ordata.com**.

From *QEX*, March/April 2005

A Study of Phased Vertical Arrays

The author describes a number of possible multi-element vertical antenna arrays, outlining their features and the differences in expected performance.

By Al Christman, K3LC

Introduction

The 4-square phased-vertical array is widely used by DXers and contesters for operation on the low bands from 40 through 160 meters. There are a number of different permutations of this array that are useful to hams, including variations in element spacing and current phasing. Several larger arrays are available, using from six to nine elements, and these also deserve consideration. This article reviews the performance of the typical 4-square array in terms of forward gain, front-to-back ratio, front-to-rear ratio, and azimuthal beamwidth. A variety of modified 4-squares are also included, along with the 6-hex and 9-circle array designs.

Element Description

A frequency of 3.65 MHz in the 80-meter band was selected for the computer analysis, which was performed using EZNEC-4.[1] A full-size quarter-wave vertical element composed of # 12 AWG copper wire was placed over average soil with a conductivity of 5 millisiemens per meter and a dielectric constant of 13. The segment lengths for the vertical monopole were tapered in accordance with the most conservative NEC guidelines, with the shortest (feed-point) segment having a length of 6 inches. I selected # 16 AWG copper wire for the radials, since this wire is widely available at reasonable cost. The inner segment of each buried radial slopes downward into the soil from the base of the vertical element, to a depth of 3 inches, with the major portion of the radial extending horizontally at this same depth.

I decided to utilize 5000 feet of wire in the radial ground-screen for each element, and conducted a series of tests to determine the optimum number and length for these buried wires. Simulation on the computer indicated that I could get the same gain using anywhere from 46 to 52 radials, whose corresponding lengths spanned the range from 108.7 to 96.2 feet. Because I planned to use arrays containing either 4, 6, or 8 symmetrically-placed elements (the 9-circle array positions one element in the center), I decided to utilize 48 radials per element, with each one having a maximum length of about 104.2 feet.

Modelling Considerations

For all of the arrays, I configured the radials in a fashion similar to what would be found at an AM-broadcast antenna site. Fig 1 illustrates this technique, and shows the computer-generated model for the conventional 4-square. Copper wires (# 12 AWG) serve as "bus-bars" to which radials from adjoining elements are soldered where they intersect. This methodology prevents any overlap of one element's ground-screen with another, and reduces the total amount of wire needed for the radial system. For consistency, I oriented each array so that its main beam would fire at an azimuth of 45°, common practice here in the north-eastern part of the US.

Classical 4-Square

The first array to be examined was the popular 4-square, which has now been in use for several decades, since it was popularized in the well-known 1976 *QST* article.[2] In this configuration, the four elements are placed at the corners of a square whose side-length is 0.25 λ. Normally the feed system is set up so the array fires through the corners of the square. This can be accomplished by driving all four elements with equal-amplitude currents. The usual current phase-angles are 0° for the rear element, –90° for the two side elements, and –180° for the front element. EZNEC predicts a peak forward gain of 6.16 dBi at a take-off angle (TOA) of 23°, with a front-to-back ratio (FBR) of 16.58 dB in the elevation plane. In the azimuthal plane the FBR is 22.71 dB, while the half-power beamwidth (HPBW) is 100.6°.

Fig 2 shows the elevation-plane radiation pattern for the array, and also illustrates the pattern that results when firing through the sides of the square. To do this, all four elements are driven with equal-amplitude currents, as before, but now the two rear elements are driven at a phase-angle of 0°, while the two front elements are both driven at –90°. In this case the maximum forward gain is 5.18 dBi at 25° TOA, and the FBR is 12.87 dB. In the azimuthal plane the FBR is 21.97 dB, while the HPBW is 130.8 degrees.

[1]Notes appear on page 62.

Grove City College
100 Campus Drive
Grove City, PA 16127-2104

Fig 3 displays both azimuthal-plane radiation patterns together on the same plot, showing how the secondary feed configuration helps to fill in those regions where the primary beams are at their half-power points.

4-Square with W8JI Feed

As Tom, W8JI, points out,[3] the phase-angles of the currents that are used to drive the typical 4-square array are far from optimal, and should be much larger. W8JI suggests that the two middle elements be driven with currents whose phase angle is –120° with respect to the rear element, while the current in the front element should have a phase-angle of –240°. This strategy yields a peak forward gain of 6.61 dBi at a TOA of 22°, with a front-to-back ratio of 25.23 dB in the elevation plane. In the azimuthal plane, the FBR is 31.50 dB and the HPBW is 80.2°. The radiation patterns are shown in Fig 4 and 5. Notice that almost half a decibel of extra gain has been obtained from the same physical array as before, simply by changing the phase-angles of the base currents. The new pattern is much sharper, as evidenced by a significant reduction in the azimuthal beam-width, and the FBR is very high, although there are some fairly-large side lobes.

Since the HPBW has been reduced from 100.6 to 80.2°, the four traditional directions of fire (through the corners) no longer provide good coverage of all sectors of the compass. Thus, it is advantageous to design the feed system of this array so it can fire through the sides of the square as well as through the corners. To do this, we could simply repeat the strategy used earlier with the classic 4-square, feeding the two rear elements at 0° and the two front elements at –90°, which would produce the same values of gain, FBR, etc, as were quoted above. However, by driving the two front elements at –110°, we can increase the maximum forward gain to 5.57 dBi at 24° TOA, while the FBR improves to 16.51 dB. In the azimuthal plane, the FBR is now 18.36 dB, and the HPBW is 116.6°. These radiation patterns are also displayed in Fig 4 and 5.

Expanded 4-Square

Another way to get more gain from

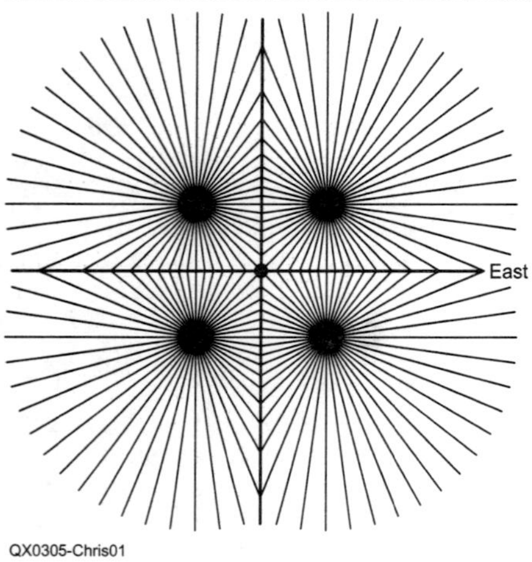

Fig 1—A bird's-eye view of a traditional 4-square phased-array, showing details of the broadcast-style radial ground-screen. Each element has 48 radials, whose maximum length is 104.2 feet.

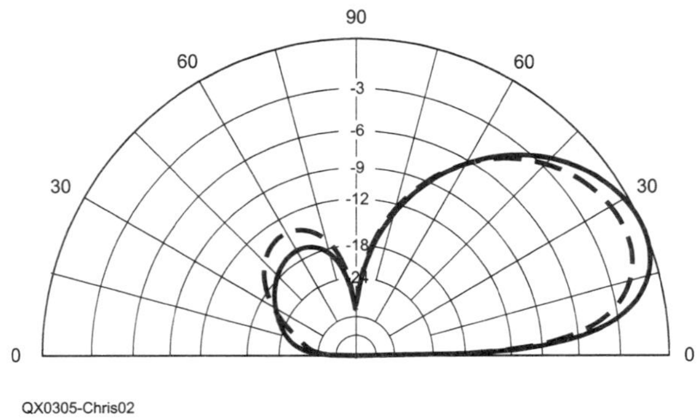

Fig 2—Elevation-plane radiation patterns for the conventional 4-square array. Solid trace: firing through the corner; gain = 6.16 dBi at 23° take-off angle, front-to-back ratio = 16.58 dB. Dashed trace: firing through the side; gain = 5.18 dBi at 25° take-off angle, front-to-back ratio = 12.87 dB.

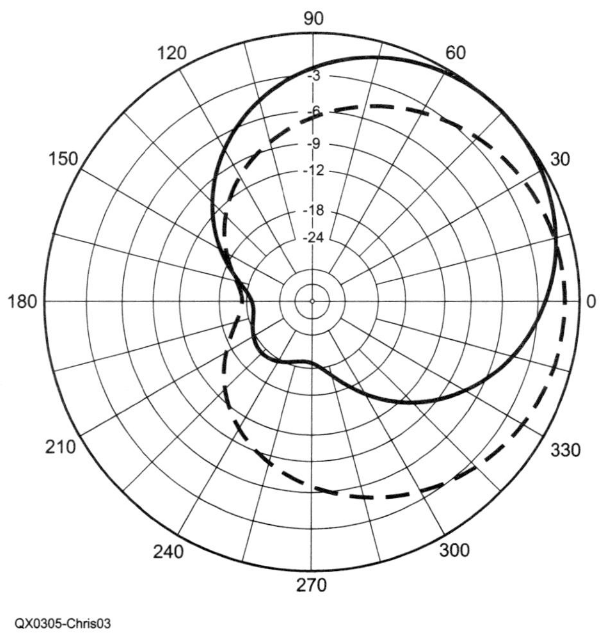

Fig 3—Azimuthal-plane radiation patterns for the conventional 4-square array. Solid trace: firing through the corner; front-to-back ratio = 22.71 dB, half-power beamwidth = 100.6°. Dashed trace: firing through the side; front-to-back ratio = 21.97 dB, half-power beamwidth = 130.8°.

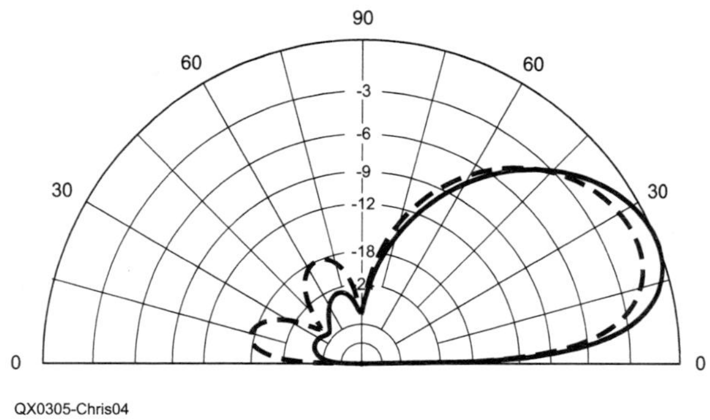

Fig 4—Elevation-plane radiation patterns for the 4-square array, with W8JI-style feed. Solid trace: firing through the corner; gain = 6.61 dBi at 22° take-off angle, front-to-back ratio = 25.23 dB. Dashed trace: firing through the side; gain = 5.57 dBi at 24° take-off angle, front-to-back ratio = 16.51 dB.

a 4-square is to make it larger by spacing the elements farther apart. If we want to use the same phase-angles as in the traditional array (0, –90, –90, and –180°), then a diagonal spacing of 0.5 λ is suitable,[3] which means the side-length of the square expands from 0.25 λ to about 0.354 λ. The peak forward gain rises to 6.84 dBi at a TOA of 24°, and the FBR is 20.98 dB in the elevation plane. In the azimuthal plane, the FBR is 35.26 dB (although there are large side-lobes) and the HPBW is 87.0°. Fig 6 and 7 illustrate the radiation patterns.

As was true with the W8JI-style feed, the azimuthal-plane beam-width of this antenna is less than 90°, so we need to make the array fire through the sides of the square as well. We can use the traditional phase-shifts of 0 and –90°, which yield a peak forward gain of 6.11 dBi at 27° TOA, and a FBR

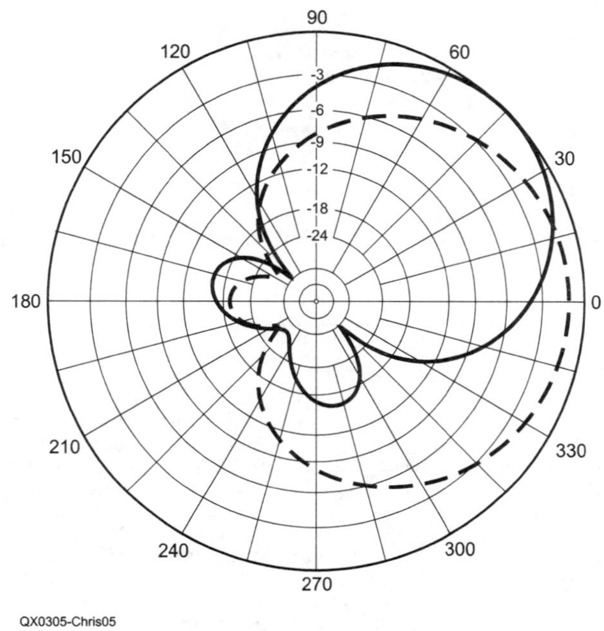

Fig 5—Azimuthal-plane radiation patterns for the 4-square array, with W8JI-style feed. Solid trace: firing through the corner; front-to-back ratio = 31.50 dB, half-power beamwidth = 80.2°. Dashed trace: firing through the side; front-to-back ratio = 18.36 dB, half-power beamwidth = 116.6°.

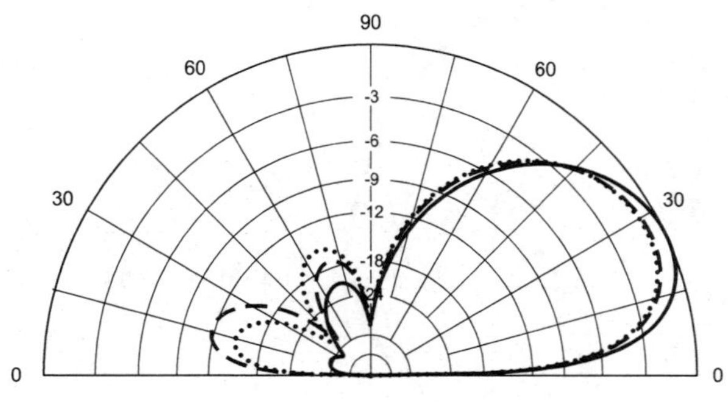

Fig 6—Elevation-plane radiation patterns for the expanded 4-square array. Solid trace: firing through the corner; gain = 6.84 dBi at 24° take-off angle, front-to-back ratio = 20.98 dB. Dashed trace: firing through the side, current phase-angles of 0 and –90°; gain = 6.11 dBi at 27° take-off angle, front-to-back ratio = 11.01 dB. Dotted trace: firing through the side, current phase-angles of 0 and –80°; gain = 5.98 dBi at 27° take-off angle, front-to-back ratio = 13.71 dB.

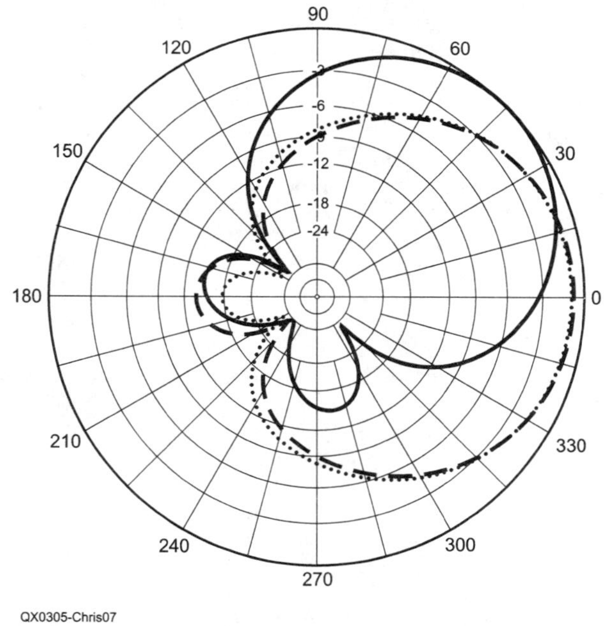

Fig 7—Azimuthal-plane radiation patterns for the expanded 4-square array. Solid trace: firing through the corner; front-to-back ratio = 35.26 dB, half-power beamwidth = 87.0°. Dashed trace: firing through the side, current phase-angles of 0 and –90°; front-to-back ratio = 12.94 dB, half-power beamwidth = 105.4°. Dotted trace: firing through the side, current phase-angles of 0 and –80°; front-to-back ratio = 16.98 dB, half-power beamwidth = 111.0°.

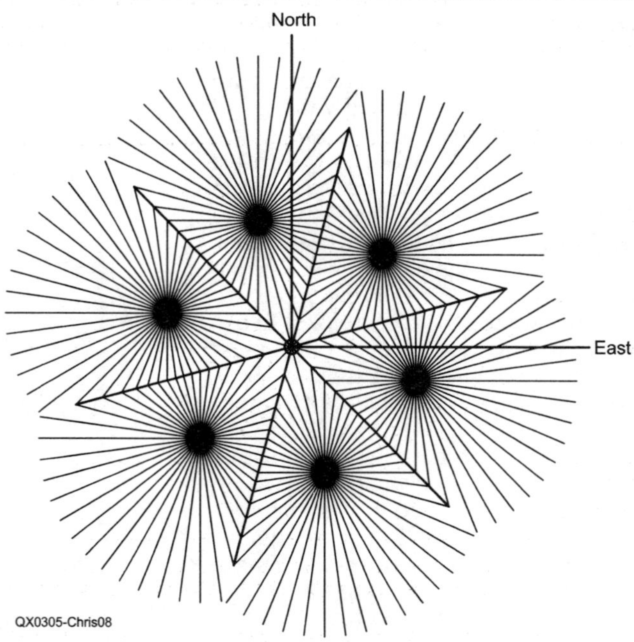

Fig 8—A bird's-eye view of the K3YA 6-element hexagonal array, showing details of the broadcast-style radial ground-screen. Each element has 48 radials, whose maximum length is 104.2 feet.

of just 11.01 dB in the elevation plane. There are two minor lobes in the rear portion of the radiation pattern, but the lower one is quite large. In the azimuthal plane, the FBR is 12.94 dB and the HPBW is 105.4°.

Reducing the phase-angle of the driving-point currents for the two front elements from –90 to –80° improves the performance by equalizing the size of the two rear lobes. Now the maximum forward gain falls a bit, to 5.98 dBi at a TOA of 27°, but the FBR rises to 13.71 dB in the elevation plane. In the azimuthal plane, the FBR increases to 16.98 dB, with a HPBW of 111.0°. All of these additional patterns are included in Fig 6 and 7.

K3YA 6-Hex Array

In the early part of 2004, I had the opportunity to visit Joe Johnson, K3RR, at his new QTH near Gettysburg, PA. While there, Joe told me of his plans to construct a 6-element array of short verticals that would be used exclusively for receiving purposes on 80 meters. As K3RR described the antenna design, which he credits to K3YA, I realized that it would make an excellent transmit array as well, and Joe agreed. Fig 8 is a computer rendition of the antenna with its ground screen, when viewed from directly overhead.

The design is very clever, and consists of six elements which are equally spaced to form a hexagonal shape. The spacing between adjacent elements is equal to the distance from any element to the center of the array. Initially, I performed some computer analysis of the array over a very simple MiniNEC-style ground in order to optimize the element spacing, and found that Joe's suggested value of 70 feet worked well at an operating frequency of 3.65 MHz. If the inter-element spacing is made larger, the forward gain will rise, but the azimuthal-plane beamwidth decreases.

Only four of the elements are active at any given time, while the other two are open-circuited at their bases. To fire northeast (see Fig 8), both the northeast and southwest elements are idle, while the other four are all driven with equal-amplitude base currents. The two elements in the northwest quadrant form an end-fire (cardioid) pair, as do the two elements in the southeast quadrant. Together, these two 2-element sub-arrays are then fed as a broadside array that beams to the northeast. The two rear elements have a current phase-angle of 0°, while the two front elements are fed at –105°. A phase-angle of –90° can also be used to drive the two front elements, but the gain and FBR in the elevation plane are not as good. Fig 9 and 10 display the radiation patterns. The peak forward gain is 7.53 dBi at a TOA of 24°, and the FBR is 16.69 dB in the elevation plane. In the azimuthal plane, the FBR is 19.24 dB, with a half-power beamwidth of 71.4°.

In the K3YA array, we can see that the element spacing of the two end-fire pairs is greater than what would typically be used, while the broadside spacing is somewhat smaller than normal. However, despite the compromises in spacing, this 6-element configuration provides the low-band operator with the equivalent of three separate four-element reversible arrays, using a system which requires only 6 elements.

While discussing phased vertical arrays with John, WØUN, he pointed me to the Web-site of Greg, W8WWV.[4] Greg has done a lot of work with 6-element hexagonal arrays, and has constructed a multi-band antenna at his QTH using an inter-element spacing of just 40 feet. The K3YA-style 6-hex fires through the corners, but W8WWV's array fires through the sides, and all six elements are driven at all times. In essence, Greg's antenna functions as two 3-element arrays which are parallel to one another, except that the two middle elements have been moved outward so they are farther apart than the front and rear pairs (which produces the hexagonal shape).

I chose to keep the hexagon's element spacing fixed at 70 feet, as before, and then adapted Greg's current phase-angles to achieve a half-power beamwidth which was similar to that of the K3YA version. In W8WWV's array, the two center elements receive twice as much current as the front and rear elements, so I did the same with mine. In the final version here, the two center

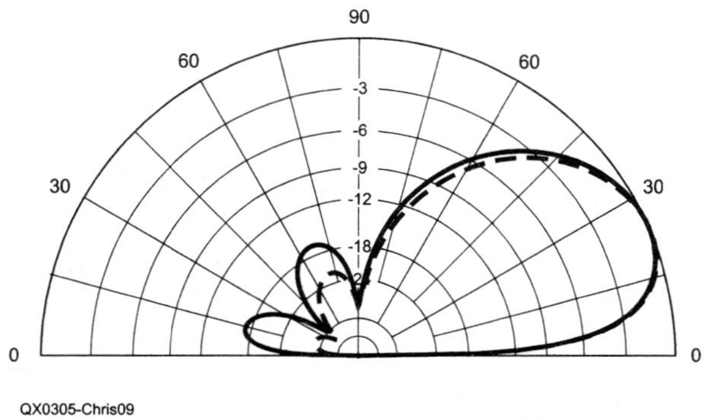

Fig 9—Elevation-plane radiation pattern for the two 6-element hexagonal arrays. Solid trace—K3YA version: gain = 7.53 dBi at 24° take-off angle, front-to-back ratio = 16.69 dB. Dashed trace—W8WWV version: gain = 7.54 dBi at 23° take-off angle, front-to-back ratio = 21.94 dB.

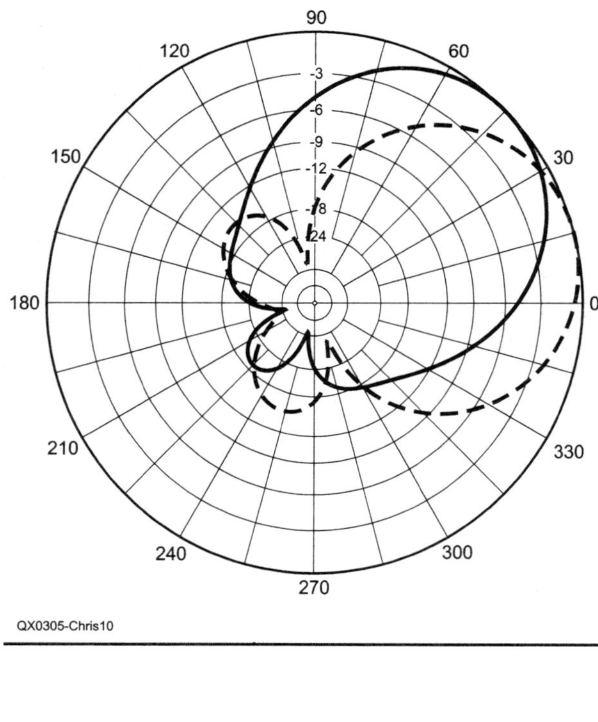

Fig 10—Azimuthal-plane radiation pattern for the two 6-element hexagonal arrays. Solid trace—K3YA version: front-to-back ratio = 19.24 dB, half-power beamwidth = 71.4°. Dashed trace—W8WWV version: front-to-back ratio = 33.67 dB, half-power beamwidth = 71.2°.

elements have the reference current phase-angle of 0°, while the two front and two rear elements are fed at –100° and +100° respectively. Figures 9 and 10 display the new radiation patterns, in conjunction with those of the K3YA array. The W8WWV-style antenna has a peak forward gain of 7.54 dBi at a TOA of 23°, and the FBR is 21.94 dB in the elevation plane. In the azimuthal plane, the FBR is 33.67 dB (although there are significant side-lobes), and the HPBW is 71.2°. Comparing the two versions of the 6-hex, we can see that both yield similar gains, to within 0.01 dBi, although the W8WWV design has a slightly-lower take-off angle. The azimuthal-plane beam-widths are also extremely close, which was my intention. The two arrays differ mainly in their ability to reject unwanted signals arriving from the back, with the W8WWV antenna being the clear winner. Theoretically, it would be possible to combine both versions together into a single antenna (using a lot of networks and switching circuitry) to create a 6-element hexagonal array with 12 directions of fire.

WØUN 9-Circle Array

The final antenna to be investigated is the 9-circle array of John Brosnahan, WØUN.[5] This antenna is formidable in design and construction, and the feed network is rather complex, but it pays big dividends, both in terms of gain and front-to-back ratio. One element is installed at the center of a circle whose radius is 0.39 λ at the frequency of interest, while the other 8 elements are placed at equal intervals around this circle. At 3.65 MHz, the radius of the circle is 105.1 feet. There are 8 directions of fire, directly outward through each of the 8 outer elements, and all 9 elements are active at all times.

When beaming to the northeast, the extreme front and rear (NE and SW) elements are fed with equal-amplitude (1.0 A relative) currents. The phase-angle of the current into the NE element is –180°, while that of the SW element is +180°. The center element and the two elements located directly to its northwest and southeast are all fed with currents whose phase-angle is 0°; the central element has a relative current amplitude of 3.0, while the two out-riggers have unity (1.0) current amplitude. The two intermediate front/side elements both have base currents of 1.64 at a phase-angle of –90°. (The suggested current amplitude of 1.66 was adjusted slightly to provide better performance.) As already stated, the rear-most element is driven with a

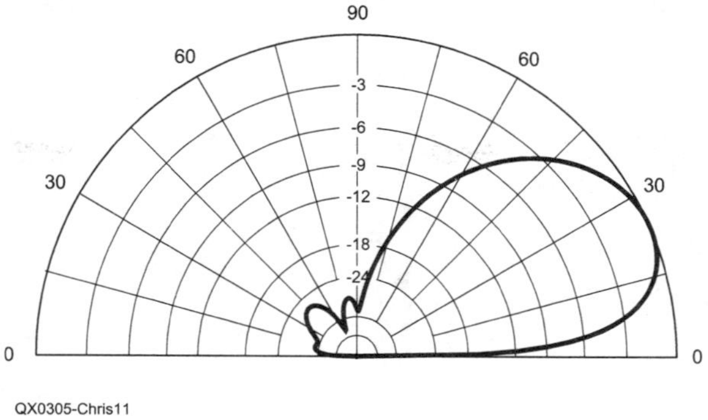

Fig 11—Elevation-plane radiation pattern for the WØUN 9-element circular array. Gain = 8.75 dBi at 23° take-off angle, front-to-back ratio = 26.72 dB.

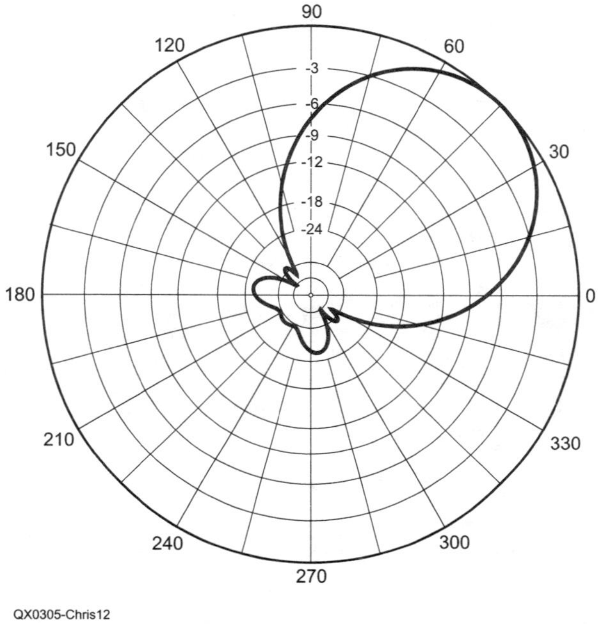

Fig 12—Azimuthal-plane radiation pattern for the WØUN 9-element circular array. Front-to-back ratio = 33.81 dB, half-power beamwidth = 60.4°.

Table 1—Performance Data for the Phased Vertical Arrays discussed herein. Elevation-plane data includes the gain and take-off angle (TOA) of the main lobe, the front-to-back ratio (FBR) and the front-to-rear ratio (FRR). Azimuthal-plane data includes the front-to-back ratio (FBR) and front-to-rear ratio (FRR), along with the half-power beamwidth (HPBW). The three 4-square arrays all have two lines of data; the first line is for the main beam when firing through the diagonal of the square, and the second line is when firing through the side of the square.

	Elevation-Plane Data			*Azimuthal-Plane Data*		
	Gain & TOA	*FBR*	*FRR*	*FBR*	*FRR*	*HPBW*
Array	*(dBi & deg)*	*(dB)*	*(dB)*	*(dB)*	*(dB)*	*(deg)*
9-Circle	8.75 @ 23	26.72	32.10	33.81	32.51	60.4
6-Hex (W8WWV)	7.54 @ 23	21.94	30.79	33.67	20.84	71.2
6-Hex (K3YA)	7.53 @ 24	16.69	21.63	19.24	22.17	71.4
Expanded	6.84 @ 24	20.98	30.21	35.26	21.73	87.0
4-Square	5.98 @ 27	13.71	18.93	16.98	16.65	111.0
4-Square	6.61 @ 22	25.23	30.58	31.50	22.23	80.2
W8JI-feed	5.57 @ 24	16.51	21.11	18.36	17.34	116.6
Regular	6.16 @ 23	16.58	21.66	22.71	21.22	100.6
4-Square	5.18 @ 25	12.87	20.13	21.97	12.96	130.8

current of 1.0 at an angle of +180°, while the two intermediate back/side elements both have base currents of 1.64 at a phase-angle of +90°.

The maximum forward gain of this array is 8.75 dBi at a TOA of 23°, and the FBR in the elevation plane is 26.72 dB. In the azimuthal plane, the FBR is 33.81 dB, and the HPBW is 60.4°. Plots of the radiation patterns are included as Figures 11 and 12.

Conclusions

This article has summarized the capabilities of a variety of phased-vertical arrays that are of interest to the low-band operator. Table 1 lists the important performance parameters of each array, for reference and comparison. In this table, I have listed the "FRR" (front-to-rear ratio) in addition to the typical front-to-back ratio (FBR). For each array, the customary "front- to-back ratio" data has already been given in the text and in the accom-panying figures, but these statistics fail to account for the presence or absence of multiple secondary lobes, which can be large. To perform the FRR calculations, I averaged the gain at five-degree increments throughout the rear portion of each radiation pattern, and then subtracted this value from the peak forward gain in order to obtain the FRR value.

Each of these arrays was also simulated using a ground-screen consisting of 50 100-foot radials per element. In this scenario, none of the radials were truncated in length, although many had to be buried at different depths in order to avoid unwanted "collisions" which led to software crashes. In every case, this alternative ground-screen configuration yielded slightly-higher values of forward gain, ranging from about 0.1 to 0.3 dB. Since all of the radials in these ground screens are full-length, considerably more wire is required to construct them.

Notes

[1]EZNEC is available from Roy Lewallen, W7EL, PO Box 6658, Beaverton OR 97007.

[2]D. Atchley, W1CF, H. Stinehelfer, ex W2ZRS, and J. White, PhD, "360-degree-Steerable Vertical Phased Arrays," *QST*, April 1976, pp 27-30.

[3]T. Rauch, W8JI, **www.w8ji.com/tx_four_square.htm**.

[4]G. Ordy, W8WWV, **www.seed-solutions.com/gregordy/Amateur%20Radio/Experimentation/HexArray/8040Array.htm.**

[5]J. Devoldere, ON4UN, *ON4UN's Low-Band DXing, Third Edition*, ARRL, 1999, pp 11-68—11-71.

Al Christman, K3LC, has a PhD in electrical and computer engineering from Ohio University and is currently serving as a professor in the EE department of Grove City College in Western Pennsylvania. First licensed as WA3WZD in 1974, Al is an active DXer with 328 countries confirmed on 20-m SSB. You can contact Al at Grove City College, 100 Campus Drive, Grove City, PA 16127-2104. □□

A 40-Meter Comparison: 4-Square versus Yagi

Al Christman, K3LC
Grove City College
100 Campus Dr
Grove City, PA 16127-2104

This article compares the performance of a 40-meter 4-square phased-vertical array to that of a two-element "shorty forty" Yagi, when both are installed over several different types of soil. The number of radials in the ground system of the 4-square is varied, as is the height of the Yagi (which should probably be called the "Uda-Yagi array" to give proper credit to its developers).

Background

Forty meters is the primary night-time band during many contests, and a big signal on 7 MHz is important if an operator wants to be competitive. Both contesters and DXers know that 40 meters is the lowest-frequency band where Yagis are readily available to the average ham. They can be bought with 2, 3, or 4 elements, while the elements themselves may be either full-length, or utilize coils or linear-loading wires. Towers to support these Yagis need not be gargantuan in size, and flashing beacons are not required at normal installation heights.

As an alternative, one might choose to construct a 4-square phased-vertical array for this band. Quarter-wave elements are only about 34 feet long, and can easily be made to be self-supporting. Such an antenna is relatively unobtrusive, and can be erected without violating most zoning ordinances.

But which antenna style is best? One is horizontal, the other is vertical, and both have their adherents. Let's take a closer look and see what we can learn.

Computer Simulation

Figure 1 shows a plan view of a typical 4-square array. We are looking directly down from above, so the vertical elements are not visible. A "broadcast-style" ground system is used, and none of the radials overlap. Instead, they are truncated at the points where they intersect one another, and are then bonded to bus-wires as shown. Thus, the ground screen of each element is electrically connected to all the others. The number of radials per element was varied from 15 to 30 to 60 to 120, and the maximum radial length was fixed at 0.25 wavelength (WL). The inner end of the first segment of each radial is located at the air/soil interface (H = 0), and the segment slopes downward to a "final" burial depth of 3 inches. All four of the monopoles have a physical height of 0.25 WL, or about 34.39 feet. All elements are fed with equal-amplitude currents,

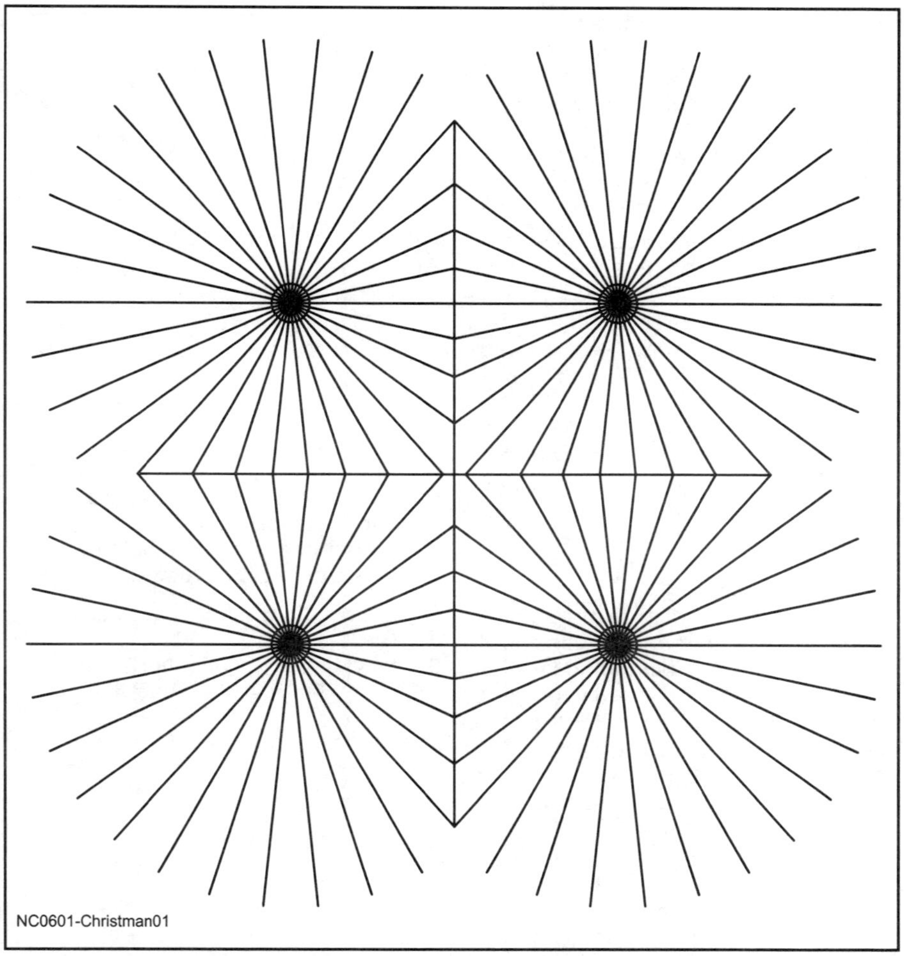

Figure 1—Computer-generated drawing (plan view) of a 40-meter four-square array using a "broadcast-style" ground screen. Here, each element has 30 buried radials (depth = 3 inches), whose maximum length is 0.25 WL.

Table 1
Gain of the 4-square phased-vertical array, at various take-off angles, versus the number of radials per element. A "broadcast-style" ground screen is used, and the maximum radial length is 0.25 WL (average soil).

Take-off Angle (deg)	*Gain (dBi) as a function of Number of radials per element* N = 15	*N = 30*	*N = 60*	*N = 120*
2.5	-5.99	-5.36	-5.07	-4.96
5	-1.26	-0.64	-0.34	-0.23
7.5	+1.08	+1.71	+2.01	+2.11
10	2.49	3.12	3.42	3.53
12.5	3.41	4.04	4.34	4.44
15	4.02	4.65	4.95	5.06
17.5	4.42	5.06	5.36	5.47
20	4.67	5.31	5.62	5.73
22.5	4.81	5.45	5.76	5.87
25	4.84	5.49	5.80	5.91
27.5	4.80	5.45	5.76	5.88
30	4.68	5.34	5.66	5.78

using progressive 90-degree phase shifts.

Probably the most familiar 40-meter Yagi on the planet is the so-called "shorty forty," which typically employs two coil-loaded elements on a boom whose length is about 24 feet. A computer model of such an antenna was obtained from John Brosnahan, WØUN,[1] for use in this study, and its operational height was varied from 40 to 120 feet in steps of 20 feet.

The antenna models were run using the new *EZNEC Pro* software, version 4.0,[2] with a double-precision *NEC-4* calculating engine. For simplicity, it was assumed that the 4-square was constructed entirely from #12 AWG copper wire, while aluminum was used for the Yagi. The wire segment-lengths for all antenna elements were selected in accordance with the *NEC* guidelines. Three different sets of ground constants were used: "average" soil (conductivity = 0.005 Siemens per meter and dielectric constant = 13), "very good" soil (conductivity = 0.0303 S/m and dielectric constant = 20), and "very poor" soil (conductivity = 0.001 S/m and dielectric constant = 5). A frequency of 7.15 MHz (mid-band) was selected for the analysis, and the terrain is assumed to be flat in all cases.

If our goal is to optimize the DX performance of our antenna system on 40 meters, then it is important to know the range of take-off angles which must be covered. Dean Straw, N6BV, has expended a considerable amount of time and effort to answer this question using a number of software tools, and is an acknowledged expert in this area. Dean has stated that "the vast majority of the time the take-off angles for DX are less than 30 degrees."[3] Therefore, we need to determine which antenna configurations generate the most gain within this span of elevation angles.

Results for Average Soil

Table 1 lists the detailed results for the

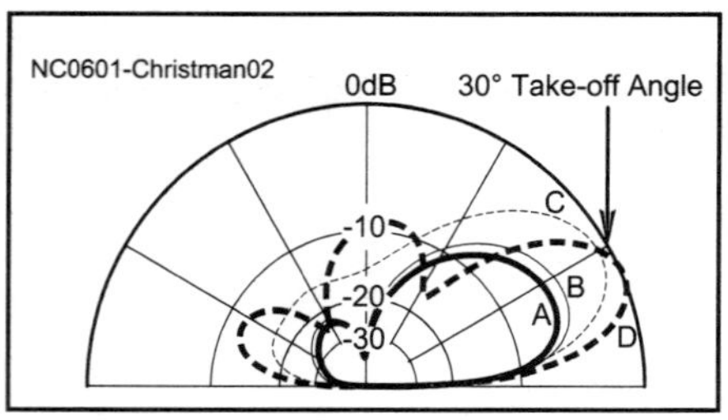

Figure 2—Elevation-plane radiation patterns for several 40-meter antennas, using "average" soil.
A = 4-square with 15 radials per element
B = 4-square with 120 radials per element
C = 2-el Yagi up 60 feet
D = 2-el Yagi up 80 feet.

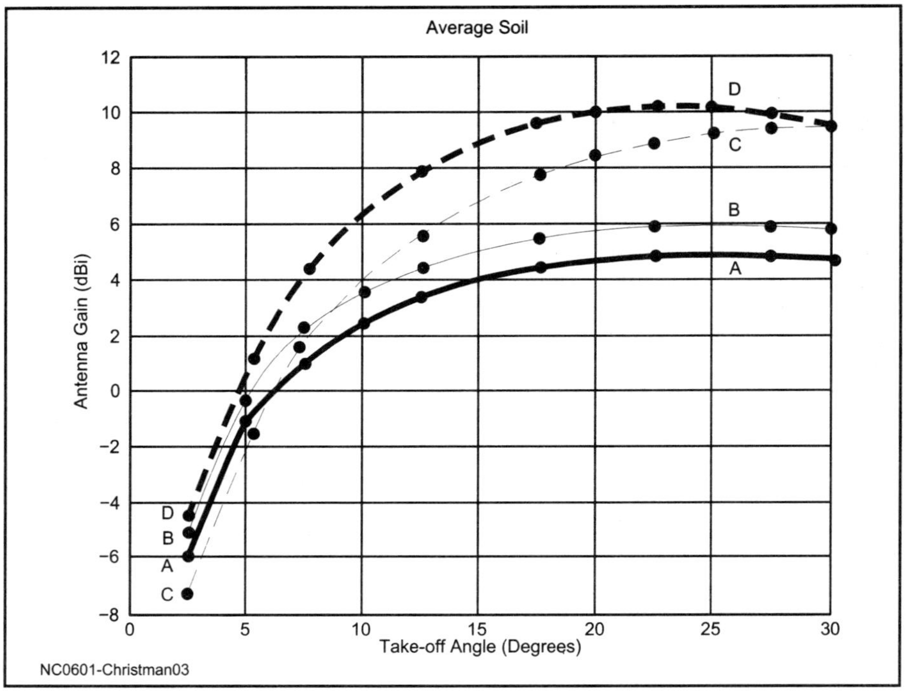

Figure 3—Elevation-plane radiation patterns for several 40-meter antennas, using "average" soil.
A = 4-square with 15 radials per element
B = 4-square with 120 radials per element
C = 2-el Yagi up 60 feet
D = 2-el Yagi up 80 feet.

Table 2
Gain of the 2-element "shorty-forty" Yagi, at various take-off angles, versus the height of the antenna, over average soil.

Take-off Angle (deg)	*Gain (dBi) as a function of antenna height*				
	H = 40'	*H = 60'*	*H = 80'*	*H = 100'*	*H = 120'*
2.5	-10.92	-7.35	-4.65	-3.10	-1.61
5	-5.04	-1.49	+1.15	+2.64	+4.05
7.5	-1.68	+1.81	4.37	5.75	7.02
10	+0.63	4.03	6.48	7.70	8.78
12.5	2.34	5.64	7.94	8.95	9.76
15	3.66	6.83	8.95	9.70	10.17
17.5	4.72	7.73	9.62	10.05	10.08
20	5.56	8.40	10.03	10.05	9.51
22.5	6.24	8.88	10.19	9.73	8.43
25	6.78	9.20	10.15	9.09	6.77
27.5	7.20	9.38	9.92	8.12	4.32
30	7.53	9.43	9.50	6.78	0.63

Table 3
Gain of the 4-square phased-vertical array, at various take-off angles, versus the number of radials per element. A "broadcast-style" ground screen is used, and the maximum radial length is 0.25 WL (very good soil).

Take-off Angle (deg)	*Gain (dBi) as a function of Number of radials per element*			
	N = 15	*N = 30*	*N = 60*	*N = 120*
2.5	-0.87	-0.56	-0.34	-0.24
5	+3.05	+3.36	+3.58	+3.68
7.5	4.78	5.10	5.31	5.41
10	5.72	6.04	6.25	6.35
12.5	6.27	6.58	6.80	6.90
15	6.58	6.90	7.11	7.22
17.5	6.74	7.06	7.27	7.38
20	6.78	7.10	7.32	7.43
22.5	6.74	7.07	7.28	7.39
25	6.63	6.95	7.17	7.28
27.5	6.45	6.78	7.00	7.11
30	6.22	6.55	6.77	6.89

4-square array, for take-off angles from 2.5 to 30 degrees, as the number of radials is varied between 15 and 120 per element. We can see that the maximum gain produced by any of these vertical antennas is less than 6 dBi. In a similar fashion, Table 2 displays the outcome for the horizontally-polarized "shorty-forty" Yagi, at heights ranging from 40 to 120 feet. Here, a peak gain of more than 10 dBi is possible. Figure 2 is a complete polar plot of the elevation-plane radiation patterns for several selected antennas of both polarizations. Figure 3 presents the same information in rectangular form, covering only those elevation angles which are important for DX propagation. Note that the "shorty-forty" at a height of 80 feet yields more gain than any of the 4-squares, throughout the range of take-off angles from 2 to 30 degrees. However, if the Yagi height is set at 60 feet, then the 4-square is better at very low elevation angles, even with a sparse groundscreen.

Results for Very Good Soil

Increasing the conductivity of the soil allows the 4-square array to generate more gain, as is evident from an examination of Table 3. Now the peak gain obtainable from the phased verticals approaches 7.5 dBi, an improvement of slightly more than 1.5 dB when compared to average soil. Table 4 lists the results for the "shorty-forty," and reveals that the gain of the Yagi actually climbs by a few tenths of a decibel when placed over very good soil, reaching a maximum of just over 10.7 dBi. Several examples of the elevation-plane radiation patterns (for both antenna types) are shown in Figure 4. At very low take-off angles, the 4-square is superior, even with only 15 radials per element. To match its performance at elevation angles below about 5 degrees, the Yagi would have to be placed more than 120 feet in the air, although this reduces its gain at take-off angles in the 20-to-30 degree region.

Results for Very Poor Soil

Conversely, a decrease in the soil conductivity diminishes the gain of the 4-square, as indicated by the data in Table 5. Here the maximum gain produced by the phased verticals is just under 5 dBi, roughly one decibel less than for average soil. Table 6 lists the results for the "shorty-forty," and we see (as expected) that the gain of the Yagi now falls a bit when installed over very poor soil, peaking at slightly less than 9.8 dBi. Figure 5 includes the radiation-pattern plots for a number of antennas, and it can be seen that the "shorty-forty" easily out-performs the 4-square when the Yagi is mounted at a height of just 60 feet.

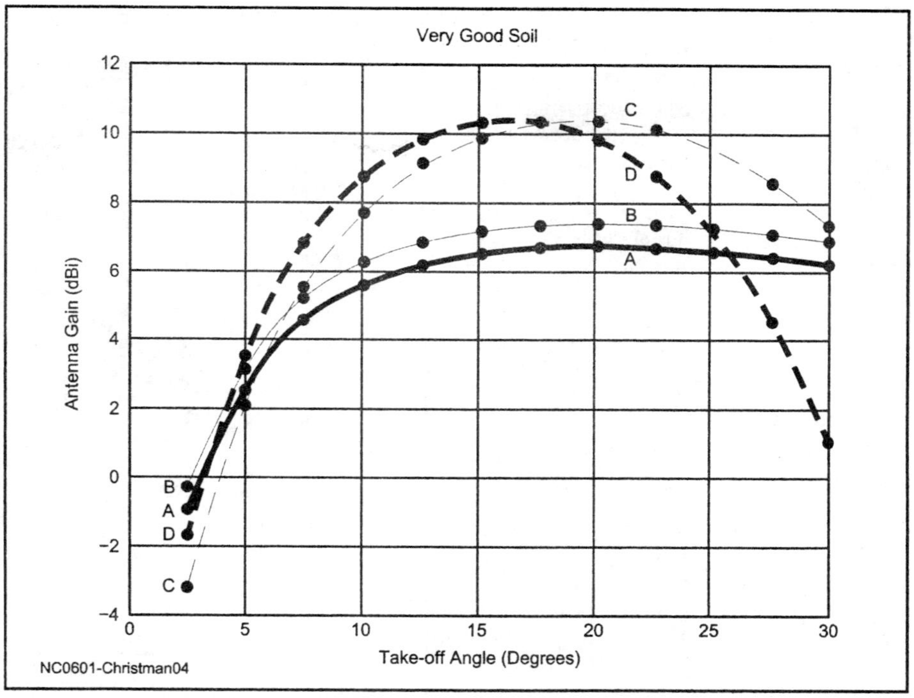

Figure 4—Elevation-plane radiation patterns for several 40-meter antennas, using "very good" soil.
A = 4-square with 15 radials per element **C = 2-el Yagi up 100 feet**
B = 4-square with 120 radials per element **D = 2-el Yagi up 120 feet.**

Table 4
Gain of the 2-element "shorty-forty" Yagi, at various take-off angles, versus the height of the antenna, over very good soil.

Take-off Angle (deg)	*Gain (dBi) as a function of antenna height*				
	H = 40'	*H = 60'*	*H = 80'*	*H = 100'*	*H = 120'*
2.5	-11.11	-7.44	-4.58	-3.11	-1.64
5	-5.17	-1.53	+1.28	+2.68	+4.07
7.5	-1.76	+1.82	4.55	5.84	7.10
10	+0.59	4.09	6.71	7.84	8.91
12.5	2.35	5.75	8.22	9.14	9.95
15	3.73	7.00	9.29	9.95	10.41
17.5	4.82	7.95	10.02	10.35	10.37
20	5.72	8.66	10.47	10.41	9.86
22.5	6.44	9.19	10.70	10.14	8.84
25	7.03	9.56	10.71	9.56	7.24
27.5	7.50	9.78	10.53	8.65	4.83
30	7.87	9.89	10.16	7.35	1.09

Table 5
Gain of the 4-square phased-vertical array, at various take-off angles, versus the number of radials per element. A "broadcast-style" ground screen is used, and the maximum radial length is 0.25 WL (very poor soil).

Take-off Angle (deg)	*Gain (dBi) as a function of Number of radials per element*			
	N = 15	*N = 30*	*N = 60*	*N = 120*
2.5	-9.06	-8.41	-8.19	-8.10
5	-3.91	-3.27	-3.05	-2.96
7.5	-1.23	-0.58	-0.36	-0.27
10	+0.47	+1.11	+1.33	+1.43
12.5	1.62	2.26	2.49	2.58
15	2.43	3.08	3.31	3.40
17.5	3.01	3.66	3.89	3.99
20	3.42	4.07	4.30	4.40
22.5	3.69	4.35	4.58	4.68
25	3.84	4.51	4.74	4.84
27.5	3.91	4.58	4.82	4.92
30	3.90	4.58	4.82	4.92

Conclusions

Computer analysis indicates that, under most conditions, a two-element "shorty-forty" Yagi can generate more gain in the range of take-off angles that are useful for DX communications than a conventional 4-square array with an extensive ground system. Over "average" soil, the Yagi must be placed about 80 feet above the ground to achieve superiority at most important elevation angles. For "very poor" soil the required height is roughly 60 feet, while about 120 feet is necessary over "very good" soil.

Limitations

Computer simulation is a great tool for antenna analysis, but the models are limited in scope, and by no means do they include all of the factors which are present in a real-world environment. For instance, if the "shorty-forty" Yagi is installed at a typical operating height, it is likely to be "in the clear," (relatively speaking) with respect to its surroundings. On the other hand, a four-square phased-vertical array will normally be installed either at or near the earth's surface, making it more difficult for the antenna to "see out," which can be a significant handicap. In addition, this article gives no consideration to wave-propagation phenomena. Because it is immersed in a magnetic field, the ionosphere can give preferential treatment to waves of a specific polarization, but that issue was not addressed here.[4]

Acknowledgements

The author would like to thank several individuals for their helpful reviews of this article while it was being prepared. Advice, comments, and suggestions were solicited and received from John Brosnahan, WØUN, Carl Luetzelschwab, K9LA, and Dean Straw, N6BV. Thank you gentlemen!

Notes:

[1]Private communication with John Brosnahan, WØUN.

[2]Several versions of the *EZNEC* antenna-modeling software are available from Roy Lewallen, W7EL, PO Box 6658, Beaverton, OR 97007.

[3]Private communication with Dean Straw, N6BV.

[4]See the article by K9LA in the Propagation column in this issue.

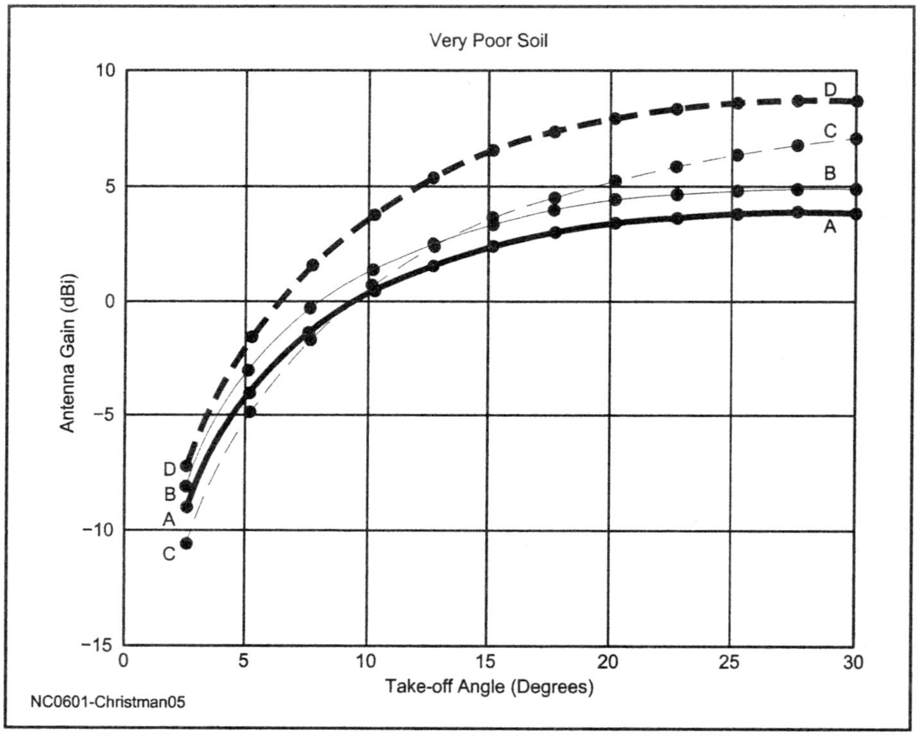

Figure 5—Elevation-plane radiation patterns for several 40-meter antennas, using "very poor" soil.
A = 4-square with 15 radials per element
B = 4-square with 120 radials per element
C = 2-el Yagi up 40 feet
D = 2-el Yagi up 60 feet.

Table 6
Gain of the 2-element "shorty-forty" Yagi, at various take-off angles, versus the height of the antenna, over very poor soil.

Take-off Angle (deg)	*Gain (dBi) as a function of antenna height*				
	H = 40'	*H = 60'*	*H = 80'*	*H = 100'*	*H = 120'*
2.5	-10.60	-7.22	-4.71	-3.07	-1.58
5	-4.80	-1.45	+1.01	+2.58	+3.99
7.5	-1.52	+1.77	4.14	5.61	6.88
10	+0.70	3.91	6.16	7.47	8.55
12.5	2.33	5.43	7.53	8.63	9.45
15	3.58	6.55	8.46	9.30	9.77
17.5	4.55	7.37	9.05	9.56	9.59
20	5.32	7.96	9.37	9.49	8.95
22.5	5.92	8.36	9.46	9.09	7.81
25	6.39	8.60	9.35	8.39	6.12
27.5	6.75	8.71	9.04	7.37	3.75
30	7.02	8.71	8.56	6.02	0.54

NCJ

A Study of Elevated Radial Ground Systems for Vertical Antennas — Part 1

Al Christman, K3LC
Grove City College
100 Campus Drive
Grove City, PA 16127-2104

Part 1 of this article compares the gain developed by elevated 0.25 WL vertical monopole antennas when they are installed over a variety of elevated ground screens. All of these ground systems utilize symmetrically placed elevated horizontal radials whose length is also 0.25 WL. The number and elevation height of the radials is varied, along with the electrical characteristics of the soil beneath the antenna. Part 2 will compare these results for elevated antennas with those for ground-mounted antennas with large numbers of buried radials.

Overview

Many low-band operators use ground mounted verticals in combination with extensive buried-radial ground systems. An alternative to this practice is to install an elevated vertical monopole with an elevated ground screen comprised of just a few radials. The purpose of this study is to determine how many radials are needed, and how high they must be placed above the ground. All three of the "low bands" will be investigated, using three different types of soil on each band. The number of elevated radials will be varied from 3 to 6 (uniformly spaced around the base of the vertical element), and several different elevation heights will be examined.

I have access to the new *EZNEC Pro* software, version 4.0[1], with a double precision NEC 4 calculating engine, which enables me to analyze ground systems using either buried or elevated radials. For simplicity, all computer simulations were performed using #12 AWG copper wire for both the vertical element and the elevated radials. The length of all wires was fixed at 0.25 WL, which is easy to calculate. In addition, an antenna with these dimensions generally yields a reasonable value for the input resistance, on the order of 30 to 40 Ω. No attempt was made to "prune" either the vertical element or the radials to achieve resonance. The wire segment lengths for the vertical element and the radials are all tapered in accordance with the most conservative *EZNEC* guidelines. The shortest segments, such as the one containing the feed point at the base of the vertical element, and the inner segment of each radial, have a length of about 0.0025 WL.

Results on 40 meters

A frequency of 7.15 MHz was selected for the analysis on 40 meters, leading to a length of about 34.39 feet for the vertical element and radials. The number of radials was varied in sequential order from 3 to 6, and elevation heights of 5, 7.5, 10, 12.5, and 15 feet were utilized. Figure 1 shows a version of this antenna with four radials, at a height of 7.5 feet above the ground.

Table 1 displays the results when the antenna is mounted over "average" soil with a conductivity of 0.005 Siemens/meter and a relative permittivity (or dielectric constant) of 13. Notice that when the number of radials is fixed at some particular value, increasing their height yields both higher gain and a lower takeoff angle, as one would expect. As the base height of the antenna is raised from 5 to 15 feet, the gain goes up by about 0.2 dB or so, and the takeoff angle falls by around 4 degrees. If the base height of the antenna is held constant, adding

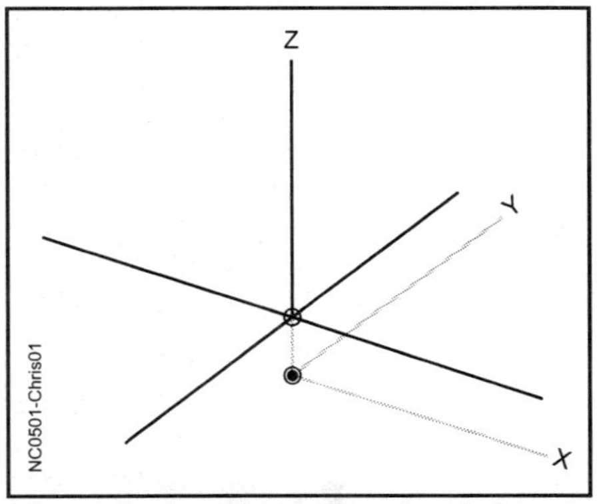

Figure 1—Drawing of an elevated vertical monopole antenna with four elevated horizontal radials installed at H = 7.5 feet. The length of both the vertical element and the radials is 0.25 WL (34.39 feet at 7.15 MHz).

Table 1
Antenna gain versus number and height of elevated horizontal radials. The frequency is 7.15 MHz, and the radials are installed over average soil (conductivity = 0.005 Siemens/meter, and dielectric constant = 13). The vertical element, and all of the radials, have a length of 0.25 WL, or about 34.39 feet.

Number Of Radials	*Height of Radials (feet)*	*Gain and Take-off Angle (dBi & degrees)*		*Pattern Non-Circularity (dB)*
3	5	-0.08	24.0	0.23
	7.5	0.02	22.5	0.24
	10	0.10	21.5	0.25
	12.5	0.15	20.5	0.26
	15	0.20	19.8	0.27
4	5	0.01	24.0	0.01
	7.5	0.08	22.8	0.02
	10	0.14	21.5	0.01
	12.5	0.18	20.8	0.01
	15	0.22	20.0	0.01
5	5	0.03	23.8	0.0
	7.5	0.10	22.8	0.0
	10	0.15	21.8	0.0
	12.5	0.20	20.8	0.01
	15	0.24	20.0	0.01
6	5	0.04	23.5	0.0
	7.5	0.11	23.0	0.0
	10	0.16	22.0	0.0
	12.5	0.20	20.5	0.0
	15	0.24	20.0	0.0

[1]Several versions of the *EZNEC* antenna modeling software are available from Roy Lewallen, W7EL, PO Box 6658, Beaverton, OR 97007.

Table 2

Antenna gain versus number and height of elevated horizontal radials. The frequency is 7.15 MHz, and the radials are installed over very poor soil (conductivity = 0.001 Siemens/meter, and dielectric constant = 5). The vertical element, and all of the radials, have a length of 0.25 WL, or about 34.39 feet.

Number Of Radials	*Height of Radials (feet)*	*Gain and Take-off Angle (dBi & degrees)*		*Pattern Non-Circularity (dB)*
3	5	-0.98	27.0	0.37
	7.5	-0.80	26.0	0.38
	10	-0.64	25.2	0.37
	12.5	-0.48	24.2	0.37
	15	-0.34	23.2	0.38
4	5	-0.90	26.8	0.01
	7.5	-0.74	25.8	0.0
	10	-0.58	25.0	0.01
	12.5	-0.44	24.0	0.01
	15	-0.31	23.2	0.02
5	5	-0.88	27.0	0.01
	7.5	-0.72	26.0	0.0
	10	-0.58	25.2	0.01
	12.5	-0.44	24.2	0.01
	15	-0.30	23.2	0.01
6	5	-0.87	26.5	0.0
	7.5	-0.71	26.0	0.0
	10	-0.56	25.0	0.0
	12.5	-0.42	24.0	0.0
	15	-0.29	23.5	0.0

Table 3

Antenna gain versus number and height of elevated horizontal radials. The frequency is 7.15 MHz, and the radials are installed over very good soil (conductivity = 0.0303 Siemens/meter, and dielectric constant = 20). The vertical element, and all of the radials, have a length of 0.25 WL, or about 34.39 feet.

Number Of Radials	*Height of Radials* (feet)	*Gain and Take-off Angle* (dBi & degrees)		*Pattern Non-Circularity* (dB)
3	5	1.72	19.5	0.09
	7.5	1.81	18.5	0.10
	10	1.89	18.0	0.10
	12.5	1.95	17.0	0.10
	15	2.00	16.2	0.11
4	5	1.77	19.5	0.02
	7.5	1.85	18.5	0.02
	10	1.91	17.5	0.02
	12.5	1.97	17.0	0.02
	15	2.02	16.2	0.01
5	5	1.79	19.5	0.0
	7.5	1.86	18.5	0.0
	10	1.92	17.5	0.0
	12.5	1.98	17.0	0.0
	15	2.02	16.2	0.0
6	5	1.80	19.5	0.0
	7.5	1.87	18.5	0.0
	10	1.93	17.5	0.0
	12.5	1.98	17.0	0.0
	15	2.03	16.0	0.0

Table 4

Antenna gain versus number and height of elevated horizontal radials. The frequency is 3.75 MHz, and the radials are installed over average soil (conductivity = 0.005 Siemens/meter, and dielectric constant = 13). The vertical element, and all of the radials, have a length of 0.25 WL, or about 65.57 feet.

Number Of Radials	*Height of Radials (feet)*	*Gain and Take-off Angle (dBi & degrees)*		*Pattern Non-Circularity (dB)*
3	5	0.08	23.8	0.17
	10	0.23	22.5	0.18
	15	0.31	21.5	0.20
	22.5	0.38	20.0	0.21
	30	0.40	18.8	0.23
4	5	0.20	23.5	0.01
	10	0.30	22.5	0.01
	15	0.36	21.5	0.02
	22.5	0.40	19.5	0.02
	30	0.42	18.5	0.02
5	5	0.26	23.5	0.0
	10	0.33	22.5	0.0
	15	0.38	21.5	0.01
	22.5	0.41	19.8	0.01
	30	0.42	18.5	0.01
6	5	0.28	23.5	0.0
	10	0.34	22.5	0.0
	15	0.38	21.5	0.0
	22.5	0.42	20.0	0.0
	30	0.43	18.5	0.0

Table 5

Antenna gain versus number and height of elevated horizontal radials. The frequency is 3.75 MHz, and the radials are installed over very poor soil (conductivity = 0.001 Siemens/meter, and dielectric constant = 5). The vertical element, and all of the radials, have a length of 0.25 WL, or about 65.57 feet.

Number Of Radials	*Height of Radials (feet)*	*Gain and Take-off Angle (dBi & degrees)*		*Pattern Non-Circularity (dB)*
3	5	-1.62	27.5	0.36
	10	-1.38	26.5	0.37
	15	-1.22	25.2	0.38
	22.5	-1.03	23.8	0.40
	30	-0.86	22.2	0.41
4	5	-1.47	27.5	0.0
	10	-1.29	26.5	0.0
	15	-1.16	25.5	0.0
	22.5	-0.99	23.5	0.0
	30	-0.83	22.2	0.0
5	5	-1.42	27.5	0.01
	10	-1.26	26.5	0.0
	15	-1.14	25.0	0.01
	22.5	-0.98	23.5	0.01
	30	-0.82	22.2	0.0
6	5	-1.39	27.5	0.0
	10	-1.24	26.0	0.0
	15	-1.13	25.5	0.0
	22.5	-0.97	23.5	0.0
	30	-0.81	22.0	0.0

Table 6

Antenna gain versus number and height of elevated horizontal radials. The frequency is 3.75 MHz, and the radials are installed over very good soil (conductivity = 0.0303 Siemens/meter, and dielectric constant = 20). The vertical element, and all of the radials, have a length of 0.25 WL, or about 65.57 feet.

Number Of Radials	*Height of Radials (feet)*	*Gain and Take-off Angle (dBi & degrees)*		*Pattern Non-Circularity (dB)*
3	5	2.23	18.5	0.06
	10	2.42	17.5	0.06
	15	2.54	16.5	0.07
	22.5	2.68	15.5	0.07
	30	2.80	14.5	0.07
4	5	2.32	18.2	0.01
	10	2.46	17.5	0.01
	15	2.56	16.5	0.01
	22.5	2.70	15.5	0.01
	30	2.82	14.8	0.01
5	5	2.36	18.5	0.0
	10	2.48	17.5	0.0
	15	2.58	17.0	0.0
	22.5	2.70	15.8	0.01
	30	2.82	14.5	0.0
6	5	2.39	18.0	0.0
	10	2.49	17.5	0.0
	15	2.58	16.5	0.0
	22.5	2.71	15.5	0.0
	30	2.83	14.5	0.0

Table 7

Antenna gain versus number and height of elevated horizontal radials. The frequency is 1.835 MHz, and the radials are installed over average soil (conductivity = 0.005 Siemens/meter, and dielectric constant = 13). The vertical element, and all of the radials, have a length of 0.25 WL, or about 134 feet.

Number Of Radials	*Height of Radials (feet)*	*Gain and Take-off Angle (dBi & degrees)*		*Pattern Non-Circularity (dB)*
3	10	0.86	21.5	0.12
	20	1.01	20.5	0.12
	30	1.08	19.5	0.13
	45	1.15	18.2	0.14
	60	1.18	17.0	0.15
4	10	0.98	21.5	0.01
	20	1.08	20.5	0.01
	30	1.13	19.5	0.02
	45	1.18	18.0	0.02
	60	1.21	17.0	0.02
5	10	1.02	21.5	0.01
	20	1.10	20.5	0.0
	30	1.14	19.8	0.01
	45	1.19	18.2	0.0
	60	1.22	17.0	0.0
6	10	1.05	21.5	0.0
	20	1.11	20.5	0.0
	30	1.15	19.5	0.0
	45	1.20	18.0	0.0
	60	1.22	17.0	0.0

Table 8

Antenna gain versus number and height of elevated horizontal radials. The frequency is 1.835 MHz, and the radials are installed over very poor soil (conductivity = 0.001 Siemens/meter, and dielectric constant = 20). The vertical element, and all of the radials, have a length of 0.25 WL, or about 134 feet.

Number Of Radials	*Height of Radials (feet)*	*Gain and Take-off Angle (dBi & degrees)*		*Pattern Non-Circularity (dB)*
3	10	-1.54	26.8	0.30
	20	-1.38	24.8	0.32
	30	-1.30	23.5	0.33
	45	-1.24	22.2	0.36
	60	-1.21	20.8	0.38
4	10	-1.39	26.2	0.02
	20	-1.29	24.8	0.02
	30	-1.24	23.8	0.02
	45	-1.20	22.2	0.01
	60	-1.17	21.0	0.02
5	10	-1.34	26.2	0.01
	20	-1.26	25.0	0.01
	30	-1.22	23.8	0.0
	45	-1.18	22.5	0.01
	60	-1.16	21.0	0.01
6	10	-1.31	26.5	0.0
	20	-1.24	25.0	0.0
	30	-1.21	23.5	0.0
	45	-1.17	22.0	0.0
	60	-1.15	21.0	0.0

Table 9

Antenna gain versus number and height of elevated horizontal radials. The frequency is 1.835 MHz, and the radials are installed over very good soil (conductivity = 0.0303 Siemens/meter, and dielectric constant = 20). The vertical element, and all of the radials, have a length of 0.25 WL, or about 134 feet.

Number Of Radials	*Height of Radials (feet)*	*Gain and Take-off Angle (dBi & degrees)*		*Pattern Non-Circularity (dB)*
3	10	2.86	16.0	0.04
	20	3.04	15.5	0.04
	30	3.18	15.0	0.05
	45	3.35	14.0	0.04
	60	3.51	13.0	0.04
4	10	2.94	16.2	0.01
	20	3.08	15.5	0.01
	30	3.20	15.0	0.01
	45	3.36	14.0	0.01
	60	3.52	13.0	0.01
5	10	2.97	16.5	0.0
	20	3.10	15.5	0.0
	30	3.22	15.0	0.01
	45	3.38	14.0	0.0
	60	3.54	13.0	0.0
6	10	2.99	16.5	0.0
	20	3.11	15.5	0.0
	30	3.22	15.0	0.0
	45	3.38	14.0	0.01
	60	3.54	13.5	0.0

more elevated radials yields only a slight increase in gain.

Since the typical elevated ground screen contains only a few radials, the shape of the radiation pattern in the azimuthal plane may not be circular. That is, the gain measured directly off the tips of the radials may be different than the gain measured at a point located midway between two adjacent radials. The amount of this asymmetry is listed in the table column headed "pattern non-circularity." If only three elevated radials are used, the non-circularity is about 0.25 dB, but when N = 4 or more, the patterns are almost perfectly circular, with N = 6 providing the best performance.

Table 2 shows the outcome for the same quarter wave 40-meter vertical when the soil is "very poor," with a conductivity of 0.001 Siemens/meter and a dielectric constant of 5. If the soil is "very good," (conductivity = 0.0303 Siemens/meter and dielectric constant = 20), then the results are as listed in Table 3. In both cases, the trends are the same as when the soil was "average." Notice that when N = 3, the degree of non-circularity in the radiation pattern is higher for very poor soil than for average soil, while the reverse is true for very good soil.

Results for 80 meters

On 80 meters a frequency of 3.75 MHz was utilized, so the length of the vertical element (and the radials) is about 65.57 feet. Again, the number of radials was varied from 3 to 6, but now the elevation heights are 5, 10, 15, 22.5, and 30 feet (because the wavelength is longer at this frequency). When the antenna is installed over average soil, the results are as shown in Table 4. Tables 5 and 6 list the outcomes for very poor soil and very good soil, respectively.

Results on 160 meters

A frequency of 1.835 MHz was selected for the computer simulation on top band, requiring a length of about 134 feet for the quarter-wave vertical element and the radials. As usual, the number of radials was varied from three to six, but on this band, the base heights are 10, 20, 30, 45 and 60 feet. Table 7 displays the results when the antenna is mounted over average soil, while Tables 8 and 9 give the results for top-band verticals installed over very poor and very good soil, respectively.

Comparing All Three Bands

If a circular radiation pattern is desired, then one should probably install at least four elevated radials, uniformly spaced around the base of the antenna. If N = 3, then there will be a fair amount of non-circularity in the pattern, especially if the soil is very poor. The base height of the antenna does not appear to be critical, although greater heights yield slightly more gain at lower takeoff angles. In any event, all wires need to be sufficiently high above the ground to preclude any accidental contact by humans or animals.

Conclusions

For an elevated vertical antenna with a sparse ground screen consisting of elevated horizontal radials, the gain and takeoff angle depend upon the soil type, the number of radials, and their height above the ground. Part 1 of this article supplies tables that allow the low-band operator to design an elevated system that will best meet his or her needs. Part 2 will compare the data presented here with the results from an analysis of conventional ground-mounted vertical antennas utilizing large numbers of buried radials. NCJ

A Study of Elevated-Radial Ground Systems for Vertical Antennas—Part 2

Al Christman, K3LC
Grove City College
100 Campus Dr
Grove City, PA 16127-2104

Part 1 of this article discussed the gain of elevated 0.25 wavelength vertical monopole antennas, when installed over a variety of 0.25 wavelength elevated radial ground screens. Part 2 analyzes the performance of ground mounted 0.25 wavelength antennas with large numbers of buried 0.25 wavelength radials, and then compares these results with those found earlier for elevated vertical antenna systems.

Overview

Many contesters and DXers use vertical antenna elements, either singly or in phased arrays, for operation on the low bands. When ground mounted, each element is installed over a ground screen consisting of many radials that are either shallowly buried in the soil or simply laid upon the surface of the earth. The purpose of this study is two-fold: first, to determine how the gain of the conventional antenna configuration is impacted by changing the number of buried radials; second, to compare the gain of a buried-radial antenna with that of an elevated system. All three of the low bands (40, 80, and 160 meters) will be investigated, using three different types of soil on each band. The number of buried radials will be varied from 20 to 120, in steps of 20, with all radials uniformly spaced around the base of the vertical element.

I have access to the new *EZNEC Pro* software, version 4.0[1], with a double precision *NEC-4* calculating engine, which enables me to analyze ground systems using buried radials. For simplicity, all computer simulations were performed using #12 AWG copper wire for both the vertical element and the radials. The length of all wires was fixed at 0.25 wavelength, which is easy to calculate. Further, an antenna with these dimensions generally yields a reasonable value for the input resistance, on the order of 30 to 40 Ω. No attempt was made to "prune" the vertical element to achieve resonance. The wire segment lengths for the vertical element and the radials are all tapered in accordance with the most conservative NEC guidelines. The shortest segments, such as the one containing the feedpoint at the base of the vertical element, and the first segment of every radial, have a length of about 0.0025 wavelength. The inner segment of each radial slopes downward from the bottom end of the vertical element (at H = 0) to a final depth of 3 inches, with the remainder of the radial extending outward at this depth.

[1]Several versions of the *EZNEC* antenna-modeling software are available from Roy Lewallen, W7EL, PO Box 6658, Beaverton, OR 97007.

Results for Buried Radials on 40 meters

As in Part 1, a frequency of 7.15 MHz was selected for the analysis on 40 meters, leading to a length of about 34.39 feet for both the vertical element and the buried radials. The number of radials was varied sequentially from 20 to 40 to 60 to 80 to 100 to 120. Figure 1 shows a version of this antenna with 20 radials.

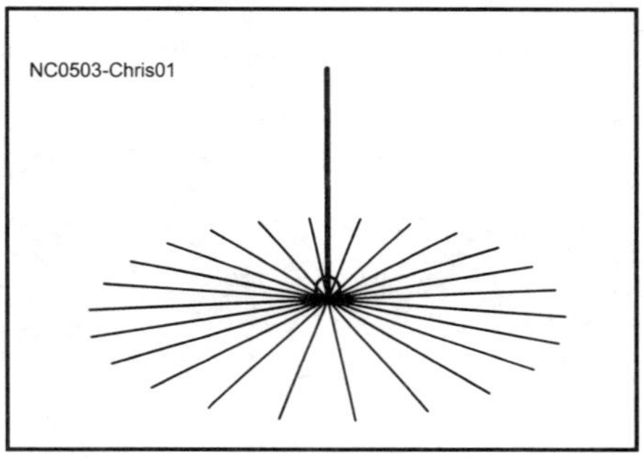

Figure 1—Drawing of a ground mounted vertical monopole antenna with 20 buried radials. The length of both the vertical element and the radials is 0.25 wavelength (34.39 feet at 7.15 MHz).

Table 1 displays the results when the antenna is mounted over average, very poor and very good soils, respectively. Average soil is assumed to have a conductivity of 0.005 Siemens per meter and a relative permittivity (or dielectric constant) of 13. For very poor soil, the conductivity is 0.001 Siemens per meter and the relative permittivity is 5, while for very good soil, the conductivity is 0.0303 Siemens/meter and the relative permittivity is 20.

As expected, both the highest gain and the lowest take-off angle are achieved when the soil conductivity is "very good." When the number of radials is held constant, the gain values for average soil are typically 0.8 dB better than those for very poor soil, while the gain values for very good soil are around 1.6 dB above those for average soil, or 2.3 dB better than those for very poor soil. The elevation angle at which maximum gain occurs is about 21 degrees for very good soil, 26 degrees for average soil, and 30 degrees for very poor soil. Of course, increasing the number of buried radials always leads to more gain, although the incremental improve-

Table 1

Antenna gain versus number of buried radials, for three different types of soil, at a frequency of 7.15 MHz. The vertical element, and all of the radials, have a length of 0.25 wavelength, or about 34.39 feet. For average soil, the conductivity is 0.005 Siemens per meter and the dielectric constant is 13; for very poor soil, the conductivity is 0.001 Siemens per meter and the dielectric constant is 5; for very good soil, the conductivity is 0.0303 Siemens per meter and the dielectric constant is 20.

Number Of Buried Radials	*Average Soil* Gain and Take-off Angle (dBi & degrees)		*Very Poor Soil* Gain and Take-off Angle (dBi & degrees)		*Very Good Soil* Gain and Take-off Angle (dBi & degrees)	
20	−0.34	26.5	−0.95	29.5	1.43	21.0
40	0.16	26.5	−0.59	29.5	1.70	20.5
60	0.31	26.5	−0.50	29.5	1.82	21.5
80	0.37	25.5	−0.46	29.5	1.88	21.5
100	0.41	26.0	−0.44	29.5	1.92	21.0
120	0.42	25.5	−0.42	29.5	1.94	21.0

ment is small when more than 60 to 80 radials are used.

Results for Buried Radials on 80 meters

On 80 meters a frequency of 3.75 MHz was utilized, as in Part 1, so the length of the vertical element and the buried radials is about 65.57 feet. Again, the number of radials was varied from 20 to 120 in steps of 20. When the antenna is installed over the three different types of soil, the results are as shown in Table 2. The typical gain values for average soil exceed those for very poor soil by almost 1.5 dB, while the gain values for very good soil are about 1.9 dB better than for average soil, or 3.4 dB above those for very poor soil. The take-off angle at which maximum gain occurs is about 19 degrees for very good soil, 25 degrees for average soil, and 29 degrees for very poor soil.

Results for Buried Radials on 160 meters

A frequency of 1.835 MHz was selected for the computer simulation on top band, requiring a length of about 134 feet for the quarter-wave vertical element and the buried radials. As usual, the number of radials was varied from 20 to 120 in increments of 20. Table 3 displays the results when the antenna is mounted over the three varieties of soil.

When the number of radials is held constant, the gain values for average soil are typically 2.1 dB better than those for very poor soil, while the gain values for very good soil are around 1.8 dB above those for average soil, or 3.9 dB better than those for very poor soil. The elevation angle at which maximum gain occurs is about 17 degrees for very good soil, 22 degrees for average soil, and 28 degrees for very poor soil.

Comparing Buried and Elevated Radial Antennas on 40 meters

Figure 2 is a plot of the gain of the buried radial antenna, for average soil, along with that of the five different elevated radial antennas, as a function of the number of radials. According to the computer analysis, if one is willing to install 50 or more buried 0.25 wavelength radials, then the conventional antenna system will generate more gain than any of the elevated radial configurations.

Figure 3 shows the situation for very poor soil. Here, an elevated radial antenna with a base height of 12 feet or more can outperform a conventional ground mounted vertical, even with an extensive system of radials. When the soil is very good, Figure 4 reveals that many of the elevated radial antennas are superior to the ground-mounted version.

Table 2

Antenna gain versus number of buried radials, for three different types of soil, at a frequency of 3.75 MHz. The vertical element, and all of the radials, have a length of 0.25 wavelength, or about 65.57 feet. For average soil, the conductivity is 0.005 Siemens per meter and the dielectric constant is 13; for very poor soil, the conductivity is 0.001 Siemens per meter and the dielectric constant is 5; for very good soil, the conductivity is 0.0303 Siemens per meter and the dielectric constant is 20.

Number Of Buried Radials	*Average Soil*		*Very Poor Soil*		*Very Good Soil*	
	Gain and Take-off Angle (dBi & degrees)		*Gain and Take-off Angle (dBi & degrees)*		*Gain and Take-off Angle (dBi & degrees)*	
20	–0.05	24.5	–1.54	29.0	2.09	19.0
40	0.38	24.5	–1.07	29.0	2.30	19.5
60	0.53	24.5	–0.94	28.5	2.40	19.5
80	0.60	25.0	–0.88	28.5	2.46	19.5
100	0.63	24.5	–0.84	29.0	2.49	18.5
120	0.66	25.0	–0.82	29.0	2.52	19.0

Table 3

Antenna gain versus number of buried radials, for three different types of soil, at a frequency of 1.835 MHz. The vertical element, and all of the radials, have a length of 0.25 wavelength, or about 134.0 feet. For average soil, the conductivity is 0.005 Siemens per meter and the dielectric constant is 13; for very poor soil, the conductivity is 0.001 Siemens per meter and the dielectric constant is 5; for very good soil, the conductivity is 0.0303 Siemens per meter and the dielectric constant is 20.

Number Of Buried Radials	*Average Soil*		*Very Poor Soil*		*Very Good Soil*	
	Gain and Take-off Angle (dBi & degrees)		*Gain and Take-off Angle (dBi & degrees)*		*Gain and Take-off Angle (dBi & degrees)*	
20	0.71	22.5	–1.56	27.5	2.70	17.5
40	1.04	22.5	–1.07	27.5	2.86	17.5
60	1.18	22.5	–0.91	27.5	2.94	17.5
80	1.25	22.5	–0.84	27.5	2.99	17.5
100	1.29	22.5	–0.80	27.5	3.02	17.5
120	1.31	22.5	–0.77	28.0	3.04	16.5

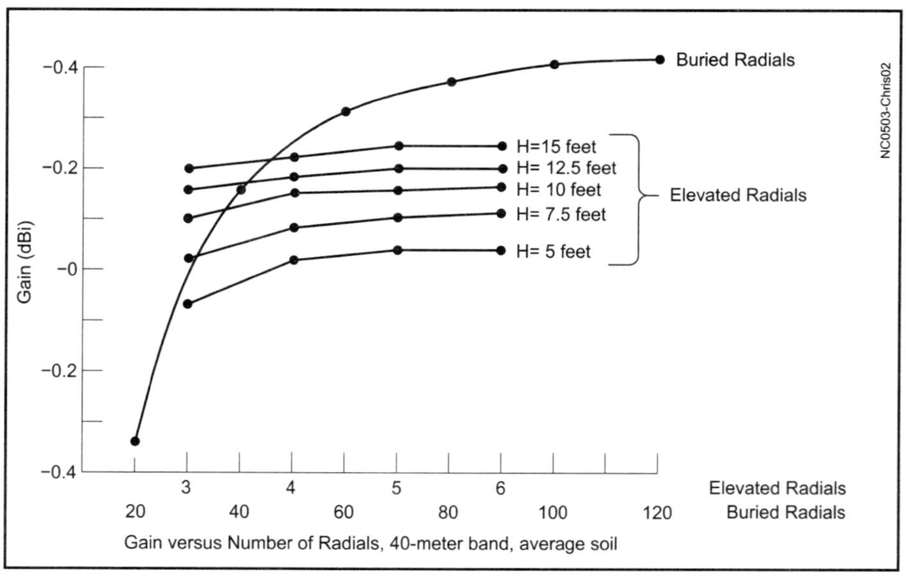

Figure 2—Gain versus Number of Radials, 40-meter band, average soil.

Comparing Buried and Elevated Radial Antennas on 80 meters

Figure 5 compares the gain of the buried radial antenna, for average soil, with that of the five different elevated radial designs, as the number of radials is varied. It is evident that a conventional antenna with 45 or more buried 0.25 wavelength radials can yield more gain than any of the elevated radial systems.

Figure 6 illustrates the outcome for very poor soil. Now, an elevated radial antenna with a base height of 22 feet or more can compete with a conventional ground mounted vertical, depending upon the number of buried radials. If the quality of the soil is very good, Figure 7 shows that most of the elevated radial antennas are better than the ground-mounted version.

Comparing Buried and Elevated Radial Antennas on 160 meters

Figure 8 plots the gain of the buried radial antenna, for

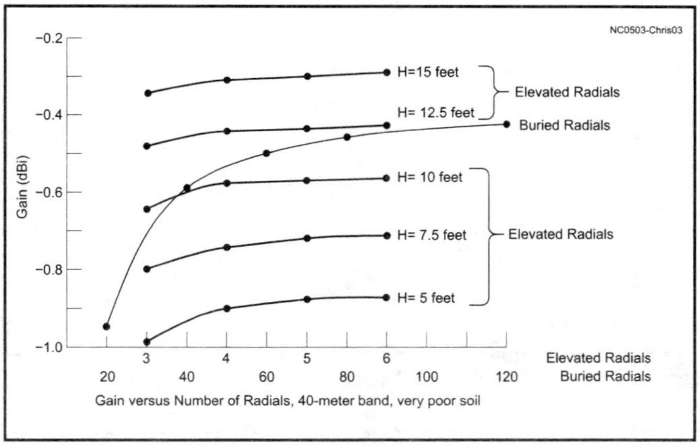

Figure 3—Gain versus Number of Radials, 40-meter band, very poor soil.

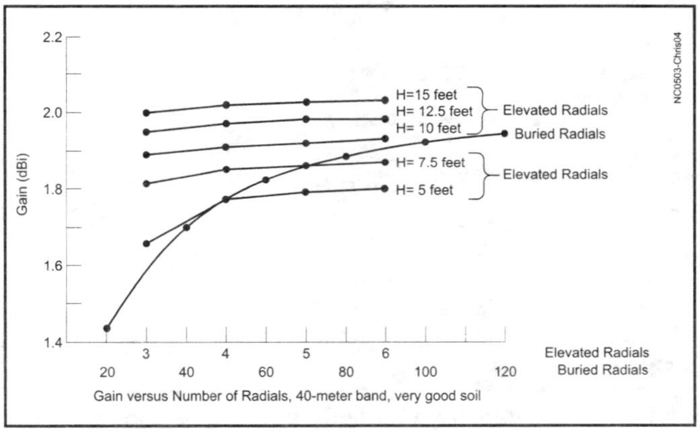

Figure 4—Gain versus Number of Radials, 40-meter band, very good soil.

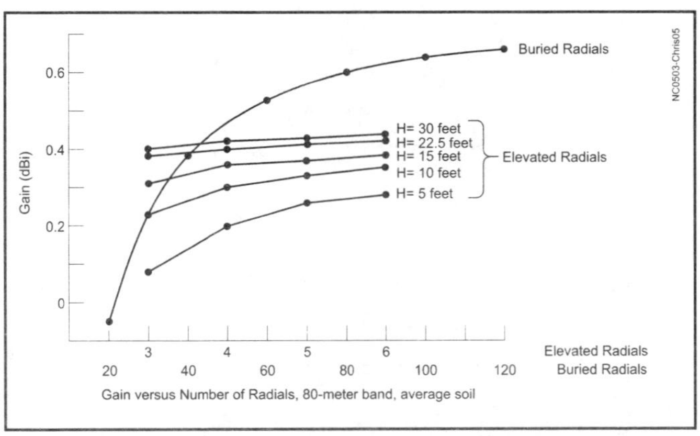

Figure 5—Gain versus Number of Radials, 80-meter band, average soil.

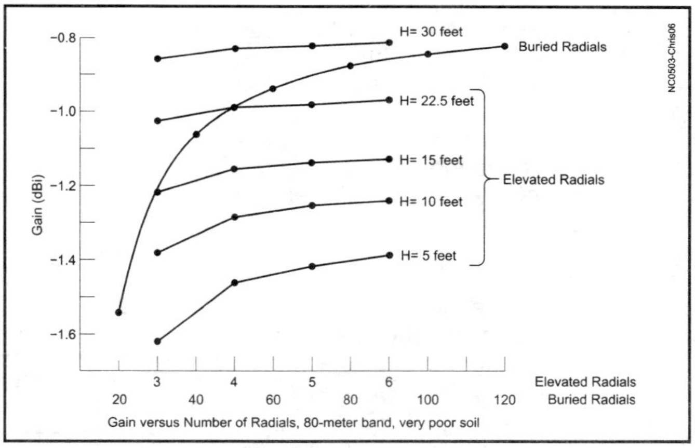

Figure 6—Gain versus Number of Radials, 80-meter band, very poor soil.

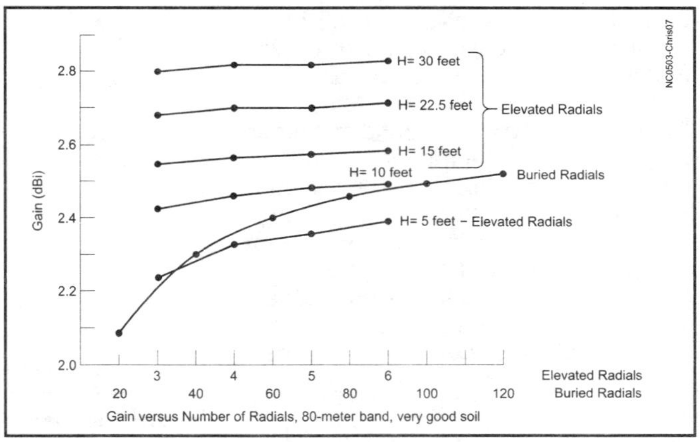

Figure 7—Gain versus Number of Radials, 80-meter band, very good soil.

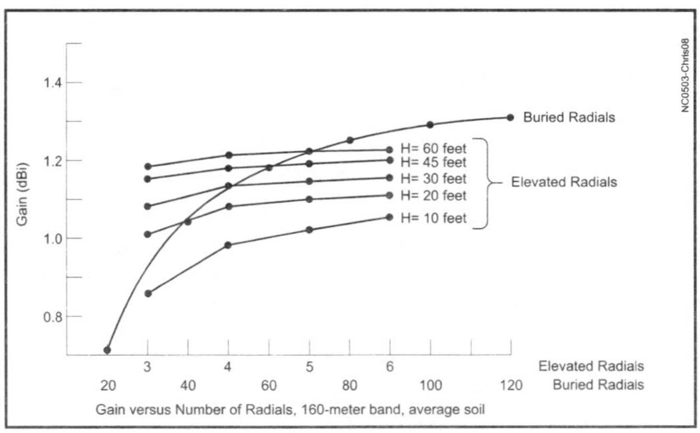

Figure 8—Gain versus Number of Radials, 160-meter band, average soil.

average soil, along with that of the five different elevated radial versions, as a function of the number of radials. If an amateur has the energy to install 70 or more buried 0.25 wavelength radials, then the conventional antenna system will generate more gain than any of the elevated radial configurations.

Figure 9 shows what happens with very poor soil. In this scenario, only about 35 buried radials are needed in order to generate more gain than can be achieved with any of the elevated radial antennas. In contrast, when the soil is very good, Figure 10 reveals that almost all of the elevated radial antennas are superior to the ground-mounted version, even when it has 120 buried radials.

Conclusions

The graphs presented here will allow an amateur to consider the relative merits of a conventional versus an elevated vertical antenna system, for a variety of soil types, elevation heights and number of radials. If the soil is of average conductivity, then a ground mounted antenna must have somewhere between 45 and 70 0.25 wavelength radials (depending upon the band of interest) in order to be competitive with all of the elevated radial configurations discussed in this article. For very poor soil, an elevated antenna must be installed relatively high above the ground to outperform a conventional antenna on 40 and 80 meters, while operation on top band virtually demands a conventional buried radial system. With very good soil, many of the elevated radial designs shown here are superior to a conventional antenna with 120 buried radials.

The statements above are correct if we can believe the results predicted by the computer software. To the author's knowledge, few measurements of far field sky wave power gain have been made on actual vertical antennas. Such a research project would be time-consuming, difficult and expensive, but the results would be of great value to all of us.

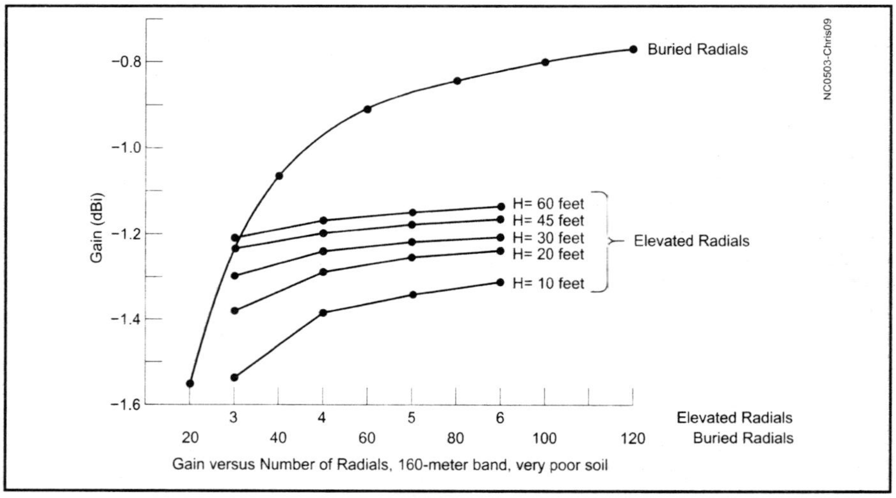

Figure 9—Gain versus Number of Radials, 160-meter band, very poor soil.

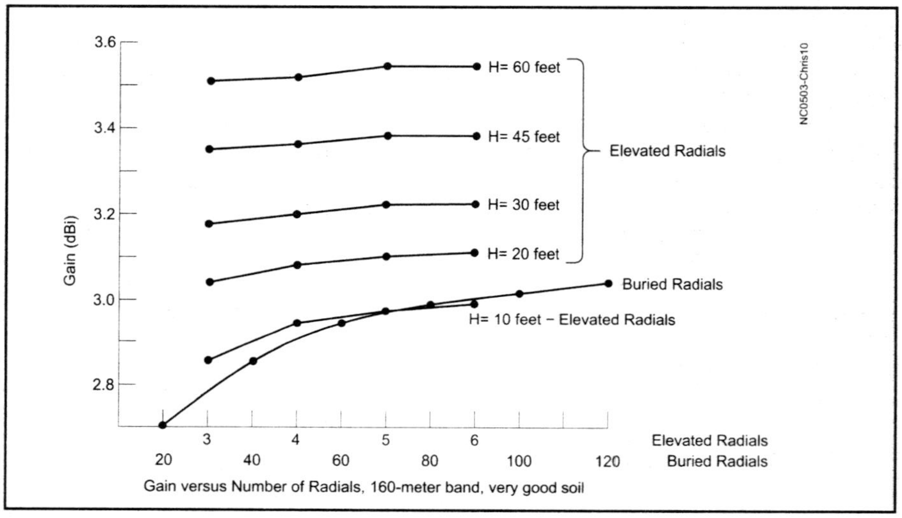

Figure 10—Gain versus Number of Radials, 160-meter band, very good soil.

NCJ

From *NCJ*, May/June 2005

A Study of Elevated-Radial Ground Systems for Vertical Antennas—Part 3

Al Christman, K3LC
Grove City College
100 Campus Dr
Grove City, PA 16127-2104

Part 3 of this article compares the gain developed by elevated 0.25 wavelength vertical monopole antennas when they are equipped with very sparse ground screens composed of either one or two radials. All of these counterpoise systems utilize elevated horizontal radials whose length is 0.25 wavelength. The elevation height of the radials is varied, along with the electrical characteristics of the soil beneath the antenna.

Overview

After Parts 1 and 2 of this article[1] were published, a number of questions were raised concerning the performance of antennas with only one or two radials in the ground screen. A two-radial antenna can be installed along a fence line or property boundary, and a vertical monopole with only a single radial may be erected at a Field Day site or on a DXpedition.

As before, all three of the low bands will be investigated, using three different types of soil on each band. The number of elevated radials will be limited to either two or one (when two are used, they are spaced 180 degrees apart in azimuth), and several different elevation heights will be examined.

I have access to the new *EZNEC Pro* software, version 4.0,[2] with a double precision *NEC4* calculating engine, enabling me to analyze ground systems using either buried or elevated radials. The computer simulations were performed using #12 AWG copper wire for both the vertical element and the elevated radials, and the length of all wires was fixed at 0.25 wavelength. No attempt was made to prune either the vertical element or the radials to achieve resonance. The wire segment lengths for the vertical element and the radials are all tapered in accordance with the most conservative *EZNEC* guidelines. The shortest segments, such as the one containing the feed point at the base of the vertical element, and the inner segment of each radial, have a length of about 0.0025 wavelength.

Details on the three soil types are as follows: average soil has an electrical conductivity of 0.005 Siemens/meter and a dielectric constant (relative permittivity) of 13; very poor soil has a conductivity of 0.001 Siemens/meter and a

Table 1

Gain of the 40-meter antenna with two elevated radials, as a function of radial height, for three different soil types. The frequency is 7.15 MHz. The vertical element and both radials have a length of 0.25 wavelength, or about 34.39 feet. "End fire" denotes the elevation plane containing the two radials, while "broadside" indicates the elevation plane that is perpendicular to the end-fire plane.

Soil Type	*Radial Height (feet)*	*Gain (dBi) & Take-off Angle (deg) end fire*		*broadside*		*Pattern Non-Circularity (dB)*
Avg	5	-0.56	23	-0.23	24	0.33
	7.5	-0.39	22	-0.02	23	0.37
	10	-0.27	21	0.12	22	0.39
	12.5	-0.19	20	0.21	22	0.40
	15	-0.12	19	0.27	21	0.39
Very Poor	5	-1.21	26	-1.44	27	0.23
	7.5	-0.94	25	-1.19	26	0.25
	10	-0.71	25	-0.99	26	0.28
	12.5	-0.51	24	-0.83	25	0.32
	15	-0.33	23	-0.69	24	0.36
Very Good	5	1.33	19	1.75	19	0.42
	7.5	1.46	18	1.93	19	0.47
	10	1.54	17	2.04	18	0.50
	12.5	1.59	16	2.13	17	0.54
	15	1.63	15	2.20	17	0.57

Table 2

Gain of the 80-meter antenna with two elevated radials, as a function of radial height, for three different soil types. The frequency is 3.75 MHz. The vertical element and both radials have a length of 0.25 wavelength, or about 65.57 feet. "End fire" denotes the elevation plane containing the two radials, while "broadside" indicates the elevation plane that is perpendicular to the end-fire plane.

Soil Type	*Radial Height (feet)*	*Gain (dBi) & Take-off Angle (deg) end fire*		*broadside*		*Pattern Non-Circularity (dB)*
Avg	5	-0.53	23	-0.08	24	0.45
	10	-0.27	21	0.24	23	0.51
	15	-0.15	20	0.41	22	0.56
	22.5	-0.05	19	0.54	20	0.59
	30	-0.02	18	0.59	19	0.61
Very Poor	5	-2.13	27	-2.04	27	0.09
	10	-1.75	25	-1.66	26	0.09
	15	-1.50	24	-1.43	25	0.07
	22.5	-1.20	23	-1.21	24	0.01
	30	-0.96	22	-1.06	23	0.10
Very Good	5	1.86	18	2.16	19	0.30
	10	2.11	17	2.46	18	0.35
	15	2.25	16	2.64	17	0.39
	22.5	2.38	15	2.82	16	0.44
	30	2.49	14	2.96	15	0.47

dielectric constant of 5; very good soil has a conductivity of 0.0303 Siemens/meter and a dielectric constant of 20.

Results for Antennas with Two Radials

A frequency of 7.15 MHz was selected for the analysis on 40 meters, leading to a length of about 34.39 feet for the vertical element and radials. Recall that the two radials are symmetrically placed, so they "point" in opposite directions. Elevation heights of 5, 7.5, 10, 12.5 and 15 feet were utilized.

Table 1 displays the results on 40 meters. The term "end fire" refers to the elevation plane containing the vertical element and the two horizontal radials, while "broadside" identifies the vertical plane that also contains the monopole, but is normal (or perpendicular) to the "end fire" plane.

As expected, when the base height of the antenna is raised, the gain increases and the take-off angle decreases, for all three types of soil. Notice that, for very poor soil, the antenna generates more gain in the plane containing the radials ("end fire"), but the opposite is true for average and very good soils, where maximum gain occurs in the "broadside" direction, normal to the radials. Since these antennas have only two radials, there is a considerable amount of non-circularity in the radiation patterns. This effect becomes progressively more pronounced as the conductivity of the soil increases.

On 80 meters a frequency of 3.75 MHz was utilized, so the length of the vertical element (and the radials) is about 65.57 feet. Now the elevation heights are 5, 10, 15, 22.5, and 30 feet (because the wavelength is longer at this frequency). Table 2 shows the outcome for all three types of soil. Peak gain occurs once again in the broadside direction, for both average and very good soils. Over very poor soil, maximum gain also occurs in the broadside direction on this band, except for those instances where the base height of the antenna is more than 15 feet. The radiation patterns are nearly circular for very poor soil, with the greatest deviation occurring over average soil.

A frequency of 1.835 MHz was selected for the computer simulation on top band, requiring a length of about 134 feet for the quarter-wave vertical element and the radials. On 160 meters, the base heights are 10, 20, 30, 45 and 60 feet. Table 3 displays the results when the antenna is mounted over three different types of soil. On top band, all of the antennas yield more gain in the plane which is perpendicular to that of the antenna ("broadside"). The azimuthal-plane radiation patterns exhibit a fair amount of non-circularity for all soil types, but especially for average soil.

Results for Antennas with One Radial

We found in Part 1 of this series that, if a circular radiation pattern is desired, then at least four elevated radials should be installed, spaced uniformly around the base of the antenna. In contrast, a single radial can deliberately be used to produce a small amount of directivity. In most cases, the tip of this radial points toward the front of the "main lobe", although there are some exceptions, which will be noted below.

Table 4 shows the outcome on 40 meters. In general, increasing the base height produces more "forward" gain at a lower take-off angle, but this is not always true in every situation. For ex-

Table 3

Gain of the 160-meter antenna with two elevated radials, as a function of radial height, for three different soil types. The frequency is 1.835 MHz. The vertical element and both radials have a length of 0.25 wavelength, or about 134 feet. "End-fire" denotes the elevation plane containing the two radials, while "broadside" indicates the elevation plane that is perpendicular to the end-fire plane.

Soil Type	*Radial Height (feet)*	*Gain (dBi) & Take-off Angle (deg)* end fire		broadside		*Pattern Non-Circularity (dB)*
Avg	10	0.32	21	0.77	22	0.45
	20	0.56	20	1.06	21	0.50
	30	0.66	19	1.22	20	0.56
	45	0.73	17	1.35	19	0.62
	60	0.75	16	1.40	18	0.65
Very Poor	10	-2.22	26	-1.72	26	0.50
	20	-1.94	24	-1.41	25	0.53
	30	-1.79	23	-1.25	24	0.54
	45	-1.66	21	-1.13	23	0.53
	60	-1.56	20	-1.11	21	0.45
Very Good	10	2.56	16	2.78	16	0.22
	20	2.80	15	3.06	15	0.26
	30	2.95	14	3.24	15	0.29
	45	3.11	13	3.44	14	0.33
	60	3.25	12	3.61	13	0.36

Table 4

Gain of the 40-meter antenna with one elevated radial, as a function of radial height, for three different soil types. The frequency is 7.15 MHz. The vertical element and the radial have a length of 0.25 wavelength, or about 34.39 feet. "Front" denotes the direction off the tip of the radial.

Soil Type	*Radial Height (feet)*	*Gain (dBi) and Take-off Angle (deg)* front		back		sides		*Front/Back Ratio (dB)*
Avg	5	-0.18	27	-4.55	21	-1.51	26	4.37
	7.5	0.11	26	-4.37	20	-1.05	26	4.48
	10	0.25	25	-4.32	18	-0.69	26	4.57
	12.5	0.30	24	-4.37	17	-0.38	27	4.67
	15	0.51	78	-4.48	16	0.27	87	4.99
Very Poor	5	-0.31	33	-7.13	22	-2.58	32	6.82
	7.5	0.02	32	-6.73	21	-2.01	32	6.75
	10	0.25	32	-6.41	20	-1.49	33	6.66
	12.5	0.40	32	-6.15	19	-1.00	34	6.55
	15	0.50	32	-5.93	18	-0.52	36	6.43
Very Good	5	1.20	20	-0.47	18	0.90	20	1.67
	7.5	1.36	19	-0.36	17	1.16	20	1.72
	10	1.38	18	-0.40	16	1.30	20	1.78
	12.5	1.31	17	-0.53	15	1.36	20	1.84
	15	1.17	17	-0.73	15	1.38	20	1.90

ample, we can see that a base height of 15 feet is too great for average soil, because of the very high take-off angle of the forward lobe at this elevation height. Further, at base heights of more than about 10 feet over very good soil, the peak forward gain actually diminishes, and more signal is beamed toward the "sides" of the antenna, rather than directly off the tip of the radial.

When the antenna is constructed over average soil, a front- to-back ratio (FBR) of 4 to 5 dB is possible, using only a single radial. For very poor soil, a FBR of 6 to 7 dB can be achieved, but only 1 or 2 dB of FBR is attainable with very good soil.

The results on 80 meters are displayed in Table 5. Again, we see a general trend toward more "forward" gain at lower take-off angles, as the base height increases. But, for both average and very poor soils, base heights of 30 feet are too large, because the peak of the forward lobe tilts upward at a very high take-off angle in these scenarios. And, at base heights of more than about 15 feet over very good soil, the peak forward gain begins to drop off, and more signal is beamed toward the "sides" of the antenna, instead of directly off the tip of the radial.

When the antenna is placed over average soil, a FBR of 3 to 5 dB is possible, using just one radial. For very poor soil, a FBR of more than 7 dB can be achieved, but only 1 dB or so of FBR is attainable with very good soil.

Table 6 illustrates the outcome on top band. As before, an increase in the base height usually generates more "forward" gain at a lower take-off angle. However, a base height of 60 feet is too great for both average and very poor soils, as evidenced by the very high take-off angle of the forward lobe at this elevation height. And, notice that the take-off angle toward the "sides" of the antenna that is mounted at a base height of 45 feet over very poor soil is 89 degrees (although the peak gain toward the front and back still occurs at low angles). Finally, at base heights of more than about 30 feet over very good soil, the peak forward gain reverses its upward trend, and more signal is beamed toward the "sides" of the antenna than toward the tip of the radial.

When the antenna is built over average soil, a modest FBR of 2 to 3 dB is possible, using only a single radial. For very poor soil, a FBR of 6 to 8 dB can be achieved, but with very good soil less than 1 dB of FBR is attainable.

Conclusions

For an elevated vertical antenna with a minimal ground screen consisting of just one or two elevated horizontal radials, the gain and take-off angle depend upon the soil type and the height of the antenna above ground. With two symmetrically placed radials, the azimuthal plane radiation pattern is somewhat noncircular, which may be undesirable. When only a single radial is used, the antenna exhibits a small amount of directivity, yielding a few decibels of front-to-back ratio. Antennas with one radial should probably not be installed at large distances above the ground, since this can lead to unwanted high-angle radiation, and peak gain may occur at azimuths other than directly off the end of the radial.

Notes:

[1]Parts 1 and 2 of this article appeared in the January/February and March/April 2005 issues of the *National Contest Journal.*

[2]Several versions of the *EZNEC* antenna modeling software are available from Roy Lewallen, W7EL, PO Box 6658, Beaverton, OR 97007.

Table 5

Gain of the 80 meter antenna with one elevated radial, as a function of radial height, for three different soil types. The frequency is 3.75 MHz. The vertical element and the radial have a length of 0.25 wavelength, or about 65.57 feet. "Front" denotes the direction off the tip of the radial.

Soil Type	*Radial Height (feet)*	*Gain (dBi) and Take-off Angle (deg)* front		back		sides		*Front/Back Ratio (dB)*
Avg	5	-0.49	25	-3.86	22	-1.49	24	3.37
	10	0.00	25	-3.48	20	-0.87	24	3.48
	15	0.21	23	-3.39	19	-0.49	25	3.60
	22.5	0.26	22	-3.53	17	-0.11	26	3.79
	30	1.06	83	-3.85	16	-0.93	88	4.91
Very Poor	5	-1.21	33	-8.29	22	-3.46	31	7.08
	10	-0.63	33	-7.77	21	-2.67	31	7.14
	15	-0.29	32	-7.47	20	-2.04	33	7.18
	22.5	0.00	32	-7.17	18	-1.19	36	7.17
	30	0.63	74	-6.96	17	0.16	87	7.59
Very Good	5	1.49	19	0.42	18	1.28	18	1.07
	10	1.90	18	0.81	17	1.77	18	1.09
	15	2.02	17	0.90	16	1.98	18	1.12
	22.5	1.96	15	0.79	15	2.07	18	1.17
	30	1.74	14	0.52	13	2.03	18	1.22

Table 6

Gain of the 160 meter antenna with one elevated radial, as a function of radial height, for three different soil types. The frequency is 1.835 MHz. The vertical element and the radial have a length of 0.25 wavelength, or about 134.0 feet. "Front" denotes the direction off the tip of the radial.

Soil Type	*Radial Height (feet)*	*Gain (dBi) and Take-off Angle (deg)* front		back		sides		*Front/Back Ratio (dB)*
Avg	10	0.13	23	-2.04	20	-0.42	22	2.17
	20	0.55	21	-1.69	19	0.12	22	2.24
	30	0.71	20	-1.62	18	0.40	22	2.33
	45	0.68	19	-1.77	16	0.61	22	2.45
	60	1.11	85	-2.11	15	1.04	87	3.22
Very Poor	10	-1.43	31	-7.47	22	-3.15	29	6.04
	20	-0.91	30	-7.14	20	-2.44	29	6.23
	30	-0.64	29	-7.08	19	-1.91	31	6.44
	45	-0.45	29	-7.17	17	-1.03	89	6.72
	60	0.79	79	-7.36	16	0.53	87	8.15
Very Good	10	2.17	16	1.49	16	2.07	17	0.68
	20	2.54	15	1.85	16	2.50	16	0.69
	30	2.65	15	1.94	15	2.67	16	0.71
	45	2.60	13	1.87	14	2.73	16	0.73
	60	2.41	13	1.65	13	2.66	16	0.76

Antenna Parameter Variations and Their Impact on Resonant Frequency

Al Christman, K3LC
Grove City College
100 Campus Dr
Grove City, PA 16127-2104

Have you ever built an antenna that you saw described in a magazine article, only to find that the actual resonant frequency was markedly different from what you expected? Or, have you sometimes found it necessary to make "substitutions" when constructing a certain antenna design, because you didn't have the exact parts that were specified, and wondered how the modifications would change the results? If so, read on...

This article examines several of the common parameters which exist in real-world ham-radio vertical antenna installations, and reviews the effects upon resonant frequency when they are varied. The items to be considered include metal composition, element diameter, element tapering, soil characteristics, and number of radials. All of the antennas were simulated using *EZNEC Pro*[1] with a double-precision *NEC-4* calculating engine.

Element Diameter and Tapering

The first antenna to be modeled is a full-size vertical monopole designed for operation at 3650 kHz (length = 0.25 wavelength [WL] = 67.3679 feet). The antenna is placed over perfectly conducting earth to eliminate the effects of losses in the ground system. The vertical element is initially assumed to be made of aluminum, but copper and zinc are also used (galvanized steel tower sections are made from steel that has been dipped in zinc to resist rust). The diameter of the monopole ranges from 0.0808 inches (#12 AWG wire) all the way up to 20 inches, in order to include vertical antennas made from wire, tubing, or tower sections. Constant-diameter monopoles are utilized, along with a tapered element whose diameter begins at 2.5 inches (at the base) and decreases in steps of 0.25 inch, ending with a tip diameter of 0.5 inches.

The outcome is summarized in Table 1. As expected, the resonant frequency of the antenna continually falls as the conductor is made "fatter." The wire element is resonant at 3554 kHz, while the 20-inch monopole resonates 160 kHz lower. Look what happens when a tapered element is used—now the resonant frequency is well above that of the antenna constructed from #12 wire (by almost 180 kHz), even though it is much larger in size. We can see that the diameter of the monopole has a significant effect upon the resonant frequency, especially when a tapered element is used.

The gain at resonance is nearly constant (at 5.14 dBi) for all versions of the antenna, decreasing by slightly more than 0.1 dB when #12 AWG wire is used instead of tubing or tower sections.

The composition of the vertical monopole was then changed from aluminum to copper, and finally from copper to zinc, with all the models being run twice more. The results for both copper and zinc were essentially identical (resonant frequency within +/- 1 kHz and gain within +/- 0.01 dBi) to those found earlier with aluminum, for all of the elements except the #12 AWG wire. The outcome changed slightly when this wire was used, and that information is included at the bottom of the

Table 1

Input impedance and resonant frequency for quarter-wave vertical elements of various diameters. The monopoles are all made of aluminum (except for the last two entries in the table), and are installed over earth that is perfectly conducting. In each case, the physical length of the element is 67.3679 feet, which is 0.25 WL at 3650 kHz.

Vertical Element Diameter	*Input Impedance at 3650 kHz (Ohms)*	*Resonant Frequency (kHz)*
#12 AWG	40.29 + *j* 23.59	3554
0.5 in.	39.84 + *j* 22.97	3534
1.0 in.	40.12 + *j* 23.14	3521
1.5 in.	40.38 + *j* 23.31	3512
2.0 in.	40.60 + *j* 23.46	3505
2.5 in.	40.80 + *j* 23.61	3498
5.0 in.	41.63 + *j* 24.20	3474
7.5 in.	42.30 + *j* 24.66	3456
10.0 in.	42.90 + *j* 25.04	3441
15.0 in.	43.97 + *j* 25.60	3415
20.0 in.	44.94 + *j* 25.95	3394
tapered from 2.5 to 0.5 inches	34.00 - *j* 13.37	3733
#12 AWG copper	39.82 + *j* 23.18	3556
#12 AWG zinc	40.61 + *j* 23.87	3552

Table 2

Input impedance and resonant frequency for vertical antennas using various numbers of buried quarter-wave radials. The monopole and radials are made from #12 AWG aluminum wire. "Average" soil, with a conductivity of 0.005 Siemens per meter and a dielectric constant of 13, is utilized. In each case, the physical length of the element is 67.3679 feet, which is 0.25 WL at 3650 kHz.

Number of Radials	*Input Impedance at 3650 kHz (Ohms)*	*Resonant Frequency (kHz)*
15	49.41 + *j* 28.55	3528
30	43.89 + *j* 26.04	3539
60	40.28 + *j* 23.72	3549
120	38.36 + *j* 22.01	3557

Table 3

Input impedance and resonant frequency for vertical antennas using various numbers of buried quarter-wave radials. The monopole and radials are made of #12 AWG aluminum wire. "Very good" soil with a conductivity of 0.0303 Siemens per meter and dielectric constant of 20 is utilized. In each case, the physical length of the element is 67.3679 feet, which is 0.25 WL at 3650 kHz.

Number of Radials	*Input Impedance at 3650 kHz (Ohms)*	*Resonant Frequency (kHz)*
15	45.48 + *j* 27.31	3534
30	43.17 + *j* 25.85	3540
60	41.29 + *j* 24.62	3545
120	39.91 + *j* 23.61	3550

Table 4

Input impedance and resonant frequency for vertical antennas using various numbers of buried quarter-wave radials. The monopole and radials are made of #12 AWG aluminum wire. "Very poor" soil with a conductivity of 0.001 Siemens per meter and a dielectric constant of 5 is utilized. In each case, the physical length of the element is 67.3679 feet, which is 0.25 WL at 3650 kHz.

Number of Radials	*Input Impedance at 3650 kHz (Ohms)*	*Resonant Frequency (kHz)*
15	48.61 + *j* 31.00	3521
30	39.89 + *j* 24.67	3547
60	36.12 + *j* 20.76	3563
120	34.50 + *j* 18.84	3571

Table 5

Resonant frequency for vertical antennas using various numbers of buried quarter-wave radials, for three different types of soil. The monopole and radials are made from #12 AWG aluminum wire. In each case the physical length of the element is 67.3679 feet, which is 0.25 WL at 3650 kHz.

	Resonant Frequency (kHz)			*Change in*
Number of Radials	*Average Soil*	*Very Good Soil*	*Very Poor Soil*	*Resonance (kHz)*
15	3528	3534	3521	13
30	3539	3540	3547	8
60	3549	3545	3563	18
120	3557	3550	3571	21

table. It is evident that the metal selected for constructing the antenna (aluminum, copper, or zinc-clad steel) is relatively unimportant.

Soil Characteristics and Number of Radials

Next, the same quarter-wave monopole was placed over "real earth" containing a ground system composed of buried quarter-wave radials whose number ranged from 15 to 120. Three different types of soil were examined: "average" soil with a conductivity of 0.005 Siemens per meter and a dielectric constant of 13; "very good" soil with a conductivity of 0.0303 Siemens per meter and a dielectric constant of 20; and "very poor" soil with a conductivity of 0.001 Siemens per meter and a dielectric constant of 5.

Table 2 displays the findings for average soil. The resonant frequency of the antenna rises continually as more radials are added, changing by nearly 30 kHz as the number of radials grows from 15 to 120. Not surprisingly, adding more radials also increases the gain, which changes from -0.31 dBi for 15 radials to 0.63 dBi for 120 radials. For very good soil, Table 3 shows that the shift in resonant frequency is not as great as before, rising by only 16 kHz when the number of radials increases from 15 to 120. The gain moves up from 1.97 dBi to 2.50 dBi at the same time. On the other hand, the increase in resonant frequency is fully 50 kHz when the antenna is installed over very poor soil (and the gain rises from -2.03 to -0.94 dBi) when the number of radials is increased from 15 to 120, as recorded in Table 4. We can see that changing the number of radials in the ground system has some impact upon the resonant frequency of the antenna, and the amount of influence becomes larger as the soil conductivity decreases.

Table 5 summarizes the variation in resonant frequency as a function of soil type, versus the number of radials. The addition of more radials always leads to a corresponding increase in the resonant frequency, for any kind of soil. When only 15 radials are utilized, changing the soil type causes the resonant frequency to vary by as much as 13 kHz, but the range decreases to 8 kHz with 30 radials. If the ground system uses 60 radials, then changing the soil type leads to a variation in resonance of up to 18 kHz, increasing still further to a maximum of 21 kHz for a system of 120 buried radials. When the number of radials is fixed, a change in the electrical character of the soil leads to a measurable shift in the resonant frequency of the antenna. However, the magnitude of this alteration is dependent upon the total number of radials.

Limitations

Several items should be mentioned with regard to the accuracy of computer simulations. These models are imperfect representations of the real world, and cannot possibly include all of the features which are actually present, such as buildings, vegetation, other conductive objects, irregularities in the terrain, non-uniformity of the ground constants, etc.

Conclusions

This article has investigated several factors which influence the resonant frequency of an antenna when it is operated in a natural environment. Parameters such as element diameter and tapering, soil characteristics, metallic composition, and number of radials have been reviewed, and the results presented in tabular form. It is hoped that this information will help us to understand why a backyard antenna may not always work "just like it does in the book."

Notes

[1] *EZNEC Pro* is available from Roy Lewallen, W7EL, PO Box 6658, Beaverton, OR 97007.

NCJ

Compact Four-Squares

Al Christman, K3LC
Grove City College, 100 Campus Dr
Grove City, PA 16127-2104

Introduction

The four-square phased vertical array is very popular for use as a low-band contest or DX antenna. Unfortunately, it requires a lot of space to install the four vertical elements and the associated ground system. This article describes several designs for compact four-squares, suitable for operation on the 80 meter DX phone band, which will fit within an area of real estate measuring less than 50 feet by 50 feet.

Background

None of these arrays have actually been constructed, but all were simulated on the computer, using *EZNEC4*[1]. The operating frequency is 3.79 MHz, and the soil is assumed to have an electrical conductivity of 0.005 Siemens per meter, with a dielectric constant of 13. In order to minimize the amount of area occupied by the antennas, each computer model utilizes a free-standing central mast, with a diameter of 8 inches. This structure could be a self-supporting tower or metal pole. These are available from a variety of manufacturers.

First Four-Square Design

Figure 1 is a computer-generated drawing of the first design, which is the smaller of the two. This antenna utilizes an 80 foot tower as the central support, with an 8 foot ground rod at the bottom. Each side of the "square" is 48.08 feet long, which places the feed-points of the 4 elements exactly 34 feet from the axis of the tower. At a height of 15 feet above the ground, four 32 foot horizontal "arms" extend outward at 90° intervals. These arms are metallic, but an additional two foot length of non-conductive material must project beyond the end of each metal arm to the actual feed-points of the elements. These four long horizontal arms may be supported by over-head guys, as well as side-guys, if needed. The individual feed-lines for the elements should be run along these arms, with a current balun or choke at each feed-point.

The elements in the computer model of the four-square are constructed from #12 AWG copper wire; each element has a single vertical radiator and two horizontal radials. The radiator wires are 69 feet long, but are not perfectly vertical; instead, they slope inward toward the tower axis, at an angle of 21.8° from the vertical. Four short non-conductive horizontal arms, each projecting outward 8 feet from the center-line of the tower, can provide anchorage for the upper ends of these vertical elements.

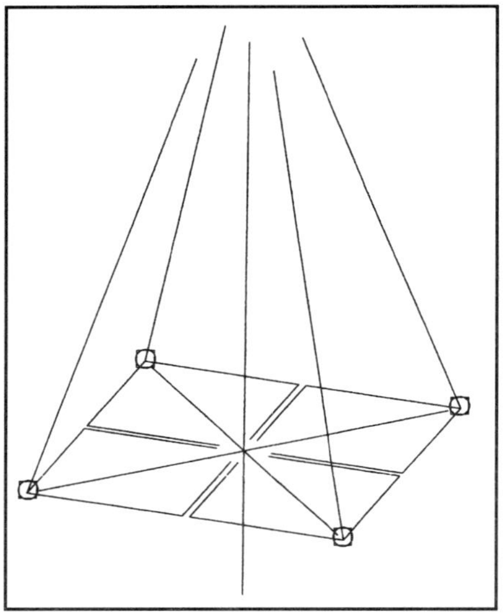

Figure 1—A computer drawing of the first compact four-square design, which utilizes an 80 foot central mast to support the antenna.

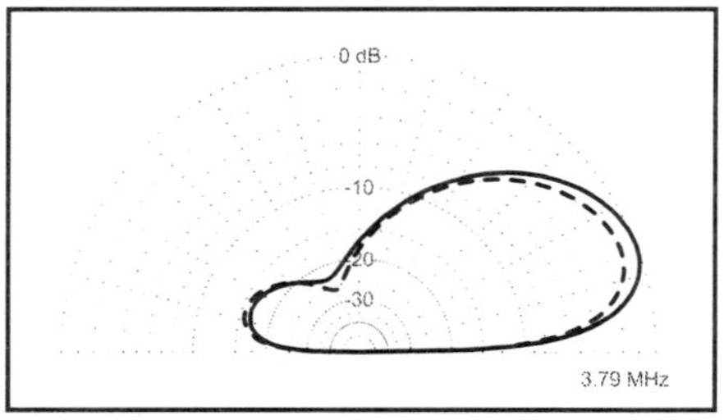

Figure 2—Elevation plane radiation patterns for the first compact four-square design—solid trace: firing through the corner of the square, using current phase-angles of 0°, -125° and -250°—dashed trace: firing through the side of the square, using current phase-angles of 0° and –125°.

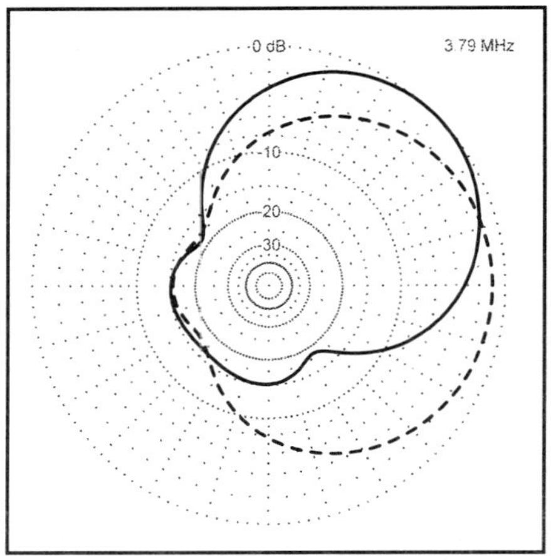

Figure 3—Azimuthal plane radiation patterns for the first compact four-square design—solid trace: firing through the corner of the square, using current phase-angles of 0°, -125°, and -250°—dashed trace: firing through the side of the square, using current phase-angles of 0° and -125°.

The radials, which make up the "sides" of the four-square, are all 43.5 feet long, but are "bent" so that two radials (one from each adjacent element) can "share" one side of the square. From each feed-point, two radials extend horizontally, perpendicular to each other, for a distance of 23.5 feet, and then bend inward at a 90° angle (directly toward the central mast) for another 20 feet. All of the radials are shorter than a full quarter-wavelength, so a loading coil must be placed at each feed-point to reduce the input reactance. In Figure 1, notice that the extreme ends of adjacent radials run parallel to each other, at a spacing of just over one foot. Thus, to avoid computational errors in the model, segment lengths of roughly one foot were used for all of the antenna wires.

Eight Directions of Fire

The parameters of the system were then adjusted to provide the best combination of gain, front-to-back ratio, beam width, and input impedance, when firing through the corners of the array. The final design utilizes equal-magnitude currents for all elements, with a phase-angle of 0° for the rear element, -125° for the two side elements, and -250° for the front element.

The resulting elevation plane radiation pattern is displayed as the solid trace in Figure 2. The peak forward gain is 4.32 dBi at a take-off angle (TOA) of 21°, and the front-to-back ratio (F/B) is 16.34 dB. In the azimuthal plane (the solid trace in Figure 3 shows the pattern) the F/B is also 16.34 dB, while the half-power beam width (HPBW) is 86.6°. Each loading coil has an inductance of 1.26 µH, or a reactance of 30Ω at 3.79 MHz. A series loss resistance of 0.3 Ω was assumed for each coil, yielding a Q of 100. The corresponding input impedances

[1]Notes appear on page 82.

are 34.3+*j*18.6 Ω for the front element, 12.9-*j*15.5 Ω for the two side elements, and -24.7-*j*10.0 Ω for the rear element.

Using these input currents, the beam width of the antenna is less than 90°, which means it is also necessary to drive the array so that it fires through the sides of the square. As before, equal-magnitude currents are used for all four elements, but now there are two "rear" elements and two "front" elements. The phase angle for the rear-element currents is 0°, and -125° is used for the front elements. Now the maximum forward gain is 3.37 dBi at 22° TOA, with an elevation plane F/B of 14.53 dB. In the azimuthal plane, the F/B is 14.57 dB and the HPBW is 122.2°. The dashed traces in Figures 2 and 3 are the corresponding radiation patterns. The driving-point impedances are 62.6-*j*23.6 Ω for the two front elements, and -11.8-*j*56.7 Ω for the rear pair of elements.

Four Directions of Fire

If the operator prefers to have an array that provides good performance with only four directions of fire (through the corners of the square), then the currents can be adjusted to produce a main lobe with less forward gain but greater beam width. To achieve this condition, the current phase angles of the "classic" four-square work well: 0° for the rear element, -90° for the two side elements, and -180° for the front element. In this configuration the loading coils should have an inductance of 2.1 µH, or 50 Ω of reactance at 3.79 MHz. The forward gain is 3.61 dBi at 22° TOA, with a F/B of 17.44 dB. In the azimuthal plane the F/B is 17.48 dB and the HPBW is 110.8°. The main radiation patterns are displayed in Figures 4 and 5. For the front element, the input impedance is 67.9+*j*28.1 Ω, with values of 27.2-*j*25.7 Ω for the two side elements and -37.8-*j*20.8 Ω for the rear element. If an even-broader main lobe is desired, the phase-angles can be decreased still further, with an accompanying reduction in gain.

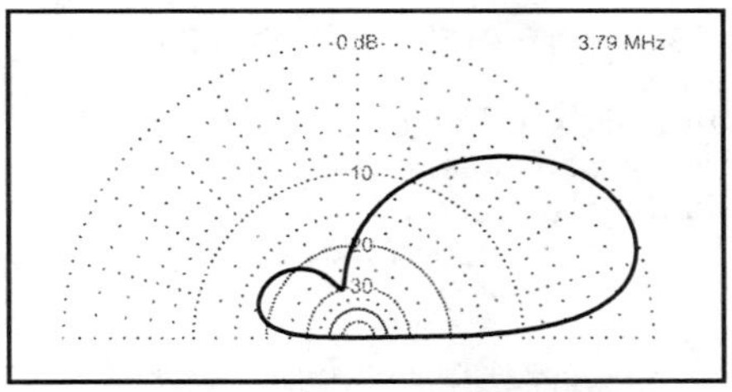

Figure 4— Elevation plane radiation pattern for the first compact four-square design, using current phase-angles of 0°, -90°, and -180° to produce a main lobe with a broad beam-width.

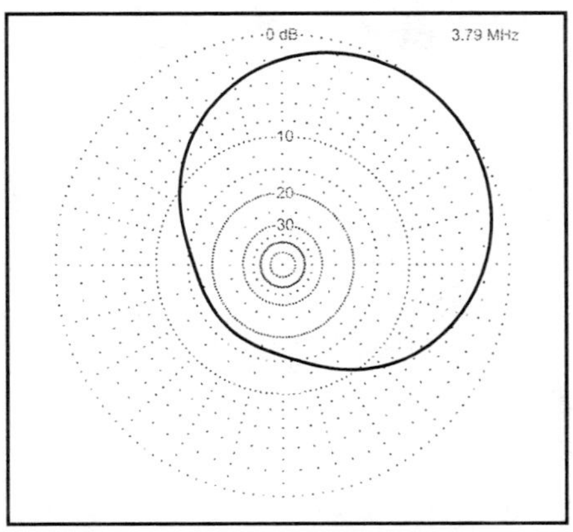

Figure 5—Azimuthal plane radiation pattern for the first compact four-square design, using current phase-angles of 0°, -90°, and -180° to produce a main lobe with a broad beam-width.

Second Four-Square Design

Inspiration for the second design came from an on-line article by L. B. Cebik, W4RNL[2]. This larger antenna (shown in Figure 6) requires a 120 foot tower as the central support, and again an 8 foot ground rod is installed at the base. As before, each side of the four-square is 48.08 feet long, placing the feed-points of the 4 elements 34 feet from the axis of the tower. In this design, full-size half-wave elements are used, with the feed-points located 64 feet above the ground. At this height, four 32 foot horizontal arms extend outward at 90-degree intervals. Again these "arms" are metallic, with an additional two foot length of non-conductive material projecting from the end of each metallic arm to the actual feed-point of the element. These arms will need to be supported by overhead guys, and probably by side-guys as well. The feed lines for the elements run along these "arms", with a current balun or choke at each feed point.

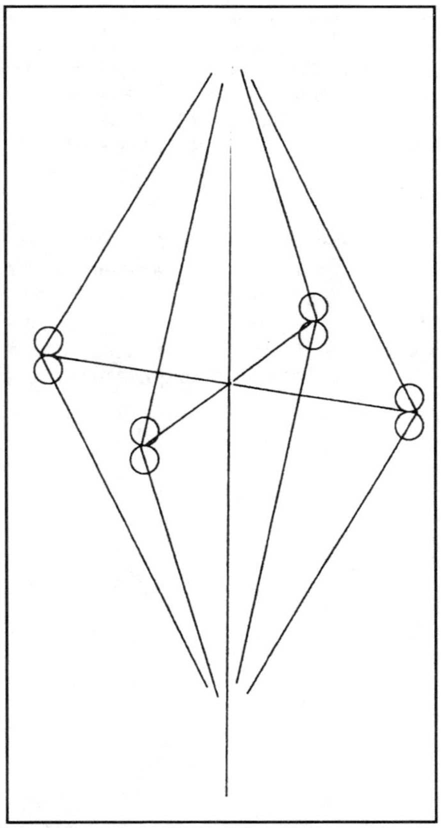

Figure 6—A computer drawing of the second compact four-square design, which utilizes a 120 foot central mast to support the antenna.

The elements of the four-square are constructed from #12 AWG copper wire. Each element is a half-wave vertical radiator that is fed in the center, and there are no radials. For this application each element-half is 61.5 feet long, but the wires are not perfectly vertical. Rather, they slope inward from the feed-point toward the tower axis, at an angle of about 29.5° from the vertical, so that the included angle between the two legs of the "vertical vee" is almost 121°. Four short non-conductive horizontal arms, each projecting outward 2.27 feet from the centerline of the tower, can provide anchorage for the upper ends of the vertical elements if installed at the apex height of 120 feet. Identical arms at a height of 8 feet will serve as tie-off points for the lower ends of the wires.

Eight Directions of Fire

The parameters of the system were then adjusted to provide the best combination of gain, F/B, beam width, and input impedance when firing through the corners of the array. The final design utilizes equal-magnitude currents for all elements, with a phase-angle of 0° for the rear element, -135° for the two side elements, and -270° for the front element. Because full-size elements are used in this larger antenna, no loading coils are needed.

The elevation plane radiation pattern for the array design is displayed as the solid trace in Figure 7. The peak forward gain is 5.48 dBi at a TOA of 18°, and the F/B is 23.54 dB. In the azimuthal plane, the F/B is 23.56 dB, while the HPBW is 86.8°. This radiation pattern is the solid trace shown in Figure 8. The input impedances are 4.8+*j*33.1 Ω for the front element, 23.0-*j*17.7 Ω for the two side elements, and -7.8-*j*16.0 Ω for the rear element.

When driven with these input currents, the beam width of the antenna is less than 90°, so it is also necessary to reconfigure this array to allow it to fire through the sides of the square. As before, equal-magnitude currents are used for all four elements. The phase-angle for the two rear element currents is 0°, with -135° used for the two front ele-

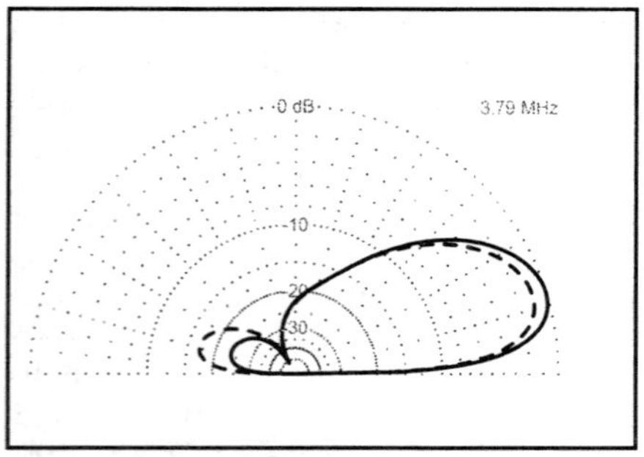

Figure 7—Elevation plane radiation patterns for the second compact four-square design—solid trace: firing through the corner of the square, using current phase-angles of 0°, -135°, and -270°—dashed trace: firing through the side of the square, using current phase-angles of 0° and -135°.

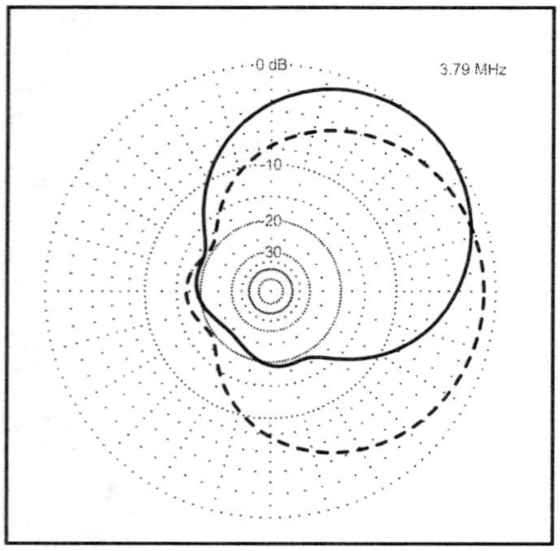

Figure 8—Azimuthal plane radiation patterns for the second compact four-square design—solid trace: firing through the corner of the square, using current phase-angles of 0°, -135°, and -270°—dashed trace: firing through the side of the square, using current phase-angles of 0° and -135°.

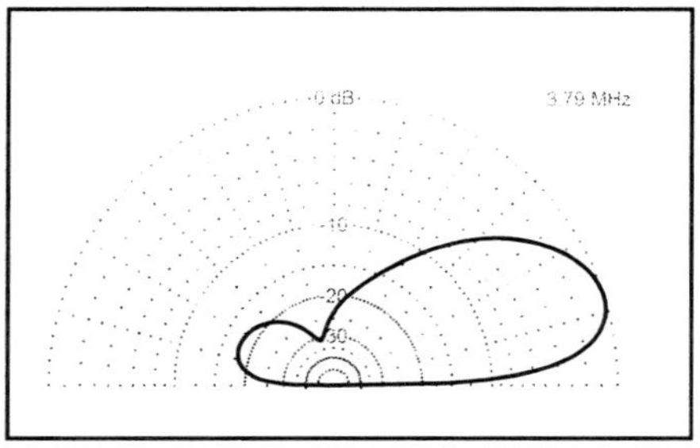

Figure 9—Elevation plane radiation pattern for the second compact four-square design, using current phase-angles of 0°, -105°, and -210° to produce a main lobe with a broad beam-width.

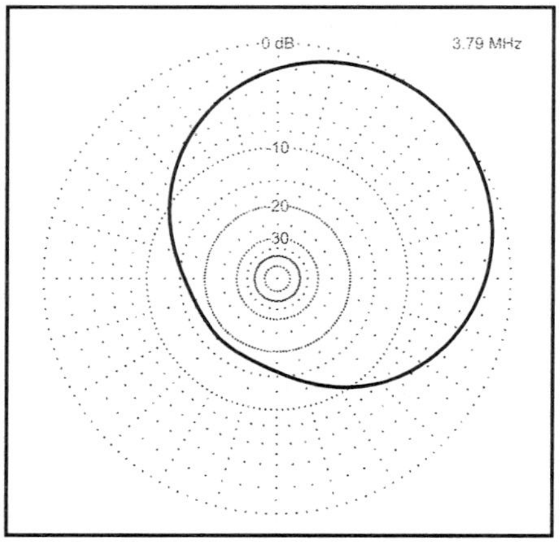

Figure 10—Azimuthal plane radiation pattern for the second compact four-square design, using current phase-angles of 0°, -105°, and -210° to produce a main lobe with a broad beam-width.

ments. Now the maximum forward gain is 4.57 dBi at 18° TOA, with an elevation plane F/B of 15.94 dB. In the azimuthal plane, the F/B is 15.98 dB and the HPBW is 122.0°. Figures 7 and 8 include the corresponding radiation patterns, shown as a dashed line. The driving-point impedances are 75.3+*j*29.7 Ω for the front pair of elements, and 5.7-*j*54.0 Ω for the two rear elements.

Four Directions of Fire

As was true with the first design, it is entirely possible to modify this larger array so it provides good performance with only four directions of fire. Of course, the resulting main lobe will have less forward gain, but a larger beam width. In this scenario, the current phase-angles are selected as 0° for the rear element, -105° for the two side elements, and -210° for the front element. The forward gain is now 4.84 dBi at 18° TOA, with an elevation plane F/B of 17.64 dB. In the azimuthal plane the F/B is 17.87 dB and the HPBW is 111.0°. The important radiation patterns are displayed in Figures 9 and 10. For the front element, the input impedance is 39.8+*j*65.4 Ω, with values of 54.1-*j*38.3 Ω for the two side elements and -22.9-*j*44.2 Ω for the rear element. Again, a broader main lobe can be produced by further decreasing the current phase-angles, although this reduces both the forward gain and the front-to-back ratio.

Conclusion

This article has shown how to construct several different compact four-square phased-vertical arrays that fit into a 50x50-foot plot of land. Their performance is inferior to that of a conventional ground-mounted antenna, but only 1/16 as much space is required.

Notes

[1] *EZNEC* is available from Roy Lewallen, W7EL, PO Box 6658, Beaverton OR 97007.

[2] L. B. Cebik, W4RNL, "Pursuing the (Nearly) Perfect Parasitic Vertical Array for 160 Meters, Part 1: A Review of Design and Modeling Techniques,"

NCJ

Vertical Antennas with Top Loading

Al Christman, K3LC
Grove City College
100 Campus Dr
Grove City, PA 16127-2104

I recently was asked by a W7 in Montana about the optimum design for a top-loaded vertical antenna. This topic seems well suited to computer analysis, so I decided to simulate a simple antenna in which the height of the vertical portion was varied, along with the down-slope angle of the top hat wires. The results indicate that the gain of the antenna is essentially constant, independent of the antenna height and the slope-angle of the wires. However, the shorter antennas have less bandwidth, as one would expect. These conclusions support the fine experimental work of Dr Jerry Sevick, W2FMI, which was performed back in the 1970s using horizontal top hats.[1-5] However, the information reported here extends his studies to include sloping top hats, which are easier to build.

Reference Antenna

A frequency of 3.65 MHz in the 80 meter band was selected for the computer analysis, which was performed using *EZNEC*-4.[6] The first antenna to be studied was a full-size quarter-wave vertical element, which serves as a standard of reference. The monopole is composed of #10 AWG wire, and is placed over "average" soil where the conductivity is 5 milliSiemens per meter and the dielectric constant is 13. The ground system consists of 120 quarter-wave radials made from #12 AWG wire and buried to a depth of 3 inches. All of these wires were assumed to be lossless (that is, perfect conductors) in order to remove this variable from the analysis. I trimmed the vertical element until resonance was achieved (zero input reactance), which occurred when the length was 65.68 feet. The initial length of this wire was 67.368 feet, or 0.25 WL, so the full-size vertical monopole has a resonant length of about 87.7°, rather than 90°. The peak gain of the antenna was 0.66 dBi at a take-off angle of 24°, and the input resistance at resonance was 35.37 Ω. Using this impedance value as a reference, the 2:1 SWR bandwidth is approximately 220 kHz.

Procedure

For the computer analysis, I varied the height of the vertical portion of the antenna in 5° increments from 35° to 80°, to cover a wide range of possible lengths. The top hat was made from four equal length #12 AWG wires that were symmetrically spaced in azimuth around the top of the vertical element. The down-slope angle of these top-loading wires was varied in 10° increments from 0° (horizontal top hat) to 60°. Figure 1 shows a 35° vertical whose top-hat wires have a 0° down-slope angle, while Figure 2 illustrates the same vertical when the down-slope angle of the top hat wires is 60°. The same ground system described earlier (120 buried quarter-wave radials) was utilized for each test. As the down-slope angle was changed, the length of the four top hat wires was always readjusted to maintain resonance at the operating frequency of 3.65 MHz. (I assumed that an input reactance value of 0.025 Ω, or less, was equivalent to resonance.)

Results

Table 1 shows the outcome when the vertical element is only 35° tall (roughly 26.2 feet). Although very short, this antenna has a peak gain that is slightly higher than that of the full size radiator. However, the difference is so small that it would be undetectable in practice. The elevation angle where peak gain occurs is also slightly higher than for the reference antenna, by about 3 degrees. The input resistance (at resonance) changes markedly as the down-slope angle of the top loading wires increases, spanning the range from 13.4 Ω when the wires are horizontal, to only 4.89 Ω when the down-slope angle is 60°. In a similar fashion, the resonant length of the top hat wires continually increases as these wires are sloped more steeply downward. Tables 2 through 10 display the results for the other cases, in which the height of the vertical monopole is varied from 40° to 80°.

Figure 1—A computer-generated view of an antenna with a 35° vertical section and a horizontal top hat (0° down-slope angle).

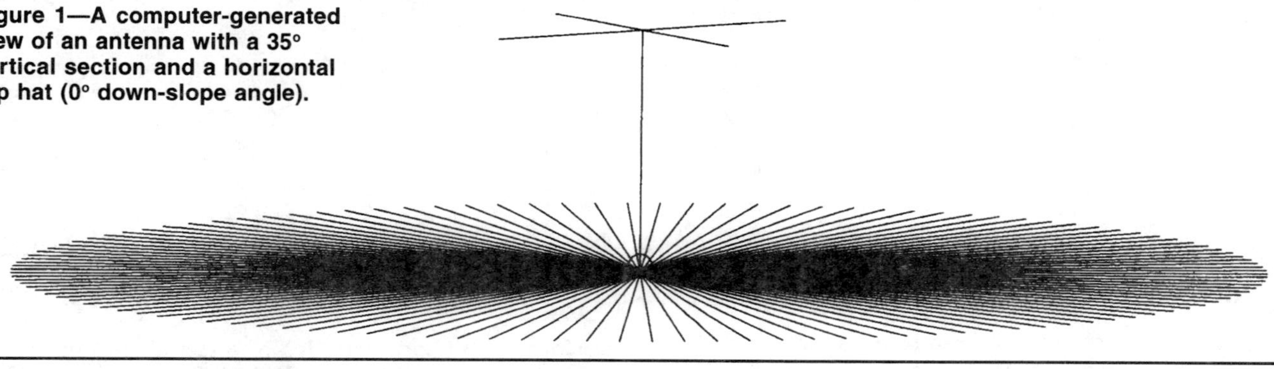

Figure 2—A computer-generated view of an antenna with a 35° vertical section and a top hat whose wires slope downward at 60°.

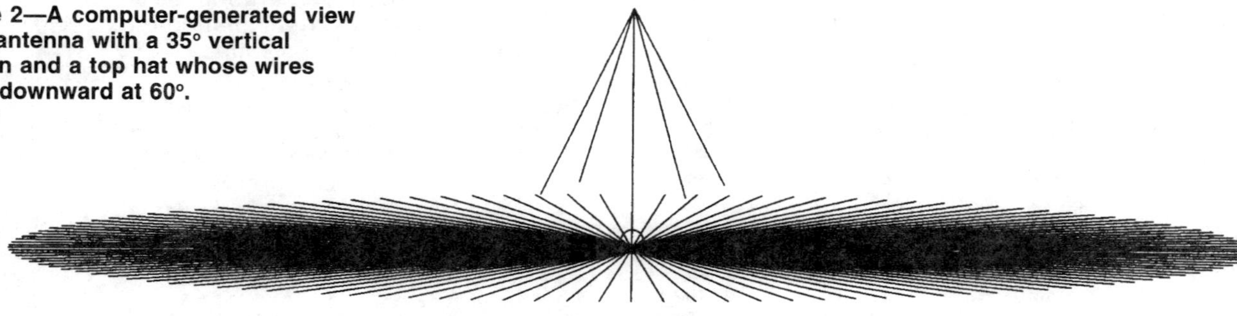

Table 11 lists the 2:1 SWR bandwidth for some of the antennas that were examined. As the height of the antenna is reduced, the bandwidth narrows in a corresponding manner. When the monopole is relatively tall, changing the down-slope angle of the top hat wires has only a small impact upon the bandwidth. However, if the vertical element is very short, the bandwidth decreases sharply as the down-slope angle increases. In every case, the broadest bandwidth was achieved with a horizontal top hat.

Conclusions

Table 12 contains the average values for the gain, takeoff angle, input resistance at resonance, and 2:1 SWR bandwidth, for the antennas that were analyzed in this study. Notice that the gain was virtually constant throughout the range of antenna heights that were simulated, although the takeoff angle did rise slightly as the antenna was shortened. Further, there was a continual reduction in the input resistance and SWR bandwidth as the height of the antenna decreased. If this loss of bandwidth is not a problem, then a short top loaded vertical monopole, equipped with a good impedance matching network and an extensive ground screen, can compete well with a full size quarter-wave antenna. To quote Jerry, W2FMI: "With low-loss matching and loading techniques, short verticals over a

Table 1

Performance characteristics of a top-loaded vertical antenna whose electrical height is 35 degrees (26.199 feet at 3.65 MHz). As the down-slope angle of the four top-loading wires is varied, their length is also adjusted to maintain resonance.

Down-Slope Angle (deg)	*Elevation-Plane Peak Gain and Takeoff Angle (dBi and degrees)*		*Input Resistance (Ohms)*	*Top-Load Wire Length (feet)*
0	0.67	27	13.40	18.027
10	0.67	27	11.90	18.474
20	0.67	27	10.41	19.032
30	0.68	27	8.95	19.723
40	0.67	27	7.60	20.453
50	0.69	27	6.24	21.589
60	0.73	27	4.89	23.141

Table 2

Performance characteristics of a top-loaded vertical antenna whose electrical height is 40 degrees (29.941 feet at 3.65 MHz). As the down-slope angle of the four top-loading wires is varied, their length is also adjusted to maintain resonance.

Down-Slope Angle (deg)	*Elevation-Plane Peak Gain and Takeoff Angle (dBi and degrees)*		*Input Resistance (Ohms)*	*Top-Load Wire Length (feet)*
0	0.67	27	16.62	15.388
10	0.67	27	15.31	15.785
20	0.67	27	13.98	16.291
30	0.67	27	12.64	16.934
40	0.68	27	11.29	17.769
50	0.70	27	9.91	18.892
60	0.68	26	8.61	20.275

Table 3

Performance characteristics of a top-loaded vertical antenna whose electrical height is 45 degrees (33.684 feet at 3.65 MHz). As the down-slope angle of the four top-loading wires is varied, their length is also adjusted to maintain resonance.

Down-Slope Angle (deg)	*Elevation-Plane Peak Gain and Takeoff Angle (dBi and degrees)*		*Input Resistance (Ohms)*	*Top-Load Wire Length (feet)*
0	0.67	26	19.84	13.21
10	0.67	26	18.72	13.56
20	0.67	26	17.58	14.01
30	0.67	26	16.41	14.59
40	0.68	26	15.20	15.35
50	0.66	26	14.11	16.102
60	0.67	26	12.78	17.562

Table 4

Performance characteristics of a top-loaded vertical antenna whose electrical height is 50 degrees (37.427 feet at 3.65 MHz). As the down-slope angle of the four top-loading wires is varied, their length is also adjusted to maintain resonance.

Down-Slope Angle (deg)	*Elevation-Plane Peak Gain and Takeoff Angle (dBi and degrees)*		*Input Resistance (Ohms)*	*Top-Load Wire Length (feet)*
0	0.66	26	22.99	11.122
10	0.66	26	22.10	11.422
20	0.66	26	21.17	11.811
30	0.66	26	20.21	12.313
40	0.66	26	19.21	12.976
50	0.67	26	18.14	13.886
60	0.66	26	17.08	15.037

Table 5

Performance characteristics of a top-loaded vertical antenna whose electrical height is 55 degrees (41.169 feet at 3.65 MHz). As the down-slope angle of the four top-loading wires is varied, their length is also adjusted to maintain resonance.

Down-Slope Angle (deg)	*Elevation-Plane Peak Gain and Takeoff Angle (dBi and degrees)*		*Input Resistance (Ohms)*	*Top-Load Wire Length (feet)*
0	0.66	26	25.88	9.419
10	0.66	26	25.19	9.678
20	0.66	26	24.46	10.016
30	0.66	26	23.71	10.454
40	0.66	25	22.91	11.031
50	0.66	26	22.17	11.621
60	0.66	26	21.23	12.756

Table 6

Performance characteristics of a top-loaded vertical antenna whose electrical height is 60 degrees (44.912 feet at 3.65 MHz). As the down-slope angle of the four top-loading wires is varied, their length is also adjusted to maintain resonance.

Down-Slope Angle (deg)	*Elevation-Plane Peak Gain and Takeoff Angle (dBi and degrees)*		*Input Resistance (Ohms)*	*Top-Load Wire Length (feet)*
0	0.66	25	28.50	7.732
10	0.66	26	27.99	7.951
20	0.66	26	27.47	8.238
30	0.66	26	26.92	8.61
40	0.66	26	26.34	9.099
50	0.66	25	25.70	9.775
60	0.67	26	24.95	10.772

large ground system suffer only in bandwidth."[7]

Notes

[1]Jerry Sevick, "The Ground-Image Vertical Antenna," *QST*, July 1971.
[2]Jerry Sevick, "The W2FMI 20-Meter Vertical Beam," *QST*, June 1972.
[3]Jerry Sevick, "The W2FMI Ground-Mounted Short Vertical," *QST*, March 1973.
[4]Jerry Sevick, "A High-Performance 20, 40, and 80-Meter Vertical System," *QST*, December 1973.
[5]Jerry Sevick, "The Constant Impedance Trap Vertical," *QST*, March 1973.
[6]*EZNEC* is available from Roy Llewellyn, W7EL, PO Box 6658, Beaverton, OR 97007.
[7]Jerry Sevick, "Short Ground-Radial Systems for Short Verticals," *QST*, April 1978, page 33.

Table 7

Performance characteristics of a top-loaded vertical antenna whose electrical height is 65 degrees (48.655 feet at 3.65 MHz). As the down-slope angle of the four top-loading wires is varied, their length is also adjusted to maintain resonance.

Down-Slope Angle (deg)	*Elevation-Plane Peak Gain and Takeoff Angle (dBi and degrees)*		*Input Resistance (Ohms)*	*Top-Load Wire Length (feet)*
0	0.65	25	30.75	6.207
10	0.65	25	30.41	6.39
20	0.65	25	30.06	6.629
30	0.65	25	29.69	6.938
40	0.65	25	29.29	7.343
50	0.66	26	28.85	7.902
60	0.66	25	28.33	8.73

Table 8

Performance characteristics of a top-loaded vertical antenna whose electrical height is 70 degrees (52.397 feet at 3.65 MHz). As the down-slope angle of the four top-loading wires is varied, their length is also adjusted to maintain resonance.

Down-Slope Angle (deg)	*Elevation-Plane Peak Gain and Takeoff Angle (dBi and degrees)*		*Input Resistance (Ohms)*	*Top-Load Wire Length (feet)*
0	0.65	24	32.59	4.966
10	0.65	24	32.37	5.122
20	0.65	24	32.15	5.329
30	0.66	25	31.90	5.592
40	0.66	25	31.64	5.922
50	0.65	25	31.44	6.096
60	0.65	25	31.12	6.769

Table 9

Performance characteristics of a top-loaded vertical antenna whose electrical height is 75 degrees (56.140 feet at 3.65 MHz). As the down-slope angle of the four top-loading wires is varied, their length is also adjusted to maintain resonance.

Down-Slope Angle (deg)	*Elevation-Plane Peak Gain and Takeoff Angle (dBi and degrees)*		*Input Resistance (Ohms)*	*Top-Load Wire Length (feet)*
0	0.66	25	33.97	3.455
10	0.66	25	33.87	3.576
20	0.65	24	33.75	3.745
30	0.65	24	33.64	3.969
40	0.65	24	33.50	4.242
50	0.65	24	33.35	4.551
60	0.66	25	33.19	4.90

Table 10

Performance characteristics of a top-loaded vertical antenna whose electrical height is 80 degrees (59.883 feet at 3.65 MHz). As the down-slope angle of the four top-loading wires is varied, their length is also adjusted to maintain resonance.

Down-Slope Angle (deg)	*Elevation-Plane Peak Gain and Takeoff Angle (dBi and degrees)*		*Input Resistance (Ohms)*	*Top-Load Wire Length (feet)*
0	0.65	24	34.93	2.062
10	0.65	24	34.89	2.137
20	0.65	24	34.85	2.255
30	0.65	25	34.81	2.429
40	0.65	25	34.76	2.663
50	0.65	25	34.70	2.934
60	0.65	25	34.63	3.192

Table 11

Approximate 2:1 SWR bandwidth for various top-loaded vertical antennas, as a function of antenna height and top hot slope angle. In each case, the reference impedance used for calculating the SWR is the value of input resistance given previously in the corresponding tables. Recall that the full-height antenna has a bandwidth of about 220 kHz.

Vertical Monopole Height (degrees)	*2:1 SWR Bandwidth (kHz)*		
	0-degree slope angle	*30-degree slope angle*	*60-degree slope angle*
35	120	80	55
40	135	105	75
45	150	125	95
50	165	145	125
55	175	160	145
60	185	180	165
65	190	185	180
70	205	200	190
75	210	205	200
80	215	212	210

Table 12

Average values of the main performance characteristics for top-loaded vertical antennas, as a function of antenna height. Values for the full-size (resonant quarter-wave) element are shown at the bottom.

Vertical Monopole Height (degrees)	Elevation-Plane Peak Gain and Take-off Angle (dBi and degrees)		Input Resistance (Ohms)	2:1 SWR Bandwidth (kHz)
35	0.683	27.0	9.06	85
40	0.677	26.9	12.62	105
45	0.670	26.0	16.38	123
50	0.661	26.0	20.13	145
55	0.660	25.9	23.65	167
60	0.661	25.7	26.84	176
65	0.653	25.1	29.63	186
70	0.653	24.6	31.89	198
75	0.654	24.4	33.61	207
80	0.650	24.6	34.80	213
full-size	0.660	24.0	35.37	220

NCJ

Verticals by the Sea—Part 1

Al Christman, K3LC
Grove City College
100 Campus Drive
Grove City, PA 16127-2104

This article analyzes the gain of vertical monopole antennas with two gull-wing elevated radials, when they are installed near the seacoast. The height of the radials is varied, along with the set-back distance between the antenna and the shoreline. Many of the amateur bands from 160 to 10 meters are reviewed.

Background

Contesters and DXers are well aware of the advantages that are inherent in the use of vertical antennas when they are placed near large bodies of salt water. Many years ago, Tom Erdmann, W7DND, had a rotatable 4-element 40-meter phased vertical array that was located at the edge of Puget Sound in Bremerton, Washington.[1] Well-known contester Rush Drake, W7RM (SK), used verticals mounted on a beach at the base of a cliff to work the world from his former QTH in Washington state.[2] More recently, "Team Vertical," composed of K2KW, N6BT and others, conducted several successful contest expeditions from the Caribbean using verticals.[3, 4] Their decision to employ bent or gull-wing elevated radials was based upon this author's earlier published work in an IEEE journal.[5]

A series of tests are described by K2KW, in which a vertical transmit antenna was erected on the beach and a receive antenna installed across the (saltwater) bay: "The team set the vertical on the land-water boundary (above) and moved the vertical away from the water in 3-foot steps. The land-water boundary was the reference point (0.0)...but when the antenna reached approximately 1/4 wavelength from the water, the signal level increased + 3 dB over the reference point! The zone of this enhancement was fairly small. As the antenna reached 1/2 wavelength from the water, the antenna was now -2 dB from the reference point! At the 3/4-wavelength point, the signal was now +2 dB. Obstacles prohibited us from going farther than 3/4 wavelength from the land-water boundary. But it was clear that the land-water boundary had a significant impact on the low-angle energy (in the pseudo-Brewster angle)."[4]

Kenny continues: "Unfortunately it appears that the gain enhancement is only in the immediate direction of the water. In directions other than the closest point of water, there is more land, thus impacting the pseudo-Brewster angle in those directions. While having 3 dB of gain is desirable, it is my conviction that the vertical should be placed as close to, or over, saltwater to ensure optimum performance in as many directions as possible."[4]

[1]Notes appear on page 89.

Computer Simulation

After considering the remarks above, it was decided to model several vertical monopole antennas, installed either at or near a land/sea boundary. Figure 1 shows a pictorial view of one antenna configuration, which is typical of those that were examined. The length of the vertical element and the two radials was fixed at exactly 0.25 wavelength at the operating frequency, while the base of the antenna was placed just one foot above the ground. The two gull-wing radials are oriented parallel to the shoreline, and they initially extend upward at a 45-degree angle from the bottom of the antenna, until reaching their final height H. This elevation height varies from 2.5 to 10 feet, depending upon the band of interest. The set-back distance from the water to the antenna is D, which ranges from 0 to 0.25 wavelengths in steps of 0.05 wavelength.

The antenna models were run using the new *EZNEC Pro* software, version 4.0,[6] with a double precision *NEC-4* calculating engine. For simplicity, it was assumed that all antennas were constructed entirely from #12 AWG copper wire. No attempt was made to prune either the vertical element or the radials to achieve resonance. The wire segment lengths for the vertical element and the radials were selected in accordance with the most conservative *NEC* guidelines. The seawater has an electrical conductivity of 5 Siemens per meter and a dielectric constant of 81. However, because the soil on most beaches is either sandy or rocky, its conductivity was set at 0.0015 Siemens/meter, with a dielectric constant of 4.

Results on 40 Meters

A frequency of 7.15 MHz was selected for the analysis on 40 meters, which corresponds to a length of about 34.39 feet for both the vertical element and the two gull-wing radials. Elevation heights

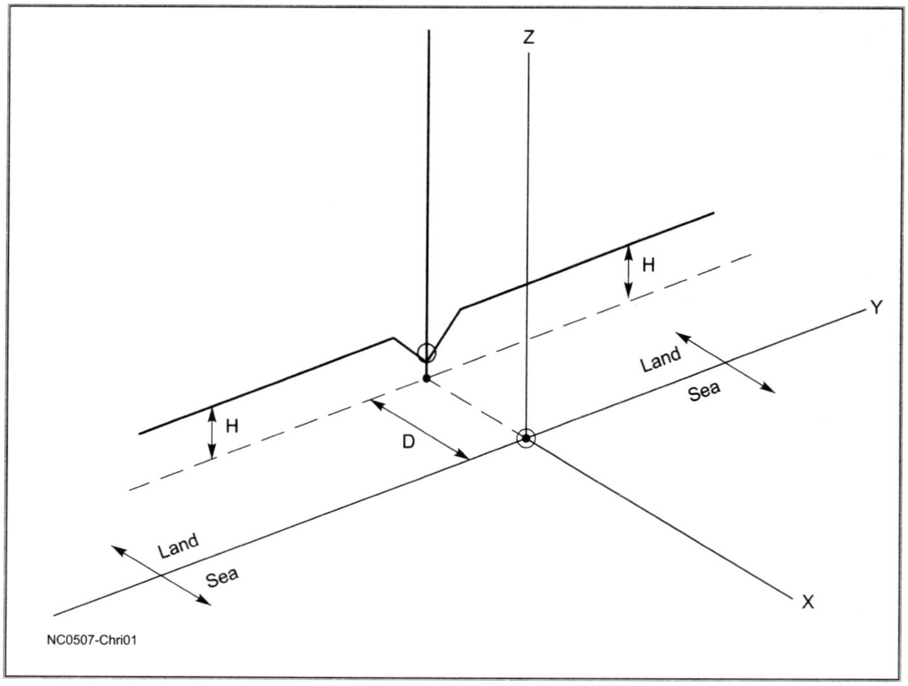

Figure 1—Drawing of an elevated vertical monopole antenna (base height = 1 foot) with two gull-wing radials. The length of both the vertical element and the radials is 0.25 wavelength (34.39 feet at 7.15 MHz). H represents the height of the horizontal portions of the radials, and D is the set-back distance from the shoreline.

of 2.5 and 5.0 feet were chosen for the horizontal portions of the radials.

Table 1 displays the resulting performance data for the gain and take-off angle, when looking either toward the front of the radiation pattern (seaward) or back toward land. Also shown are the front-to-back ratio (FBR) in the elevation plane, and the half power beamwidth (HPBW) in the azimuthal plane. We can see that increasing the height of the radials (H) always produced a small amount of extra gain and a slight improvement in the FBR, for any set-back distance (D). However, the changes were minor, with the gain remaining around 4 or 5 dBi and the FBR close to 7 dB. When the height of the radials (H) is held constant, the gain can be peaked (to a small degree) by varying the set-back distance. For either value of H, maximum gain and maximum FBR both occurred when the antenna was placed 0.1 wavelength from the edge of the water. When the height of the radials is fixed, the gain of the back lobe remains constant for any set-back distance, since the antenna always sees land in that direction.

The elevation and azimuthal plane radiation patterns for a set-back distance (D) of 0 wavelength and a radial height (H) of 2.5 feet are displayed in Figures 2 and 3, while those for D = 0.1 wavelength and H = 2.5 feet appear in Figures 4 and 5. For each plot, the sea is toward the right, and the land toward the left, as evidenced by the enhanced gain and lower take-off angles in the direction of the saltwater. Notice that the elevation plane pattern is very smooth when the antenna is mounted at the land-sea junction, but somewhat jagged and irregular when the set back distance increases. A similar effect is visible in the azimuthal plane patterns, along with a general decrease in the half power beamwidth when the antenna is moved away from the interface.

Why are some of the seawater portions of the pattern plots so odd looking? Roy Lewallen, W7EL, cautions us regarding the use of a second medium

Table 1

Antenna performance on 40 meters (f = 7.15 MHz) as a function of radial height and set-back distance. The base of the antenna is fixed at a height of one foot, and the radials are oriented parallel to the shoreline.

Radial Height H (ft)	*Set-back Distance D (WL)*	*Gain (dBi) and Take-off Angle (deg) front*		*back*		*FBR (dB)*	*Azimuthal Beamwidth (degrees)*
2.5	0.0	4.16	9	-2.65	29	6.81	179.6
	0.05	4.37	15	-2.65	28	7.02	162.4
	0.1	4.60	8	-2.65	29	7.25	157.0
	0.15	4.55	6	-2.65	28	7.20	152.6
	0.2	4.37	4	-2.65	28	7.02	156.2
	0.25	4.15	3	-2.65	28	6.80	158.0
5.0	0.0	4.52	9	-2.37	28	6.89	179.6
	0.05	4.77	17	-2.37	27	7.14	159.4
	0.1	5.17	9	-2.37	27	7.54	158.2
	0.15	5.15	6	-2.37	27	7.52	153.6
	0.2	5.01	9	-2.37	27	7.38	136.6
	0.25	5.01	7	-2.37	27	7.38	138.2

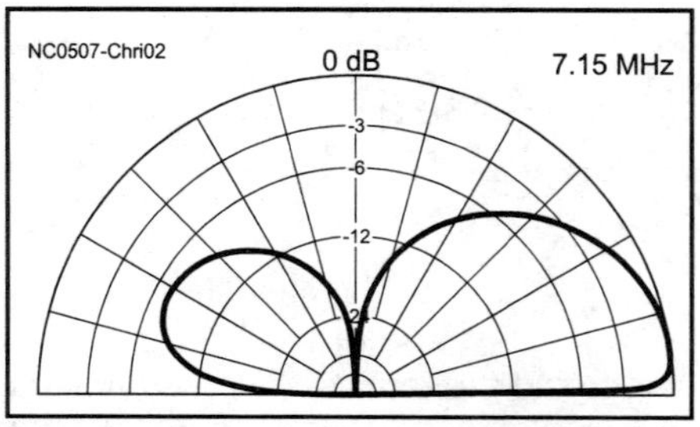

Figure 2—Elevation plane radiation pattern for the 40-meter vertical, where the set-back distance D = 0 and the height of the horizontal portion of the gull-wing radials is H = 2.5 feet. The sea is toward the right in the drawing. Gain = 4.16 dBi at 9 degrees take-off angle; front-to-back ratio = 6.81 dB.

Table 2

Antenna performance on 80 meters (f = 3.75 MHz) as a function of radial height and set-back distance. The base of the antenna is fixed at a height of one foot, and the radials are oriented parallel to the shoreline.

Radial Height H (ft)	*Set-back Distance D (WL)*	*Gain (dBi) and Take-off Angle (deg) front*		*back*		*FBR (dB)*	*Azimuthal Beamwidth (degrees)*
5.0	0.0	3.94	8	-2.49	28	6.43	179.6
	0.05	4.36	13	-2.49	28	6.85	164.2
	0.1	4.57	7	-2.49	28	7.06	159.0
	0.15	3.83	8	-2.49	28	6.32	150.8
	0.2	3.83	8	-2.49	28	6.32	140.8
	0.25	3.78	6	-2.49	28	6.27	142.2
7.5	0.0	4.20	8	-2.30	28	6.50	179.6
	0.05	4.60	12	-2.30	27	6.90	164.8
	0.1	4.75	7	-2.30	27	7.05	159.8
	0.15	4.64	4	-2.30	27	6.94	161.2
	0.2	4.49	3	-2.30	27	6.79	161.0
	0.25	4.48	6	-2.30	27	6.78	142.2
10.0	0.0	4.39	8	-2.18	28	6.57	179.6
	0.05	4.83	11	-2.18	27	7.01	167.4
	0.1	4.93	6	-2.18	27	7.11	164.4
	0.15	4.83	4	-2.18	27	7.01	164.0
	0.2	4.68	3	-2.18	27	6.86	161.8
	0.25	4.60	6	-2.18	27	6.78	143.0

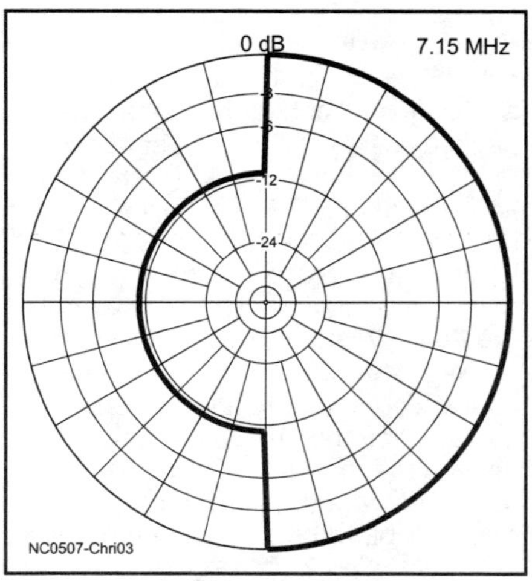

Figure 3—Azimuthal plane radiation pattern for the 40-meter vertical, where the set-back distance D= 0 and the height of the horizontal portion of the gull-wing radials is H = 2.5 feet. The sea is toward the right in the drawing. The half power beamwidth in the azimuthal plane is 179.6 degrees.

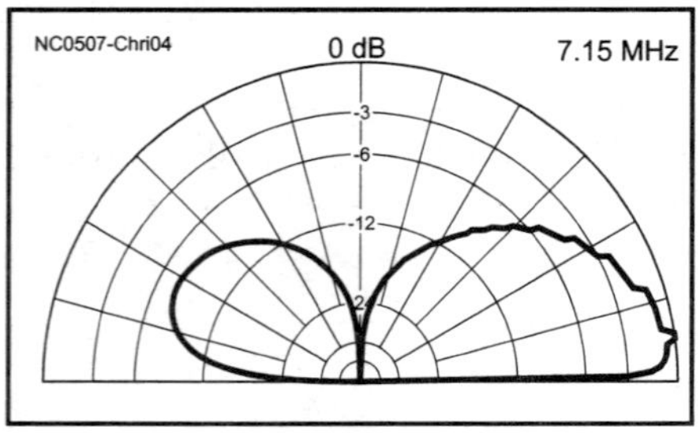

Figure 4—Elevation plane radiation pattern for the 40-meter vertical, where the set-back distance D = 0.1 wavelength and the height of the horizontal portion of the gull-wing radials is H = 2.5 feet. The sea is toward the right in the drawing. Gain = 4.60 dBi at 8 degrees take-off angle; front-to-back ratio = 7.25 dB.

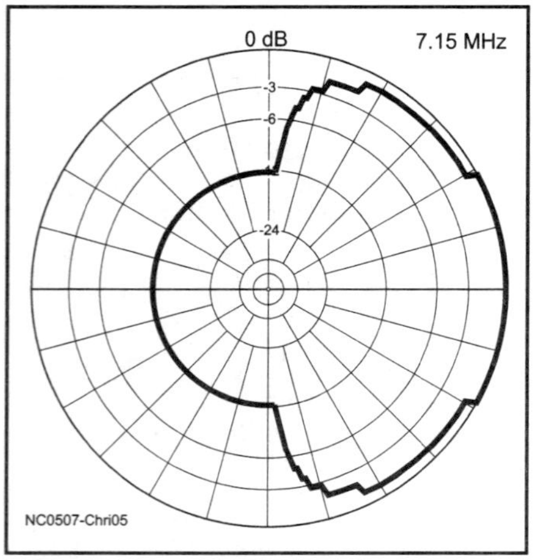

Figure 5— Figure 5. Azimuthal-plane radiation pattern for the 40m vertical, at a take-off angle of 8 degrees. The set-back distance D = 0.1 wavelength, and the height of the horizontal portion of the gull-wing radials is H = 2.5 feet. The sea is toward the right in the drawing. The half-power beamwidth in the azimuthal plane is 157.0 degrees.

Table 3

Antenna performance on 160 meters (f = 1.835 MHz) as a function of radial height and set-back distance. The base of the antenna is fixed at a height of one foot, and the radials are oriented parallel to the shoreline.

Radial Height H (ft)	*Set-back Distance D (WL)*	*Gain (dBi) and Take-off Angle (deg) front*		*back*		*FBR (dB)*	*Azimuthal Beamwidth (degrees)*
5.0	0.0	3.59	7	-1.76	26	5.35	179.6
	0.05	3.87	7	-1.76	26	5.63	170.4
	0.1	3.86	5	-1.76	26	5.62	164.0
	0.15	3.71	3	-1.76	26	5.47	165.2
	0.2	3.48	2	-1.76	26	5.24	164.8
	0.25	3.48	2	-1.76	26	5.24	162.8
7.5	0.0	3.77	7	-1.62	27	5.39	179.6
	0.05	4.14	10	-1.62	27	5.76	166.4
	0.1	4.25	6	-1.62	26	5.87	161.2
	0.15	4.21	4	-1.62	26	5.83	160.6
	0.2	4.12	3	-1.62	26	5.74	160.2
	0.25	3.89	2	-1.62	26	5.51	164.0
10.0	0.0	3.93	7	-1.49	26	5.42	179.6
	0.05	4.26	7	-1.49	26	5.75	170.6
	0.1	4.25	5	-1.49	26	5.74	166.2
	0.15	4.12	3	-1.49	26	5.61	165.4
	0.2	3.93	7	-1.49	26	5.42	143.6
	0.25	3.92	6	-1.49	26	5.41	141.2

(which in our case is seawater rather than a second type of soil) with *EZNEC*: "The effect of the second medium is taken into account only in a very simplified way. The vertical pattern is generated by tracing rays direct from the antenna and reflected from the ground. When a second medium is used, the ground reflection ray is determined by whichever medium it strikes the top of. The ray does not penetrate either medium, and diffraction or similar effects aren't considered."[7] All of the radiation patterns shown here were computed using one-degree increments in elevation angle, but the results were similar when the calculation interval was decreased to just 0.1 degree. Thus, it is likely that the information shown here is not as accurate as we would hope, although the general trends should be correct.

Results for 80 Meters

On 80 meters a frequency of 3.75 MHz was utilized, so the length of the vertical element (and the radials) is about 65.57 feet. Elevation heights of 5, 7.5, and 10 feet were chosen for the horizontal portions of the radials.

The results of the analysis are given in Table 2. As was true on 40 meters, raising the height of the radials (H) yielded a small amount of extra gain and (with two exceptions) a modest increase in the FBR, for any set back distance (D). Once again, the changes were small, with the gain staying between about 4 and 5 dBi, and the FBR close to 7 dB. When H is held constant, the gain (as before) can be peaked to a small extent by optimizing the set-back distance. For all three values of H, both maximum gain and maximum FBR occurred when the antenna was placed 0.1 wavelength from the edge of the water. And, if the height of the radials is fixed, the gain of the back lobe remains constant for any set-back distance.

The elevation and azimuthal plane radiation patterns are not shown for this band, but they closely resemble those presented for 40 meters.

Results on 160 Meters

A frequency of 1.835 MHz was selected for the computer simulation on top band, requiring a length of about 134 feet for the quarter wave vertical element and the radials. As on 80 meters, elevation heights of 5, 7.5, and 10 feet were utilized for the horizontal portions of the radials.

Table 3 lists the performance data for Top Band. We can see that increasing the height of the radials (H) usually generated a small amount of extra gain and a slight improvement in the FBR, although there were some exceptions when H = 10 feet. However, the variations are minor, with the gain remaining close to 4 dBi and the FBR between 5 and 6 dB. When the height of the radials (H) is held constant, the gain can be peaked a bit by varying the set-back distance. If H = 5 or 10 feet, maximum gain and maximum FBR both occurred when the antenna was placed 0.05 wavelength from the edge of the water, while D = 0.1 wavelength worked best at H = 7.5 feet. When the height of the radials is fixed, the gain of the back lobe remains constant for any set-back distance, since the soil characteristics are unchanged in that direction.

The elevation and azimuthal plane radiation patterns are omitted, but they are very similar to those given previously for 40-meter operation.

Results for the High Bands

After discussions with Carl, K9LA, it was decided to expand the coverage of this article to include the 20 and 10-meter bands as well. (*Results for 17, 15 and 12 meters can easily be interpolated from the*

20 and 10-meter results—Ed).

A frequency of 14.175 MHz was selected for the analysis on 20 meters, which corresponds to a length of about 17.35 feet for both the vertical element and the two-gull-wing radials. Elevation heights of 2.5 and 5.0 feet were chosen for the horizontal portions of the radials.

Table 4 displays the resulting performance data for this band. We can see that increasing the height of the radials (H) always produced a small amount of extra gain and a slight improvement in the FBR, for any set-back distance (D). However, the changes were minor, with the gain remaining around 5 dBi or so, and the FBR close to 7 dB. When the height of the radials (H) is kept fixed, the gain can be peaked (by a small amount) by varying the set back distance. For H = 2.5 feet, maximum gain and maximum FBR both occurred when the antenna was placed 0.15 wavelength from the edge of the water, while D = 0.1 wavelength worked best for H = 5 feet. Notice that, when H = 5 feet, both the gain and FBR oscillate up and down as D is made larger. If the height of the radials is held constant, the gain of the back lobe remains constant for any set-back distance, since the antenna always sees land in that direction.

The elevation and azimuthal plane radiation patterns are not shown for this band, but they closely resemble those presented for 40 meters.

On 10 meters a frequency of 28.3 MHz was utilized, so the length of the vertical element (and the radials) is about 8.69 feet. Elevation heights of 2.5 and 4 feet were chosen for the horizontal portions of the radials.

The results of the analysis on 10 meters are given in Table 5. As was true on 20 meters, raising the height of the radials (H) yielded a small amount of extra gain and (with some exceptions) a modest increase in the FBR, for any set back distance (D). Once again, the changes were small, with the gain staying between 5 and 6 dBi, and the FBR around 6 or 7 dB. When H is held constant, the gain (as before) can be peaked slightly by optimizing the set-back distance. When H = 2.5 feet, both maximum gain and maximum FBR occurred when the antenna was placed 0.15 wavelength from the shoreline, although D = 0.2 wavelength is optimal when H = 4 feet. Much as was seen on 20 meters, if H = 2.5 feet then both the gain and FBR oscillate up and down as D is made larger. On this band, the gain of the back lobe remained essentially the same for both heights, and at all set-back distances.

The elevation and azimuthal plane radiation patterns are omitted here, but they closely resemble those given previously for 40 meters.

Table 4

Antenna performance on 20 meters (f = 14.175 MHz) as a function of radial height and set-back distance. The base of the antenna is fixed at a height of one foot, and the radials are oriented parallel to the shoreline.

Radial Height H (ft)	*Set-back Distance D (WL)*	*Gain (dBi) and Take-off Angle (deg) front*		*back*		*FBR (dB)*	*Azimuthal Beamwidth (degrees)*
2.5	0.0	4.72	9	-1.97	29	6.69	179.6
	0.05	4.72	9	-1.97	28	6.69	170.0
	0.1	5.13	12	-1.97	28	7.10	154.6
	0.15	5.18	8	-1.97	28	7.15	149.8
	0.2	5.08	6	-1.97	28	7.05	149.4
	0.25	4.98	5	-1.97	28	6.95	147.2
5.0	0.0	5.00	9	-1.74	28	6.74	179.6
	0.05	5.00	9	-1.74	28	6.74	171.2
	0.1	5.68	13	-1.74	28	7.42	156.8
	0.15	5.43	14	-1.74	28	7.17	141.4
	0.2	5.56	11	-1.74	28	7.30	139.0
	0.25	5.66	5	-1.74	28	7.40	153.6

Table 5

Antenna performance on 10 meters (f = 28.3 MHz) as a function of radial height and set-back distance. The base of the antenna is fixed at a height of one foot, and the radials are oriented parallel to the shoreline.

Radial Height H (ft)	*Set-back Distance D (WL)*	*Gain (dBi) and Take-off Angle (deg) front*		*back*		*FBR (dB)*	*Azimuthal Beamwidth (degrees)*
2.5	0.0	5.01	10	-1.33	28	6.34	179.6
	0.05	5.01	10	-1.33	27	6.34	171.0
	0.1	5.14	20	-1.33	27	6.47	148.2
	0.15	5.59	14	-1.33	27	6.92	143.4
	0.2	5.30	15	-1.33	27	6.63	132.0
	0.25	5.43	12	-1.33	27	6.76	125.6
4.0	0.0	5.02	10	-1.32	27	6.34	179.6
	0.05	5.02	10	-1.32	26	6.34	172.8
	0.1	5.18	20	-1.32	26	6.50	151.8
	0.15	5.71	14	-1.32	26	7.03	151.2
	0.2	5.88	10	-1.32	26	7.20	150.6
	0.25	5.87	8	-1.32	26	7.19	150.2

Conclusions

An elevated vertical monopole with two gull-wing radials makes an effective DX antenna when installed at or near a large body of salt water. Compared to land, the high conductivity of this liquid medium yields enhanced signal strength at low take-off angles, with a resulting forward lobe that is broad in both the elevation and azimuthal planes. The height of the horizontal portions of the radials need not be great (typically on the order of a few feet), and the set-back distance from shoreline to antenna can range from zero to 0.25 wavelength, with little change in performance. This is good news for those who operate in locations with significant tidal action.

Notes

[1]Jerrold Swank, W8HXR, "The S-Meter Bender: W7DND's magic antenna," *73* magazine, June 1978.

[2]Information on the W7RM and W7DND antennas was obtained from private communications with Chip Margelli, K7JA, Ward Silver, NØAX, and Paul Kiesel, K7CW.

[3]Kenny Silverman, K2KW, and Tom Schiller, N6BT, "Verticals for Contest Expeditions: Learnings from the 6Y4A CQ WW CW Contest," *CQ Contest*, March 1998. This article can also be found on the Web at **www.k2kw.com/verticals/learning.html**.

[4]Kenny Silverman, K2KW, data posted on the Web at **www.k2kw.com/verticals/tests.html**.

[5]Al Christman and R. Paul Zeineddin, with Roger Radcliff and Jim Breakall, "Using Elevated Radials in Conjunction with Deteriorated Buried-Radial Ground Systems," *IEEE Transactions on Broadcasting*, Volume 39, Number 2, June 1993.

[6]Several versions of the *EZNEC* antenna modeling software are available from Roy Lewallen, W7EL, PO Box 6658, Beaverton, OR 97007.

[7]Roy Lewallen, W7EL, *EZNEC Pro* V.4.0 User's Manual, page 68. NCJ

Verticals by the Sea—Part 2

Al Christman, K3LC
Grove City College, 100 Campus Dr
Grove City, PA 16127-2104

This article analyzes the gain of vertical-monopole antennas with two "gull-wing" elevated radials, when they are installed near the sea-coast. The height of the radials is varied, along with the set-back distance between the antenna and the shore-line. Many of the amateur bands from 160 to 10 meters are reviewed. The goal of this section is to determine how far the antenna can be moved away from the sea before the gain-enhancement effect of the salt water is lost.

Background

Contesters and DXers are well aware of the advantages which are inherent in the use of vertical antennas when they are placed near large bodies of salt water. However, it isn't always possible to place the antenna directly at the shore-line, or even on the beach itself. Some operators feel that it's only necessary for the antenna to "see the water" in order to reap the benefits of this highly-conductive image plane, but we now have at our disposal the means to check the validity of this belief.

Computer Simulation

In the original article[1] it was assumed that the antenna was mounted either directly at the shore-line, or within a quarter-wavelength of the water. Now we will look at what happens as the antenna is moved farther inland, up to a distance of several wavelengths (WL). Figure 1 shows a pictorial view of one antenna configuration which is typical of those that were examined. In all cases, the length of the vertical element and the two radials was fixed at exactly 0.25 WL at the operating frequency, while the base of the antenna was placed just one foot above the ground. The two gull-wing radials are oriented parallel to the shore-line, and they initially extend upward at a 45-degree angle from the bottom of the antenna, until reaching their final height, which is designated as "H." This elevation height varies from 2.5 to 10 feet, depending upon the band of interest. The set-back distance from the water to the antenna is "D," which ranges from 0.25 to 3.25 WL, in steps of 0.25 WL.

The antenna models were run using the new *EZNEC Pro* software, version 4.0[2], with a double-precision *NEC-4* calculating engine. For simplicity, it was assumed that all antennas were constructed entirely from #12 AWG copper wire. No attempt was made to "prune" either the vertical element or the radials to achieve resonance.

The wire segment-lengths for the vertical element and the radials were selected in accordance with the most conservative *NEC* guidelines. The sea-water has an electrical conductivity of 5 Siemens per meter and a dielectric constant of 81. Three different sets of ground constants were used: "average" soil (conductivity = 0.005 Siemens per meter and dielectric constant = 13), "very good" soil (conductivity = 0.0303 S/m & dielectric constant = 20), and "very poor" soil (conductivity = 0.001 S/m & dielectric constant = 5).

Results on 40 meters

A frequency of 7.15 MHz was selected for the analysis on 40 meters, corresponding to a length of about 34.39 feet for both the vertical element and the two gull-wing radials. Elevation heights (H) of 2.5 and 5.0 feet were chosen for the horizontal portions of the radials. Table 1 lists the gain predicted by *EZNEC*, in both the "sea-ward" and "land-ward" directions from the antenna, at take-off angles of 5 and 10 degrees, as a function of the set-

NC0509-Chrst01

Figure 1—Drawing of an elevated vertical-monopole antenna (base height = 1 foot) with two gull-wing radials. The length of both the vertical element and the radials is 0.25 WL (34.39 feet at 7.15 MHz). "H" represents the height of the horizontal portions of the radials, and "D" is the set-back distance from the shore-line.

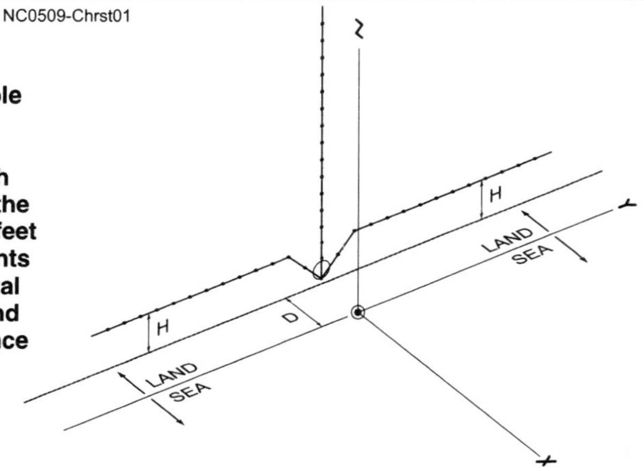

Table 1
Antenna gain versus set-back distance (D) for a 40-meter vertical antenna installed over "average" soil. The horizontal portions of the two gull-wing radials are mounted at a height (H) of either 2.5 or 5.0 feet.

Radial Height H (ft)	*Set-back Distance D (WL)*	*Antenna Gain (dBi)* Seaward 5 deg TOA	Seaward 10 deg TOA	Landward 5 deg TOA	Landward 10 deg TOA
2.5	0.25	3.49	3.10	-7.06	-3.30
	0.5	2.71	1.13	-7.06	-3.30
	1.0	1.00	-1.53	-7.06	-3.30
	1.25	-0.91	-2.97	-7.06	-3.30
	1.5	-1.92	-3.30	-7.06	-3.30
	2.0	-3.95	-3.30	-7.06	-3.30
	2.5	-6.44	-3.30	-7.06	-3.30
	2.75	-6.89	-3.30	-7.06	-3.30
	3.0	-7.06	-3.30	-7.06	-3.30
	3.25	-7.06	-3.30	-7.06	-3.30
	3.5	-7.06	-3.30	-7.06	-3.30
5.0	0.25	3.95	4.00	-6.76	-3.00
	0.5	3.78	1.89	-6.76	-3.00
	1.0	1.98	-1.03	-6.76	-3.00
	1.25	-0.04	-2.63	-6.76	-3.00
	1.5	-1.12	-3.00	-6.76	-3.00
	2.0	-3.31	-3.00	-6.76	-3.00
	2.5	-6.05	-3.00	-6.76	-3.00
	2.75	-6.57	-3.00	-6.76	-3.00
	3.0	-6.76	-3.00	-6.76	-3.00
	3.25	-6.76	-3.00	-6.76	-3.00
	3.5	-6.76	-3.00	-6.76	-3.00

back distance (D), when the vertical is mounted over "average" soil.

On the sea-ward side the low-angle gain falls continually as the antenna is moved away from the shore. On the land-ward side the gain is always the same at a specific take-off angle, since the vertical sees only land in that direction. Note that, by the time the set-back distance (D) has reached 1.5 WL, the sea-ward antenna gain at 10 degrees TOA has fallen to the same value that exists on the land-ward side. At 5 degrees TOA, the "break-even" set-back distance is 3 WL.

Figure 2 displays the table data in graphical form. We can see that the sea-ward antenna gain at low take-off angles falls very quickly as the vertical is moved away from the edge of the water, although the effect is more pronounced at 10 degrees TOA than at 5 degrees. In all cases, the sea-ward gain eventually diminishes to a value which is equal to that on the landward side of the antenna.

Figure 3 shows what happens to the shape of the elevation-plane radiation pattern as the set-back distance (D) is increased from 0.5 to 1.0 to 1.5 to 3.0 WL. The shore-line is toward the right on the drawing, and here the radial height (H) is 2.5 feet. As the vertical is moved farther from the sea, the extra low-angle gain provided by the salt water progressively disappears, although a small "tongue" remains at a take-off angle of 1 to 2 degrees, even when D = 3 WL.

When the soil is changed to "very good" and then "very poor," the results are as shown in Figures 4 and 5 respectively. Comparing these graphs to Figure 2, we notice that all of them are similar in shape. (Of course, the ultimate gain values are higher for very good soil, and lower for very poor soil.) The main point to be drawn from our work on 40 meters is that, for both radial heights and all three soil types, the set-back distance (D) at which the sea-ward gain becomes equal to that on the land-ward side is always close to 1.5 WL at 10 degrees take-off angle, and about 3 WL at 5 degrees TOA.

Outcome on the Other Bands

The results on the other bands which were investigated (160, 80, 20, and 10 meters) were amazingly consistent with what was found on 40 meters, and all of the information is summarized in Table 2.

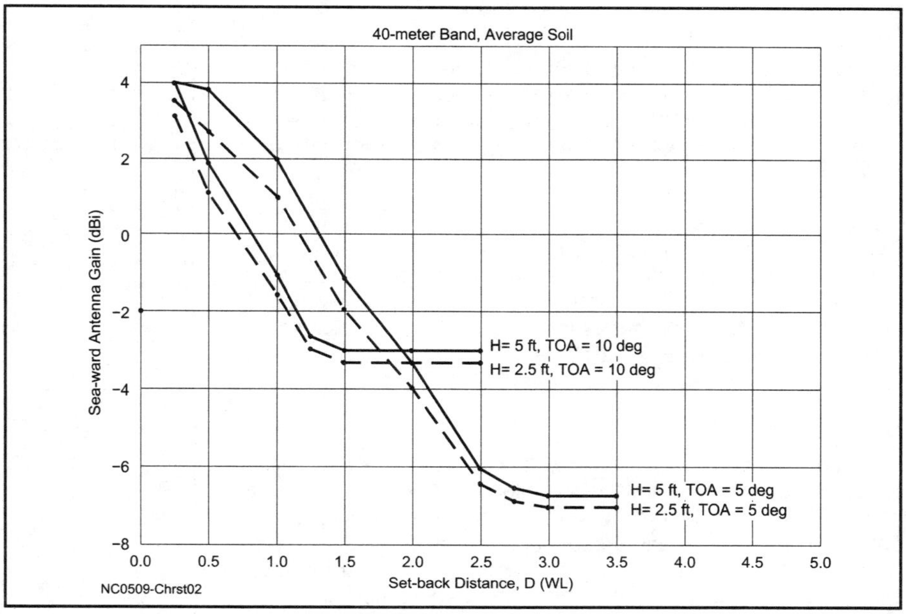

Figure 2—Sea-ward gain versus set-back distance for a 40-meter vertical antenna at take-off angles of 5 and 10 degrees. The height of the radials is either 2.5 or 5.0 feet, and the antenna is installed over "average" soil.

Table 2
Break-even set-back distance (D) for take-off angles of 5 and 10 degrees, on the bands of interest from 160 to 10 meters. A variety of radial heights (H) were used, and the antenna was mounted over three different types of soil.

Band (m)	*Frequency (MHz)*	*Radial Height H (ft)*	*Soil Type*	*Break-even Set-back Distance, D (WL) 5 deg TOA*	*10 deg TOA*
160	1.835	5.0	Average	3	1.5
			Very good	3	1.5
			Very poor	3	1.5
		7.5	Average	3	1.5
			Very good	3	1.5
			Very poor	3	1.5
		10.0	Average	3	1.5
			Very good	3	1.5
			Very poor	3	1.5
80	3.75	5.0	Average	3	1.5
			Very good	3	1.5
			Very poo	3	1.5
		7.5	Average	3	1.5
			Very good	3	1.5
			Very poor	3	1.5
		10.0	Average	3	1.5
			Very good	3	1.5
			Very poor	3	1.5
40	7.15	2.5	Average	3	1.5
			Very good	3	1.5
			Very poor	3	1.5
		5.0	Average	3	1.5
			Very good	3	1.5
			Very poor	3	1.5
20	14.175	2.5	Average	3	1.5
			Very good	3	1.5
			Very poor	3	1.5
		5.0	Average	3	1.5
			Very good	3	1.5
			Very poor	3	1.5
10	28.3	2.5	Average	3.25	1.75
			Very good	3.25	1.75
			Very poor	3.25	1.75
		4.0	Average	3.25	1.75
			Very good	3.25	1.75
			Very poor	3.25	1.75

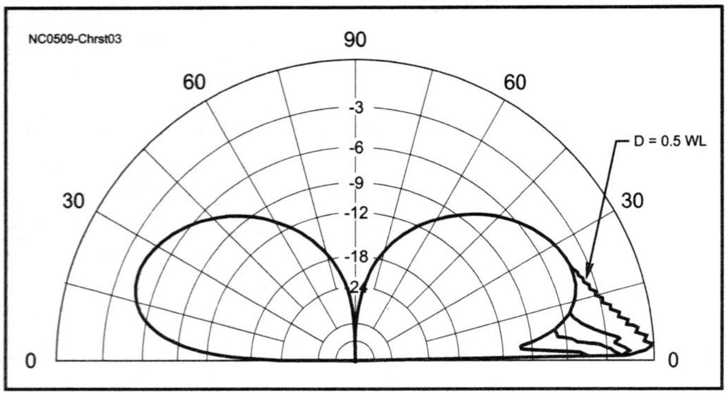

Figure 3—Elevation-plane radiation patterns for the 40-meter vertical antenna, when the set-back distance D varies from 0.5 to 1.0 to 1.5 to 3.0 WL. The sea is to the right, the radial height H is 2.5 feet, and "average" ground constants were used.

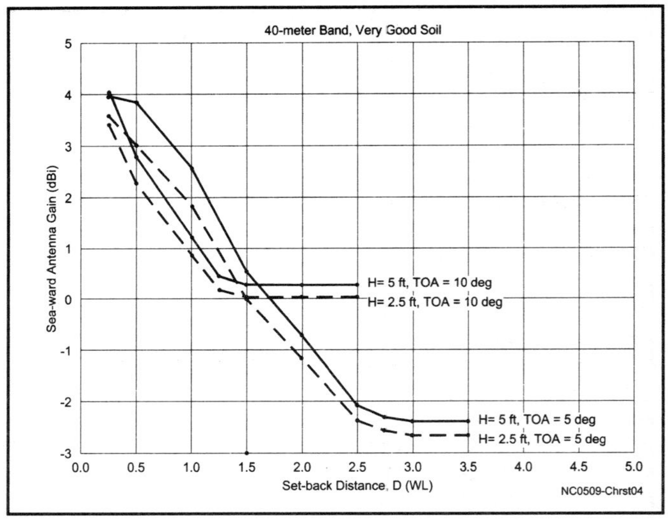

Figure 4—Sea-ward gain versus set-back distance for a 40-meter vertical antenna at take-off angles of 5 and 10 degrees. The height of the radials is either 2.5 or 5.0 feet, and the antenna is installed over "very good" soil.

Note that the "break-even" set-back distances, at take-off angles of both 5 and 10 degrees, are essentially constant over the range of frequencies from 1.8 through 30 MHz, increasing only slightly on 10 meters. These distances are independent of both the condition of the under-lying soil and the elevation height of the horizontal portions of the gull-wing radials.

Conclusions

An elevated vertical monopole with two gull-wing radials makes an effective DX antenna when installed at or near a large body of salt water. Compared to land, the high conductivity of this liquid medium yields enhanced signal strength at low take-off angles, with a resulting "forward" lobe (directed toward the water) which is broad in both the elevation and azimuthal planes. Computer analysis reveals that the enhanced gain of the vertical monopole system (in the direction of the sea-water) has completely disappeared by the time the antenna is roughly 1.5 WL away from the shore-line, for a take-off angle of 10 degrees; at 5 degrees TOA, the "break-even" distance is approximately 3 WL.

A Look Ahead to Part 3

Part 3 of this article will compare the performance of a two-element Yagi with that of a two-element phased array of gull-wing verticals, as the set-back distance from the shore-line is varied.

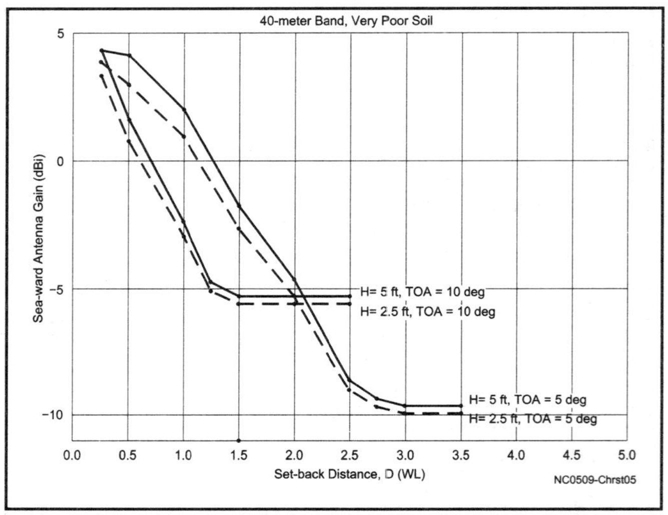

Figure 5—Sea-ward gain versus set-back distance for a 40-meter vertical antenna at take-off angles of 5 and 10 degrees. The height of the radials is either 2.5 or 5.0 feet, and the antenna is installed over "very poor" soil.

Notes

[1]Al Christman, K3LC, "Verticals by the Sea," *National Contest Journal*, Volume 33, Number 4, July/August 2005.

[2]Several versions of the *EZNEC* antenna-modeling software are available from Roy Lewallen, W7EL, PO Box 6658, Beaverton, OR 97007.

NCJ

Verticals by the Sea—Part 3

Al Christman, K3LC
Grove City College
100 Campus Dr
Grove City, PA 16127-2104

This article compares the gain of a 2-element phased-vertical array with that of a 2-element Yagi, when both are installed near the seacoast. The set-back distance between the antennas and the shoreline is varied, along with the height of the gull-wing radials on the verticals. The traditional amateur bands from 40 to 10 meters are examined. The goal of this section is to determine the range of set-back distances where each antenna works best.

Background

Contesters and DXers are well aware of the advantages that are inherent in the use of vertical antennas when they are placed near large bodies of saltwater. Part 2 of this series[1] showed that the low-angle (5 degrees) gain enhancement provided by the high conductivity of the sea essentially disappears by the time a set-back distance of three wavelengths (WL) has been reached. If this is truly the case, then is it possible to obtain better performance from a horizontal Yagi that is placed at a modest height above the earth? This is the question that will be addressed here.

Computer Simulation

After consultations with Carl, K9LA and Kenny, K2KW,[2] it was decided to use a 2-element phased array as the "reference" vertically-polarized antenna, instead of a single monopole. If an operator is serious enough to put up a 2-element Yagi, then he/she would probably also go to the trouble to install a 2-element vertical array. Figure 1 shows a pictorial view of a typical 2-element vertical array, as configured for this study. The Y-axis represents the shoreline, and the entire region in the +X direction is seawater. The monopoles are spaced 0.5-WL apart and fed in phase with equal-amplitude currents, producing a "broadside" radiation pattern that is perpendicular to the plane of the elements. The length of both vertical elements and all four radials was fixed at exactly 0.25 WL at the operating frequency, while the base of each monopole was placed just one foot above the ground. The gull-wing radials are oriented parallel to the shoreline, and they initially extend upward at a 45-degree angle from the bottom of the antenna, until reaching their final height, which is designated as "H." This elevation height varies from 2.5 to 5.0 feet, depending upon the band of interest. The set-back distance from the water to the array is "D," which ranges from 0 to 10 WL, in steps that are as small as 0.25 WL in some cases.

The Yagi which was selected to serve as the horizontally polarized "reference" antenna is the 2-element Moxon-style design introduced by Dr David Leeson, W6NL.[3] This beam incorporates unique end-loaded elements which help it produce good gain, front-to-back ratio, and SWR over the entire 40-meter band. In the analysis here, this same Yagi was also re-scaled for operation on 20, 15 and 10 meters, which is a simple task with *EZNEC*. Kenny, K2KW, suggested placing the antenna at a height of 25 feet on all bands, and his advice was followed. A computer-generated drawing of the beam is displayed in Figure 2.

The antenna models were run using the new *EZNEC Pro* software, version 4.0,[4] with a double-precision *NEC-4* calculating engine. For simplicity, it was assumed that the verticals were constructed entirely from #12 AWG copper wire, while aluminum was used for the Yagi. No attempt was made to "prune" either the vertical monopoles or the radials to achieve resonance. The wire segment-lengths for all antenna elements were selected in accordance with the most conservative *NEC* guide-lines. The seawater has an electrical conductivity of 5 Siemens per meter, and a dielectric constant of 81.

Three different sets of ground constants were used: "average" soil (conductivity = 0.005 Siemens per meter and dielectric constant = 13), "very good" soil (conductivity = 0.0303 S/m and dielectric constant = 20), and "very poor" soil (conductivity = 0.001 S/m and dielectric constant = 5). The ground is flat in all cases.

Results on 40 Meters

A frequency of 7.15 MHz was selected for the simulation on 40 meters, corresponding to a length of about 34.39 feet for both the vertical monopoles and the gull-wing radials. Elevation heights (H) of 2.5 and 5.0 feet were chosen for the horizontal portions of the radials, and (as mentioned previously) the Yagi was placed 25 feet above the ground.

Table 1 lists the seaward gain predicted by *EZNEC*, for the Yagi and for both versions of the vertical array, at take-off angles of 5 and 10 degrees, as a function of the set-back distance, when the antennas are mounted over "average" soil. No-

Table 1

Seaward antenna gain versus set-back distance (D) for a 40-meter 2-element vertical array and a 2-element horizontal Yagi installed over "average" soil. The horizontal portions of the gull-wing radials are mounted at a height (H) of either 2.5 or 5.0 feet, and the Yagi is placed at a height of 25 feet.

	2-element Phased-Vertical Array				*2-element Yagi*	
	H=2.5 feet		*H=5.0 feet*		*H=25 feet*	
Set-back Distance D (WL)	*Seaward Gain (dBi) at 5 deg*	*10 deg*	*Seaward Gain (dBi) at 5 deg*	*10 deg*	*Seaward Gain (dBi) at 5 deg*	*10 deg*
0.0	7.38	7.53	7.72	7.86	-8.56	-2.64
0.25	7.09	6.70	7.55	7.60	-8.56	-2.64
0.5	6.31	4.72	7.39	5.48	-8.56	-2.64
0.75	5.47	3.35	6.51	3.99	-8.56	-2.64
1.0	4.58	2.06	5.58	2.57	-8.56	-2.21
1.25	2.67	0.64	3.55	0.98	-8.56	-2.20
1.5	1.66	0.31	2.47	0.61	-8.56	-2.20
1.75	0.64	0.31	1.38	0.61	-8.56	-2.20
2.0	-0.36	0.31	0.29	0.61	-8.08	-2.20
2.25	-2.16	0.31	-1.69	0.61	-7.94	-2.20
2.5	-2.84	0.31	-2.45	0.61	-7.94	-2.20
2.75	-3.29	0.31	-2.96	0.61	-7.94	-2.20
3.0	-3.46	0.31	-3.15	0.61	-7.94	-2.20
4.0	-3.46	0.31	-3.15	0.61	-7.94	-2.20
5.0	-3.46	0.31	-3.15	0.61	-7.94	-2.20
6.0	-3.46	0.31	-3.15	0.61	-7.94	-2.20
7.0	-3.46	0.31	-3.15	0.61	-7.94	-2.20
8.0	-3.46	0.31	-3.15	0.61	-7.94	-2.20
9.0	-3.46	0.31	-3.15	0.61	-7.94	-2.20
10.0	-3.46	0.31	-3.15	0.61	-7.94	-2.20

tice that the horizontal Yagi is never competitive with the 2-el vertical array. A height of 25 feet for the Yagi is simply insufficient to allow it to develop good low-angle gain on the 7 MHz band. Figure 3 reveals the elevation-plane radiation patterns for both antennas when D = 0.5 WL (H = 2.5 feet for the radials) over average soil. The Yagi is inferior at all take-off angles below about 19 degrees.

The *EZNEC* results were also the same for both "very good" and "very poor" soils—the Yagi always loses. A closer examination of the listings (Table 1) for the vertical array indicates that, just as we found in part 2 of this series, the augmented gain provided by the seawater disappears at a set-back distance (D) of 1.5 WL at 10 degrees take-off angle (TOA); the "break-even" value of D is 3 WL for a TOA of 5 degrees. In contrast, the gain of the Yagi remains almost constant throughout the range of set-back distances from zero to 10 WL, and actually rises slightly as the antenna is moved away from the shoreline.

Since the Yagi is always worse than the 2-element vertical array on 40 meters, how does it fare against a single vertical element? The data for an isolated monopole is available from our study in Part 2,

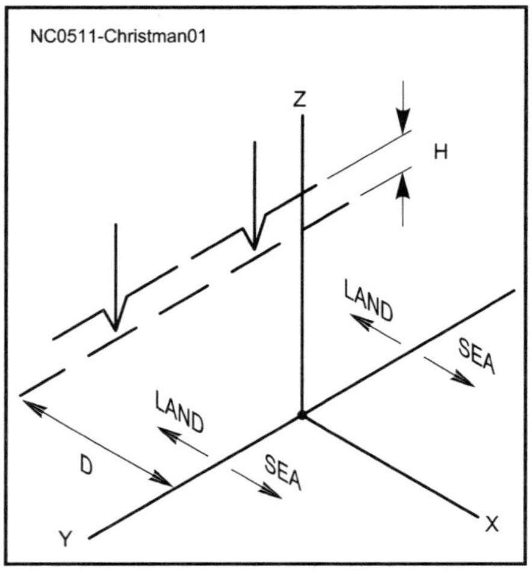

Figure 1—Drawing of a 2-element 40-meter phased array of elevated vertical-monopole antennas (base height = 1 foot), each with two gull-wing radials. The length of each radial and both vertical elements is 0.25 WL. Here the set-back distance (D) is 0.25 WL, and the height of the radials (H) is 5.0 feet. Both elements are fed with equal-amplitude, in-phase currents, producing a broadside radiation pattern. The shoreline is coincident with the Y-axis, and the seawater lies in the +X direction.

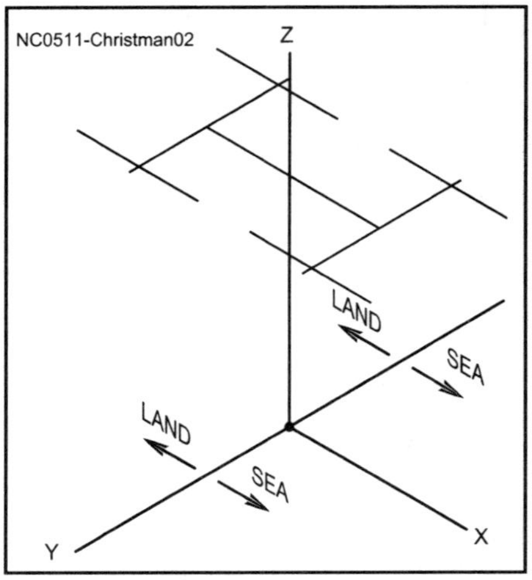

Figure 2—Drawing of a 2-element W6NL-design Moxon-style 40-meter Yagi antenna, mounted at a height of 25 feet. The shoreline is coincident with the Y- axis, and the seawater lies in the +X direction. Here, the set-back distance (D) is zero.

Table 2

On 40 meters, a single vertical monopole has more seaward gain than the 2-element horizontal Yagi until the set-back distance (D) exceeds the values shown below. Take-off angles of 5 and 10 degrees are included, while radial heights (H) of both 2.5 and 5.0 feet were used. The antennas were erected over three different types of soil.

Soil Type	*Radial Height H (feet)*	*Take-off Angle (degrees)*	*Break-even Set-back Distance D (WL)*
Average	2.5	5	infinity (Yagi always worse)
		10	1.12
	5.0	5	infinity (Yagi always worse)
		10	1.18
Very Good	2.5	5	infinity (Yagi always worse)
		10	infinity (Yagi always worse)
	5.0	5	infinity (Yagi always worse)
		10	infinity (Yagi always worse)
Very Poor	2.5	5	2.23
		10	0.92
	5.0	5	2.32
		10	0.98

Table 3

On 20 meters, a 2-element phased vertical array has more seaward gain than the 2-element horizontal Yagi until the set-back distance (D) exceeds the values shown below. Take-off angles of 5 and 10 degrees are included, and radial heights (H) of both 2.5 and 5.0 feet were used. The antennas were installed over 3 different types of soil.

Soil Type	*Radial Height H (feet)*	*Take-off Angle (degrees)*	*Break-even Set-back Distance D (WL)*
Average	2.5	5	2.73
		10	0.95
	5.0	5	4.25
		10	1.07
Very Good	2.5	5	infinity (Yagi always worse)
		10	infinity (Yagi always worse)
	5.0	5	infinity (Yagi always worse)
		10	infinity (Yagi always worse)
Very Poor	2.5	5	2.375
		10	0.82
	5.0	5	2.52
		10	0.95

so a comparison was made, with the outcome shown in Table 2. For average soil, the Yagi always loses at 5 degrees TOA, but at 10 degrees TOA it can out-perform the lone vertical if "D" is sufficiently large. Figure 4 is a graph showing the seaward gain at 10 degrees TOA for the Yagi and for the single vertical (with radial-heights of H = 2.5 and H = 5.0 feet). The "break-even" setback distances given in the table were obtained from the graph.

Over very good soil, the single vertical still wins every time. However, over very poor soil the Yagi can be competitive, if the set-back distance (D) is great enough to offset the low-angle gain enhancement provided by the saltwater.

Results on Other Bands

The outcomes on 20, 15 and 10 meters are displayed in Tables 3 through 5, respectively. As the frequency rises, the Yagi's height (in terms of wavelength) continually increases, which boosts its low-angle performance. Therefore, the Yagi was compared to a 2-element phased vertical array (as in Figure 1) on these bands, rather than to a single monopole.

On 20 meters (see Table 3) the Yagi is inferior to the array of phased verticals

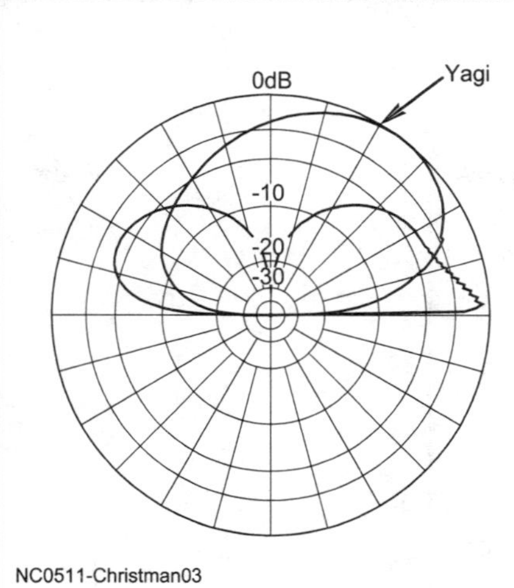

Figure 3—Elevation-plane radiation patterns for the 40-meter Yagi and the phased-vertical array, when the set-back distance D = 0.5 WL. The sea is to the right, the radial height H is 2.5 feet, and "average" ground constants were used. Here the vertical array has more seaward gain than the Yagi at all take-off angles up to about 19 degrees.

when the soil is highly conductive. For other soil types, the Yagi stands a fighting chance, but only if the antennas must be placed relatively far from the edge of the water. Table 4 displays the results on 15 meters. Notice that here the two elevation heights for the gull-wing radials are 2.5 feet (as before) and 4 feet (rather than 5). No matter what the soil type, the Yagi is now able to match the low-angle gain of the 2-element vertical array, if the set-back distance is large enough. However, "D" must be greater than 2 WL if the Yagi is to be competitive at a TOA of 5 degrees.

On 10 meters, refer to Table 5 for the information. We finally see one entry in this table where the Yagi is actually the better antenna at all set-back distances! This occurs when the take-off angle is 10 degrees, H = 2.5 feet, and the soil is "very good." If the soil is "average," the Yagi works almost as well; it is superior to the phased vertical array as long as "D" is greater than or equal to 0.04 WL (about 1.4 feet).

Conclusions

A 2-element phased-vertical array, utilizing gull-wing-style radials, is an extremely effective DX antenna when in-

Table 4

On 15 meters, a 2-element phased vertical array has more seaward gain than the 2-element horizontal Yagi, until the set-back distance (D) exceeds the values shown below. Take-off angles of 5 and 10 degrees are included, and radial heights (H) of both 2.5 and 4.0 feet were used. The antennas were mounted over three different types of soil.

Soil Type	*Radial Height H (feet)*	*Take-off Angle (degrees)*	*Break-even Set-back Distance D (WL)*
Average	2.5	5	2.04
		10	0.55
	4.0	5	2.15
		10	0.69
Very Good	2.5	5	2.21
		10	0.57
	4.0	5	2.37
		10	0.71
Very Poor	2.5	5	1.91
		10	0.56
	4.0	5	2.07
		10	0.68

Table 5

On 10 meters, a 2-element phased vertical array has more seaward gain than the 2-element horizontal Yagi until the set-back distance (D) exceeds the values shown below. Take-off angles of 5 and 10 degrees are included, and radial heights (H) of both 2.5 and 4.0 feet were used. The antennas were installed over 3 different types of soil.

Soil Type	*Radial Height H (feet)*	*Take-off Angle (degrees)*	*Break-even Set-back Distance D (WL)*
Average	2.5	5	1.82
		10	0.04
	4.0	5	1.97
		10	0.58
Very Good	2.5	5	1.88
		10	zero (Yagi always better)
	4.0	5	2.05
		10	0.54
Very Poor	2.5	5	1.78
		10	0.33
	4.0	5	1.94
		10	0.64

stalled at or near the seacoast. Under many conditions, this array generates more seaward low-angle gain than a 2-element Yagi mounted at a height of 25 feet. On 40 meters, even a single vertical monopole with two gull-wing radials is usually better than a Yagi raised to a modest height.

References

[1]Al Christman, K3LC, "Verticals by the Sea—Part 2" *National Contest Journal*, Volume 33, Number 5, September/October 2005.

[2]Private communications with Carl Luetzelschwab, K9LA, and Kenny Silverman, K2KW.

[3]David Leeson, W6NL, "W6NL 40-meter Moxon Yagi," presented at the Antenna Forum at the 2004 Dayton Hamvention.

[4]Several versions of the *EZNEC* antenna-modeling software are available from Roy Lewallen, W7EL, PO Box 6658, Beaverton, OR 97007.

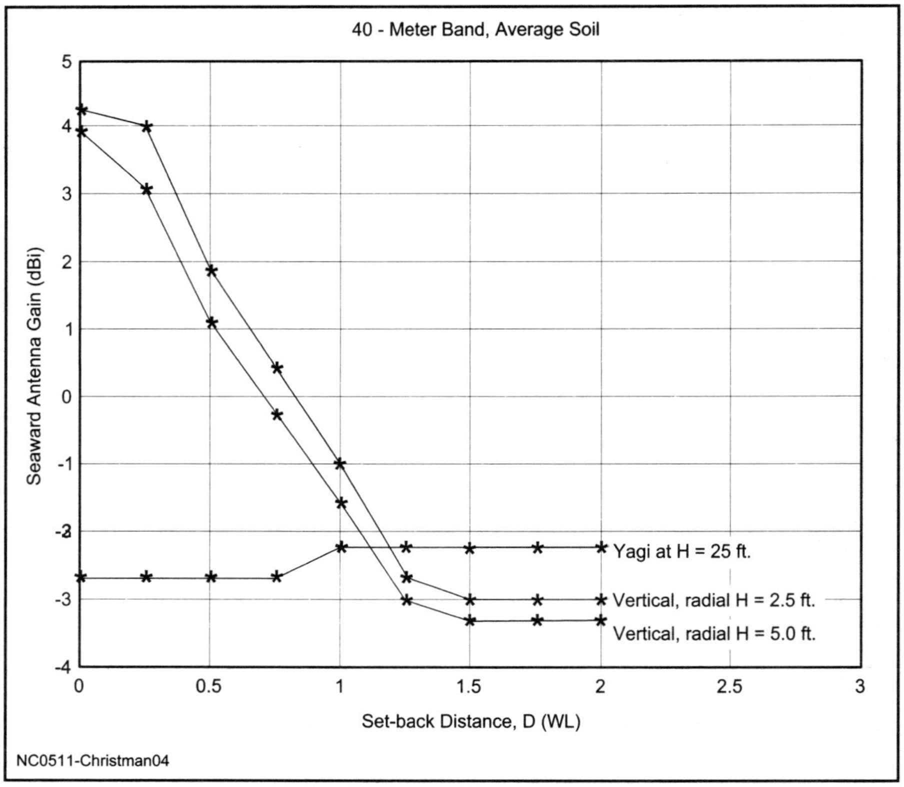

Figure 4—Seaward antenna gain at 10 degrees TOA versus set-back distance, for a single vertical monopole and for the W6NL Yagi. The height of the elevated radials is either 2.5 or 5.0 feet, and the Yagi is placed 25 feet above the ground. The band is 40 meters and the soil is "average."

A 75/80 Meter Full-Size $^1/_4$-λ Vertical

By Guy Hamblen, AA7QZ/2
16 Dongan Lane
Newfoundland, NJ 07435
e-mail: tel1gah@is.ups.com

AA7QZ/2 gives details on the design and installation of his killer vertical for 75 and 80 meters.

Verticals have always been fascinating to me. For a ground-mounted antenna, their low angle of radiation and omnidirectional radiation patterns can result in an effective DX antenna. This is particularily true for the 80 and 160-meter bands, since an 80-meter horizontal dipole would have to be higher than a wavelength to perform as well as a ground-mounted vertical. I don't know of anyone with two 250 foot towers!

In low-sunspot years with little activity on the higher frequencies, the 40, 80 and 160-meter bands provide unique opportunities for hams to explore fascinating aspects of propagation. A whole new group of low-band enthusiasts is emerging. Prior to my relocating to the East Coast, city and neighborhood restrictions constrained my antenna efforts to low-profile installations. I tried the GAP Challenger vertical and was impressed with its multiband effectiveness. However, I wondered what a full-size quarter-wave 80-meter vertical might offer in comparison. Boy, was I surprised!

Some History

The *Low Band DXing* series of books by John Develdore, ON4UN,[1] have become the standard text for 80 and 160-meter antenna enthusiasts. Chapter 9 of John's latest book provides a wealth of design data for verticals, but it can be intimidating to the newcomer. Orr and Cowan's *All About Vertical Antennas*[2] is short on details on the design and construction of simple verticals for the newcomer.

If you have tall trees available, wire verticals are easy to erect. Inverted-L antennas enjoy a devoted following. Choosing a configuration for the top part of a vertical antenna is the easy part. The more challenging part is at the bottom: making a simple, effective matching network and putting in the wire radial system required for efficiency. Elevated-radial systems have been popularized by Christman, KB8I,[3] and Russell, N4KG,[4] in recent years. However, I still recommend use of a traditional radial system for your first low-band vertical. [After all, not every 80-meter enthusiast has the space, or inclination, to install four 65-foot long radials elevated 10 feet off the ground and in plain sight of the neighbors and your own family! —*Ed.*]

Verticals require a good RF grounding system. Concrete or asphalt are not good conductors of RF currents and will definitely diminish the overall effectiveness of a vertical. Sea water is best. If you live near the ocean, verticals are ideal antennas. Seasonally wet, farm-like and tillable soil is above average for RF conductivity and will give good results with a minimal wire radial system. An excellent discussion of ground losses, radiation resistance and antenna efficiencies may be found on pages 9-1 through 9-5 of *Low Band DXing*. Paul Lee, N6PL, in Chapter 9 of his *Vertical Antenna Handbook*[5] has an excellent discussion on the effects of the earth on the efficiency of vertical antennas.

Design Goals

I wanted to construct a full-size, high-efficiency quarter-wave vertical for 80/75 meters, using aluminum tubing and non-conducting guys made of $^1/_8$-inch Philystran. I also wanted to design a simple, but easy-to-tune matching network, without bulky variable capacitors. It had to be capable of handling a kW. I planned to use a relay to change resonance from the phone to the CW portion of the band and I intended to use a low-power SWR tester from MFJ or Autek to tune the system in my backyard.

I also wanted to design the vertical so it could easily be taken down during the late spring and summer months, when static levels and daytime absorption make the 80-meter band nearly unusable. Part of this was for aesthetic reasons, but I also needed an easy-to-install portable antenna for Field Day or portable contest operations.

I planned to use insulated stranded wire radials because of their flexibility, ease of installation and because they could be rolled up in the springtime. Cost of the materials was a close secondary requirement to a design that is simple, modular, and easy to install and disassemble.

Construction Details

I followed Chapter 9 of ON4UN's book closely. Even though most of his construction details are oriented to using a 60-plus foot tower on 80 or 160 meters, I found with some interpolation that his design formulas are surprisingly accurate for smaller diameter aluminum tubing. Orr and Cowan covered matching networks in some detail in Chapter 3 of their book.

I used six telescoping lengths of 12-foot long aluminum tubing[6] with 0.058-inch wall thickness. My base section was 1.500 inch OD aluminum tubing, with sub-

Table 1

Telescoping Aluminum Tubing, Lengths for AA7QZ Vertical

Tubing OD (Inches)	*Cut Length (Feet)*	*Inst. Length (Feet)*	*Cumulative Height (Feet)*	*Hole for Guy Plate (Inches)*
0.875	5.5	4	58	
1.000	12	10.5	54	0.875
1.125	12	10.5	43.5	
1.250	12	10.5	33	
1.375	12	10.5	22.5	1.250
1.500	12	12	12	1.500

Note: overlap of telescoping joints is 18 inches.

sequent sections with ODs of 1.375, 1.250, 1.125, 1.000 and finally 0.875 inches. See **Table 1** for details.

The 0.058-inch wall thickness for each section allows for a tight fit between sections. The top section had 48 inches sticking out above the joint, to give an overall height of 58 feet. Although smaller-diameter sections could be used at the expense of some rigidity in the overall vertical, during installation you'll have lower anxiety levels with larger tubing. It always amazes me how elastic a 58-foot length of aluminum tubing is!

Most aluminum tubing has some very small burrs on the inside. When inserting smaller sections, these burrs can catch and gall and generally stymie the "ease of installation" requirement. I borrowed an 18-inch electric drill extension from a neighbor and used a small fiber grinding attachment. In five minutes I had removed all burrs on the inside of the tubing. Using the same grinding tool, I rounded all tubing edges to minimize any roughness. A 400-grit wet-grade sandpaper on the 18-inch extention finished off the mechanical portion of the vertical. I inserted each section and drilled a single 0.125-inch hole one inch down from the top of each lower section for a pinning bolt. I used stainless steel bolts, nuts and star lockwashers for each section in order to ensure easy disassembly in the springtime.

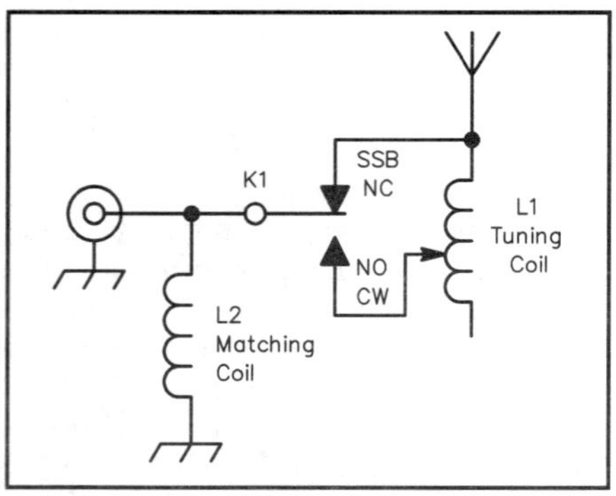

Fig 1—Schematic of tuning network, showing relay switching required to change from phone to CW bands. One setting of L2 provides a match for both phone and CW.

The overall length of the monopole can be figured from the formula:

$$L(\text{feet}) = \frac{245.9}{f(\text{MHz})} \times P \qquad \text{(Eq 1)}$$

where P can be interpolated from Figure 9-43 on page 9-34 of *Low Band DXing*. I used a P of 91%, giving a length of 58 feet at a resonant frequency of 3.850 MHz. The feedpoint impedance will be capacitive if the vertical element is shorter than the target operating frequency. The series capacitive reactance of the antenna and a shunt inductance to ground form an L-network that transforms the nominal 36-Ω feedpoint impedance to 50 Ω.

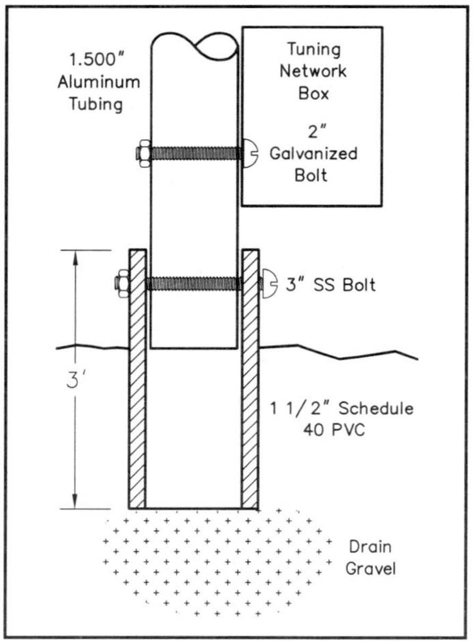

Fig 2—Details of base mount for AA7QZ/2 vertical. The 1.5-inch schedule 40 PVC pipe is placed in the hole on a bed of gravel for drainage. The 1.500-inch tubing is bolted to the PVC pipe with a 3-inch long stainless steel bolt. The tuning network is mounted to and electrically connected to the 1.500-inch tubing with a 2-inch long galvanized bolt to minimize electrolytic corrosion.

My target operating frequencies were 3.800 MHz for phone and 3.550 MHz for CW. Adding a coil in series with the vertical element allows me to tune the vertical system to the CW band. This is much easier and less expensive than using a vertical that is longer than the target operating frequencies and then tuning it with a series variable capacitor. The mode change can be done with a simple RF-rated relay.[7] See **Fig 1** for the schematic of my tuning/matching network.

I then assembled and bolted all the sections together. I also marked each section at the joint for proper bolt alignment during installation. I placed a small PVC endcap on the top section to seal it off. This was installed by carefully heating the tubing with a small propane torch and "sweating" it onto the PVC piece. This will keep rain out and will prevent wind from turning your vertical into a large, very annoying whistle.

I guyed the vertical at the 12, 33 and 54 foot levels, corresponding to tubing section changes. Three 4-inch round, $^1/_{16}$-inch thick, aluminum plates were drilled with center holes of 1.500, 1.125 and 0.875-inch diameter. Smaller holes were drilled on the periphery of the plates to attach guylines.

After disassembly, I painted all the sections the same bark-brown as the tree trunks in my backyard. This turned out so well that the vertical is nearly indistinguishable from other trees in the backyard and certainly adds to the neighborhood aesthetic appeal. Be sure to use an enamel or lacquer-based paint.

For a base, I used a 3-foot piece of 1.5-inch schedule 40 PVC pipe placed 2 feet into the ground. See **Fig 2** for construction details of the base. The pipe has a nominal ID of 1.625 inches. This insulates the vertical element from ground and provides sufficient stability. An aluminum radial mounting plate drilled with a clearance hole was mounted over the PVC base. I drilled equally spaced holes at its periphery to attach the radial wires. Stainless steel self-tapping metal screws facilitated the installation of radials.

The matching network relay and coils were housed in a heavy-duty Rubbermaid container. The price, assortment of various sizes, and the waterproof nature of these containers make them an ideal, cost-effective choice. The container was bolted to the base of the vertical, close to the radial plate. See the photo in **Fig 3**.

I eventually ended up with 40 radials, 64-feet long and made of #14 stranded and insulated wire because of easy availability at any electrical supply store. Smaller wire may be used also. Radial wires operate in the near field of the vertical antenna system and provide a path for return RF currents. They contribute to the overall efficiency of the antenna. The number required is depen-

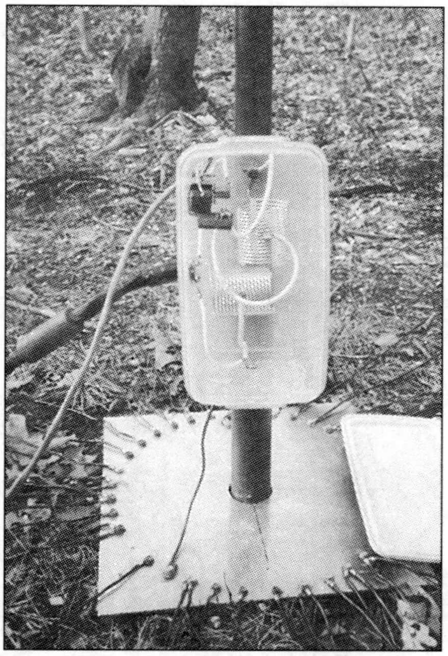

Fig 3—Photo of inside of tuning network box, along with aluminum plate used to mount the ground radials. Note the ferrite-core choke balun on the feed coax to help keep RF off the coax shield back to the shack.

dent on your ground conductivity. Lengths and required number of radials have been controversial topics of discussion over the years; see ON4UN's Chapter 9 for an excellent synopsis on this topic.

Installation and Tuning Details

In the next steps I describe the actual installation of my antenna. I cannot emphasize strongly enough that you must make *absolutely sure* you have adequate clearance from powerlines in case the vertical should fall for any reason—either during installation or later!

I set the radial mounting plate over the PVC base insulator, drove a ground rod (for lightning protection) and connected it to the radial plate. Install the bottom 1.500-inch diameter aluminum tubing first and guy it in place with the lower set of guys. I premeasured and cut the guy-line lengths. A 12-foot stepladder was placed standing next to the bottom section.

I assembled the top three sections (the 1.125, 1.000 and 0.875-inch tubing), with the top set of guys attached. This made the top assembly 25 feet long. Prior to inserting each section, I coated each with a conductive grease to facilitate disassembly. The top assembly was positioned so that its bottom was next to the PVC base.

We then raised the top section to the vertical position, with one person at each of the three guy tie-down points. All guy lines were held at just enough tension to keep the mast from bending. I then pushed the top section upwards off ground and placed the bottom tubing temporarily on one of the upper stepladder steps. Standing on the stepladder, I placed the middle guy ring and lines at the bottom of the 1.125-inch tubing, temporarily holding it from sliding downwards with tape. Then, standing on the stepladder, I pushed the top assembly up until my assistant could guide it into the 1.250-inch tubing from below. I bolted the two sections together.

The middle set of guys was walked out to the three guy handlers, and the 12-foot long 1.375-inch diameter tubing was brought to the stepladder. Again, I pushed up the whole assembly until it could be slipped into and then bolted to the 1.375-inch tubing. The whole assembly was pushed upwards until it could be fit into the top of the bottom 1.500-inch bottom section and was secured with the pinning bolt.

All throughout, tension was maintained on the guy lines for stability of the vertical, but not so much that I couldn't push the sections upwards! Of course, this is easier to say than do—good communication and planning with your guy-line team makes this whole job a 30 minute effort.

All the radials were then laid in place. For safety reasons, I used a sidewalk edger garden tool to make a slit in the ground to bury each radial. Otherwise, my wife had warned me that our lawnmower and 2000 feet of radial wires above the grass would make a memorable story! This is a time-consuming effort, so plan accordingly.

From formulas in Orr and Cowan's book, I knew I needed 2.0 to 2.5 μH of shunt loading coil and a net 1700 to 1800 pF of series capacitance to form an L-network to bring the feedpoint impedance up to 50 Ω. As explained earlier, the shorter-than-resonant vertical element itself provided the series capacitance and the shunt inductance was made up of a 2-inch diameter (B&W 3026, 8 TPI, #14 wire) air-wound coil. A series inductance would be needed for the CW part of the band, as well as providing fine-tuning of the vertical for phone.

To tune the system, I used an Autek RF Analyst, connected with a short length of coax to the SO-239 connector. On the ground side of the shunt matching coil, I attached a jumper wire to an alligator clip and connected it about 3 turns from the top. I did the same with a jumper wire and an alligator clip for the series resonating coil. I connected the RF Analyst and experimentally determined the series coil tap to resonate the vertical at 3.8 MHz. The tap should be at or near the top of the series coil. Then I experimentally adjusted the matching shunt-coil's tap for lowest SWR, readjusting the resonating series-coil tap as necessary. These became my final tap points for the phone band. Then I determined the series-coil tap to resonate the vertical for 3.550 MHz to cover the CW portion of the band, using the same shunt-coil tap as was used on phone. I switch between the series-coil taps with a 12-V relay to cover most of the 75 and 80-meter bands with one vertical.

On-the-Air Results

I used a remote coax switch to give me "antenna A versus antenna B" testing ability. On-the-air testing revealed the full-length vertical was 2 to 3 S-units stronger on receive and transmit than the GAP Challenger. It was remarkable how much difference was observed in received signal strengths—stations not readible over the noise became discernible. This doesn't mean the GAP is not doing the job it was designed for. The GAP is a linear-loaded $^1/_8$-λ vertical on 75 meters; it can't possibly be better than a full-size $^1/_4$-λ vertical with an extensive ground radial system. In fairness, the $^1/_4$-λ vertical doesn't work at all on any other band, whereas the GAP is an efficient multiband antenna.

Although my sample size is small and I continue to collect field reports, I am quite encouraged by the effectiveness of this relatively simple vertical system. It is a pleasure to consistently work European DX stations and bust pileups with a minimum of calls.

Next...

I am so enthusiastic about the effectiveness of this system, I am now planning to assemble a second vertical and build a variable phase angle network to "steer" the resultant radiation pattern. Modeling shows a 4 to 5 dB gain over a single vertical and a significant 18+ dB front-to-back ratio. On 80 meters, this should be a FUN antenna system on the East Coast!

Notes and References

[1]John Devoldere, ON4UN, *Antennas and Techniques for Low-Band DXing* (Newington: American Radio Relay League, Inc, 1994)

[2]William I. Orr, W6SAI and Stuart D. Cowan, W2LX, *All About Vertical Antennas* (Lakewood, NJ: Radio Amateur Callbook, 1993)

[3]A. Christman, "Elevated Vertical Antenna Systems," *QST*, Aug 1988, pp 35-42; Feedback, *QST*, Oct 1988, p 44.

[4]T. Russell, "Simple, Effective, Elevated Ground-Plane Antennas," *QST*, Jun 1994, pp 45-46.

[5]Paul H. Lee, N6PL, *The Amateur Radio Vertical Antenna Handbook* (Hicksville, NY: CQ Publishing, 1984)

[6]Aluminum tubing is available from national suppliers who advertise in *QST*. Try to organize or pool an order with others to reduce shipping charges.

[7]I used B&W coil stock (part number 3026, 2-inch OD, 8 TPI, #14 wire) and RF-rated 12-V relays available from Nebraska Surplus Sales.

A Homebrew Seven-Band Vertical

By Paul I. Protas, WA5ABR
15709 Seattle Street
Houston, TX 77040

WA5ABR describes a versatile, efficient vertical for 40 through 10 meters. It's easy to build, costs much less than an equivalent commercial antenna, and works great!

I built this antenna to replace an aging four-band trap vertical that had been in service for about 12 years. I had looked around at various manufactured verticals but decided to build one myself. This antenna should out-perform any trap vertical and should be at least comparable to any of the "half-wave" or "elevated-feed" products on the market today. My vertical cost around $130, considerably less than most commercially available units. If you have any scrap aluminum tubing from other projects, your cost could be even less.

My design philosophy is that if a full-size $\lambda/2$ dipole is an efficient radiator, then a full-size $\lambda/4$ radiator would also be efficient, given a good ground system, of course. On 30 through 10 meters, the vertical uses six full-sized $\lambda/4$ elements. On 40 meters a loading coil is used at the 16-foot level, with a 4-foot whip above the coil. You could elect to use a full-sized 40-meter element, but this will definitely mandate the use of guy ropes.

There are no traps or matching circuits. One feed line is used, connected to all seven elements in parallel at the base. See the drawing in **Fig 1** and the photo of the assembled antenna in **Fig 2**. All elements are made using 0.058-inch-thick aluminum tubing so that they may be telescoped easily.

If you want to build a really low-cost version, you might consider making your antenna like the wire prototype I made before changing to aluminum tubing in the final design. In the prototype I used #14 wire supported by a 23-foot tall 2×2 wood support, with wooden discs at the top and bottom to separate the individual wires. I used heavy fishing line to support the shorter elements from the top disc. While it certainly was not as durable as the final aluminum version, the prototype worked fine and only cost about $20.

Construction

See **Table 1** for the Parts List to build this antenna. To start construction of the aluminum-tubing version, cut two 15-inch diameter discs out of acrylic sheet. In one disc, drill six $5/8$-inch holes at points $6\,3/4$ inches out from the center of the circle (center to center) at 60° intervals. Drill a $1\,1/2$-inch hole in the center. This will be the bottom support. In the second disc, drill six holes as above, but make two opposing holes $5/8$ inch and the other four $1/2$-inch OD. The center hole in this disc should be $1\,1/2$ inch. This will be the top support. See the photo in **Fig 3**, showing details of the bottom support disc.

From the remainder of the acrylic sheet, cut a rectangular piece $2\,1/2$ by 15 inches. Drill a $1\,3/8$-inch hole in the exact center of this piece

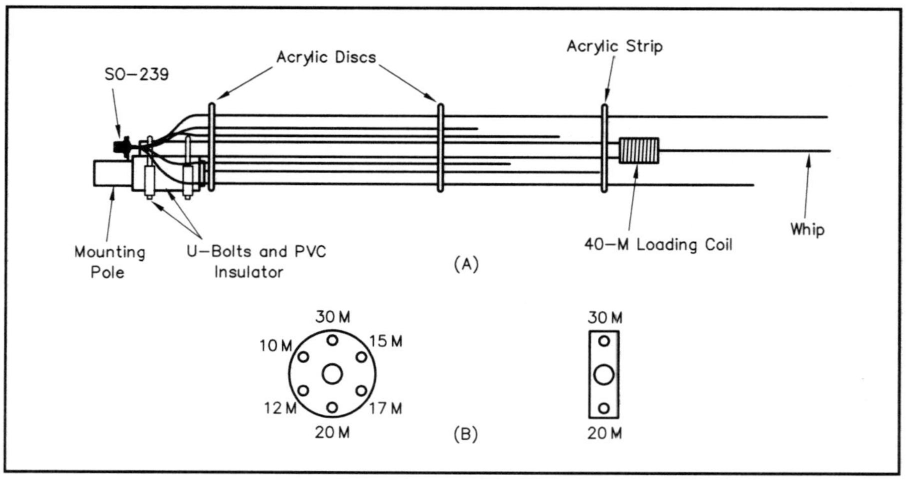

Fig 1—At A, layout of WA5ABR seven-band vertical. The acrylic discs at top and bottom support the aluminum-tubing elements. The acrylic strip at the top keeps the longer elements stable. At B, detail for the acrylic discs and strip.

Table 1

Parts List

Qty	*Description*
3	12 foot sections, 1/2-inch OD
4	12 foot sections, 5/8-inch OD
1	6 foot section, 1/2-inch OD
1	6 foot section, 1 3/8-inch OD
1	12 foot section, 1 1/2-inch OD
1	48 inch aluminum-wire whip
1	2 foot section, 1 1/4-inch Schedule-40 PVC
7	solder lugs, 1/8-inch holes
1	SO-239 connector
1	34 foot length of insulated #14 solid copper wire
1	36×18-inch sheet of 1/4-inch thick acrylic (Plexiglas)
2	2-inch U-bolt clamps
20	3/4-inch hose clamps, non-rusting
9	1 1/2-inch hose clamps, non-rusting
7	6-32×3/8-inch screws, lockwashers, and nuts

Note: all aluminum tubing is T-6061 material, with 0.058-inch-thick walls.

Fig 2—Photo of assembled antenna at WA5ABR.

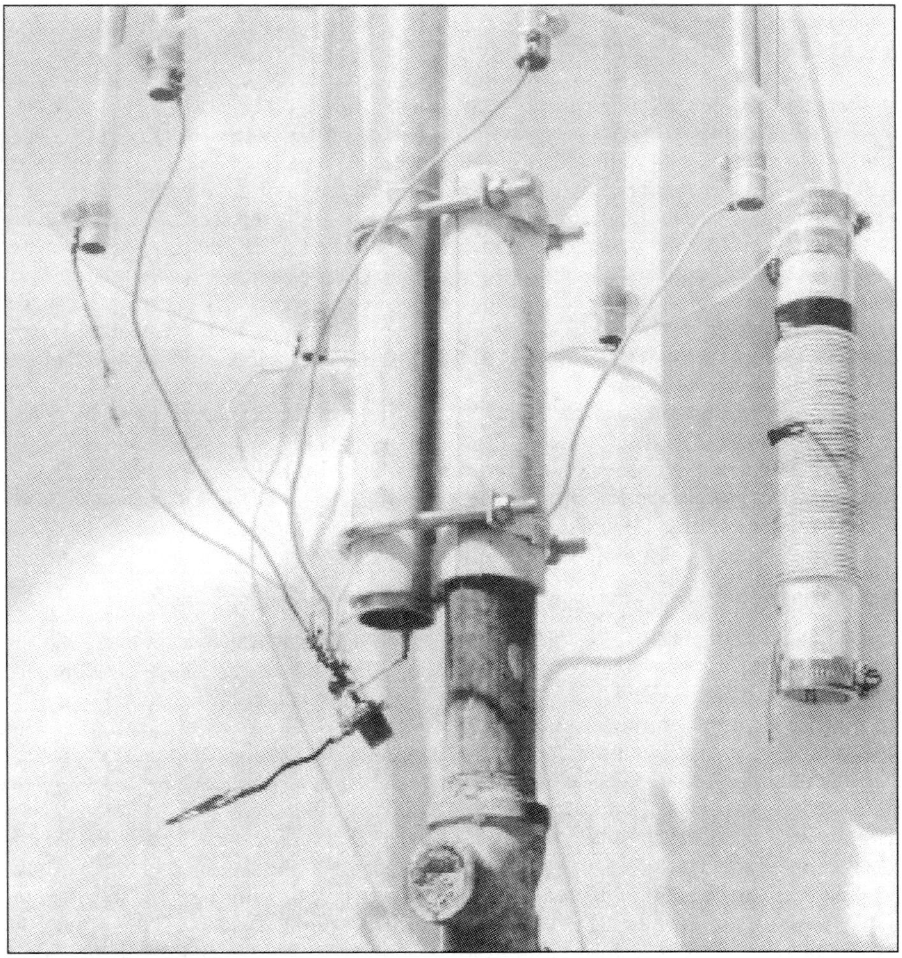

Fig 3—Photo detail of the bottom acrylic disc, with insulating "clam-shell" clamps of PVC pipe around 40-meter element and mounting pipe using U-bolts. The seven elements are connected in parallel to the SO-239 connector. In this photo an alligator clip was used to connect to the ground radials during experimentation.

and two 1/2-inch holes, 1/2 inch in from each end. This will support the top two longest elements.

Next, separate the aluminum tubing into three batches according to size. The 1 1/2 and 1 3/8-inch sections can be telescoped and set aside as one batch. The other two batches will be the 1/2-inch and the 5/8-inch OD pieces. Cut two 12-foot pieces of 5/8-inch tubing in half (each 6 feet long). Cut one piece of 1/2-inch tubing into a 4-foot and an 8-foot section. Cut another piece of 1/2-inch tubing into two 6-foot sections. The remaining two 12-foot sections of 5/8-inch tubing and the one remaining 12-foot section of 1/2-inch tubing will not be cut.

You should now have six sections of each size smaller tubing. Drill a 1/8-inch hole, 1/4 inch from the end, of each of these sections. Saw two opposing slots 1/2-inch deep into the other end of each section of 5/8-inch tubing. This will be the place where the 3/4-inch hose clamps tighten around the next telescoping section. These hose clamps can now be attached loosely. Telescope the different sections of tubing using **Table 2** as a preliminary guide. You will adjust the final length for SWR at installation. The starting lengths may seem to be short, but the length of wire attached to the SO-239 coaxial connector makes each element electrically longer.

Drill a hole, similar to the ones drilled in the 5/8-inch sections, in the end of the 12-foot section of 1 1/2-inch tubing. Also, cut two opposing slots 1/2-inch deep in the other end of this section. Similar slots need to be cut in the end of the 1 3/8-inch section. Cut six lengths of the #14 insulated copper wire about 11 inches long. Remove 1/4 inch of insulation from each end. Install solder lugs onto one end of each wire.

I made a small loop on the other end of the six wires and slid each onto a short wire soldered to the center pin of the SO-239 connector. (I crimped another solder lug on the end of this wire so the other wires didn't fall off while soldering the individual wires to the feed bus.) Also attach (either with small screws and nuts, or solder) a five-inch loop of #14 wire to the ground hole on the SO-239. This makes a bus where ground radials are attached. Use at least 8 to 12 radials, 16 feet or longer and a 4-foot or larger ground rod for lightning protection.

Now, place the bottom disc (the one with six 5/8-inch holes) about eight inches from the bottom end of the 1 1/2-inch tubing. Secure this to the tubing on either side with 1 3/4-inch hose clamps. Slide another hose clamp, the top disc, and another hose clamp after that, over the other end of the 1 1/2-inch tubing, where the 1 3/8-inch tubing is sticking out. Secure this disc about 7 1/2 feet up from the first disc by tightening the hose clamps.

Attach a seven-inch piece of Schedule-20 or 40 PVC pipe to the bottom end of the 1 1/2-inch tubing. I used a piece of 1 1/4-inch ID Schedule-40 pipe, cut in two lengthwise as the clamp-on insulator. Place the two U-bolts over the PVC at each end (so that the U-bolts don't touch the 1 1/2-inch tubing) and temporarily tighten them. Now, attach the solder lug on the SO-239 bus wire to the hole in the end of the 1 1/2-inch tubing with a 6-32×3/8-inch screw, lockwasher and nut.

Next, put the two longest sections of tubing, for 20 meters and 30 meters, through the larger holes in the uppermost disc. They should be opposite each other (at 0° and 180°). Slide a 3/4-inch hose clamp over the bottom end and insert the element into the bottom disc. Attach one of the feed wires to the element with a screw, lockwasher and nut. Keep the disc perpendicular to the larger center element, and press the tubing against the disc and tighten the hose clamp. Also, slide another hose clamp over the smaller telescoped end of the elements and tighten at the point where the element comes out of the upper disc. Do the same with the remaining elements so that they are positioned as shown in Fig 1. Once all the elements are in place, make sure that the feed wires don't cross each other.

Next, slide a 1 1/2-inch hose clamp, the 15-inch piece of acrylic, and another hose clamp over the top section of 1 3/8-inch tubing and over the 20-meter and 30-meter elements, stopping just below the top of the 20-meter element. This keeps these two elements from moving in the wind and affecting resonance. Place 3/4-inch hose clamps on these two elements above the acrylic, keeping it level, and then tighten the large hose clamps also.

If you are going to use a full-size element on 40 meters, telescope the center three sections (the top section is 1 1/4-inch OD) to an overall length of 33 feet and tighten the hose clamps. If you are using a loaded 40-meter element, as I did, you will only be using the 12-foot long 1 1/2-inch section and a 6-foot long 1 3/8-inch section. This telescopes to a total of 16 feet long.

For the 40-meter loading coil form, cut a piece of the 1 1/4-inch ID Schedule-40 PVC pipe 12 inches long. Cut two lengthwise slots about one inch long at right angles to each other on one end. Draw a line running the length of the pipe on one side. Drill two 1/4-inch holes on this line. One should be one inch from the end with the slots and the other 1 1/2 inches from the end. Drill two similar holes on the line 2 1/2 and 3 inches from the other end.

Starting at the end with the holes near the slots, insert five inches of #14 coil wire through the inside hole and back out the outer hole. This end will attach to the 1 3/8-inch tubing with a sheet metal screw. Wind 48 turns of the #14 wire tightly around the PVC and put the end of the wire through the inner hole in the far end and back out the outer hole. Leave about two inches and snip off the extra. Strip one inch of insulation off this end and make about three loops with an inside diameter the same as the whip you will be attaching to the top. Press this loop close to the PVC pipe, insert the whip into the loops, and attach the whip to the PVC pipe with two 1 1/2-inch hose clamps. Squeeze the wire loops tightly around the whip with vise grips or similar pliers.

Slip a 1 1/2-inch hose clamp over the slotted end of the PVC and install onto the 1 3/8-inch aluminum. Drill a hole in the aluminum just below where the PVC pipe is mounted and attach a small, uninsulated loop of the lower coil wire with a solder lug and sheet metal screw.

Finally, mount the antenna on its support pipe placed in the ground and attach the ground wire loop connected to the SO-239 to the ground radial system. Using low power, check the SWR on all bands. If necessary, adjust the lengths of each element by loosening the clamps at the joints and above the acrylic discs and retighten to check the SWR again. If the 40-meter SWR is too high, you may have to remove a few turns of the coil (if changing the length of the element is not sufficient). The SWR on all bands except 40 meters will typically be 1.1:1 over most of the band and less than 2:1 at the band edges. The 40-meter bandwidth will be similar if you use a full-size element. The loaded 40-meter element I used had a 2:1 SWR bandwidth of about 90 kHz.

I have been using this antenna for about one year and have worked over 60 countries during low solar-flux conditions (including ZD7, 3B8, VR6, 5N0, V6, CY9, A3, YB, and EA9, to name a few). A vertical like this is a viable option for hams with small lots, tower deed restrictions, or limited finances. It should provide years of operating enjoyment due to its simple construction.

Table 2

Preliminary Dimensions for Telescoping T-6061 Aluminum Tubing; 0.058" Wall

Band	*Overall Length*	*1 1/2" OD*	*1 3/8" OD*	*5/8" OD*	*1/2" OD*
10 meters	7' 6"	—	—	6'	4'
12 meters	8' 6"	—	—	6'	6'
15 meters	10' 3"	—	—	6'	6'
17 meters	12' 3"	—	—	6'	8'
20 meters	15' 9"	—	—	12'	6'
30 meters	22' 10"	—	—	12'	12'
40 meters	16', plus whip	12'	6'	—	—

A Close Look at the Flattop Vertical Antenna

By Thomas Kuehl, AC7A
620 S Avenida Princesa
Tucson, AZ 85748

On 80 and 160 meters, quarter-wave verticals are 66 and 134 feet tall. That may pose a problem. Thomas, AC7A, describes the flattop vertical antenna, which uses capacitive top loading to reduce the overall antenna height.

When it comes to a simple and effective antenna, the half-wave dipole installed a half wavelength above ground, is hard to beat. It offers good overall performance with the low radiation angle needed for DX communications. Many can accommodate the horizontal span of the dipole on their lot, but it is that half-wave above ground that is often a problem. This is especially true on 40, 80 and 160 meters. On 40, the half-wave height requirement is about 66 feet, and this doubles, and then doubles again as the frequency is moved down to 80 and 160 meter bands. Erecting, or even obtaining the appropriate supports, can prove difficult. For these reasons, or others such as aesthetics, many hams resort to a full-sized or shortened vertical to attain the much desired low-angle performance.

Almost everyone who experiments with, or builds their own, antennas is familiar with the quarter-wave ($\lambda/4$) vertical formula:

$$L\ (\text{ft}) = 234/f\ (\text{MHz}) \qquad \text{(Eq 1)}$$

A quarter-wave 40 meter (7 MHz) vertical, as an example, requires a length of about 33 feet—a height usually managed without too much difficulty. Aluminum tubing or TV mast can often satisfy this requirement. But when the frequency is moved down to the lower bands, 80 and 160 meters, the element lengths increase to 66 and 134 feet, respectively. Only a single tall vertical element is required, but that may still pose a problem.

An alternative is a vertical that uses capacitive top loading to reduce the overall antenna height. One such design is the flattop vertical antenna. In its simplest form it consists of a vertical element, less than a quarter wavelength long, with two opposing horizontal top wires. It looks much like the letter T and is sometimes referred to as a T-top antenna. The horizontal top wires provide capacitive loading that brings the short vertical section to resonance.

The vertical element length may be almost any convenient length; as long as sufficient horizontal space is available to accommodate the width, or span, of the top wires.

Keep in mind, as with any shortened vertical, making it progressively shorter brings along some unavoidable compromises: lower radiation resistance (R_r), and a narrower operating bandwidth. Low R_r usually isn't a problem by itself, but efficiency suffers when it is combined with high ground resistance (R_g). By using a flattop the problem is not as severe as it could be; it will always produce higher efficiency than its equally tall, inductively loaded vertical counterpart. The reasons for this will be explained later.

A variety of flattop schemes are possible but the most easily understood and built use the single-wire-T design. Multiple-wire designs are also practical and do allow for substantial reductions in the horizontal real estate requirements. Both designs will be examined.

The Single-Wire Flattop Antenna

Fig 1A shows the current distribution along a quarter-wave vertical. The vertical element is divided into 12 segments and an average current is listed for each segment. A normalized reference current of 1 A has been established at the feed point. From the figure it is evident that the current distribution is maximum at the feed point, then slowly decreases as you move upward along the element, dropping off rapidly toward the end of the antenna. If you plot these points, you'll see that the current distribution follows a cosine function along the antenna's quarter-wave (90°) length.[1]

In a similar fashion, Fig 1B shows the current distribution along the vertical and horizontal lengths for a practical flattop-antenna design. Once again the feed-point current has been normalized to 1 A. Start at the flattop's base, and compare the vertical-segment currents to those of the quarter-wave vertical and you'll see that they are nearly equal in value. At the top of the ver-

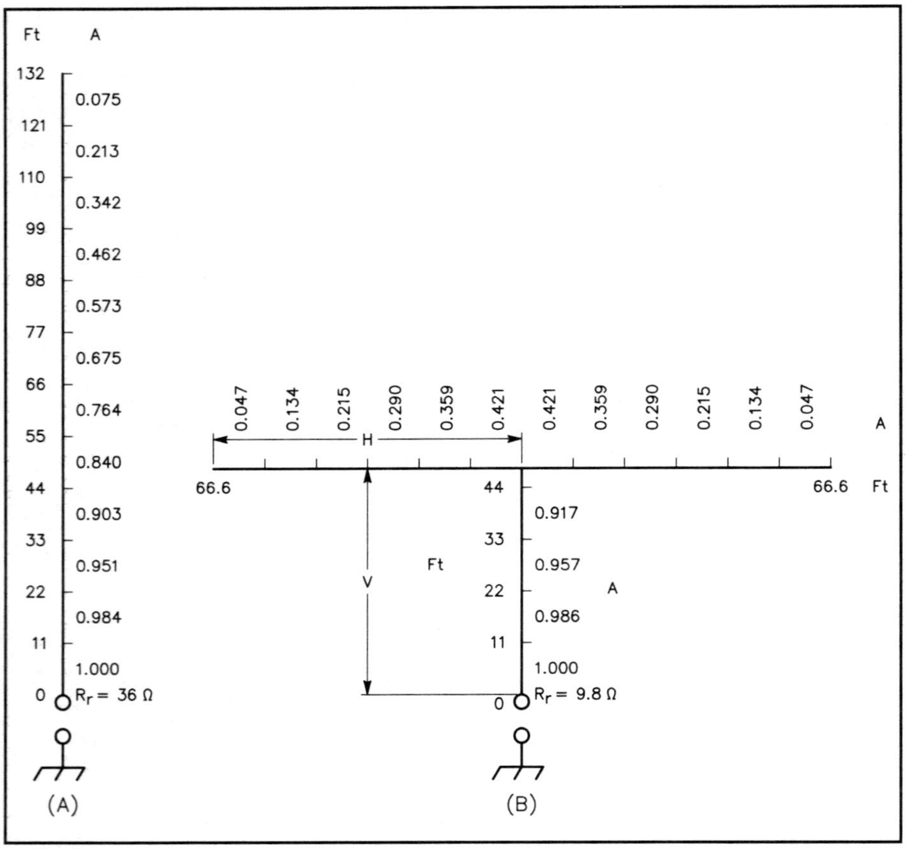

Fig 1—Comparison of current distribution in a 1/4-λ vertical (A) and a typical flattop vertical (B). Both antennas are resonant at 1.825 MHz.

tical element, where the horizontal wires join, the current splits into two equal, but opposite, components. If these two segment currents are summed, their value is very close to that of the corresponding quarter-wave-vertical segment. Thus, the current distributions produced by the two antennas are nearly identical. Because the vertical and flattop will have unequal feed-point resistance in practice, the currents will be different in magnitude, but have proportional distributions.[2]

Fig 2 shows *EZNEC* far-field elevation patterns for the quarter wave and the flattop vertical antennas. The quarter-wave vertical produces a slightly lower takeoff angle. Thus, little pattern difference exists between the two antennas, even though the flattop is much shorter!

Because of the flattop's lower R_r, its pattern is less than that of the full-sized vertical. While the quarter-wave vertical of Fig 1A has an R_r of 39.3 Ω, the flattop in Fig 1B has an R_r of 12 Ω. Assuming all other losses are 0.5 Ω, and R_g is 6 Ω, the efficiency (η) for the two antennas is 85.8% and 64.9%, respectively. Antenna efficiency is equal to:

$$\eta = 100\,[\,R_r/(R_r + R_l)\,]\% \qquad \text{(Eq 2)}$$

where R_l is the sum of the loss resistances, which is mostly R_g for each of the two antennas.[3,4]

Establishing the vertical element length is often dictated by available end supports. Taller is better, but in restricted height applications very short heights can be tolerated if a good, low resistance ground return can be provided.

Once the vertical length is decided, it is then a matter of resolving the correct top-wire lengths. Lacking more information, it has been a common practice to start with a horizontal (H) length, plus vertical (V) length, equal to a quarter-wavelength. This may serve as a starting place, but it isn't always accurate.

To understand what is involved in determining the proper top-wire lengths, it is best to visualize each top wire and the earth below it as parts of an open-wire transmission line. The top wire is easy to visualize as one conductor, but the earth return is more complex and is not as easily visualized. It consists of the earth beneath the top wire, and any conductive ground radials. Its boundaries are much more difficult to define than those of the top wire, but nonetheless do exist. The spacing of these two conductors, their effective diameters, and the dielectric characteristics of the materials between them—earth and air—define the transmission line properties; the surge impedance, Z_S (usually hundreds of ohms), and the velocity factor (VF), which is always less than unity.

The far end of this transmission line is an open circuit, or more correctly, the line is unterminated. With a length less than a quarter wavelength, the unterminated line will appear as a capacitance to the vertical element. Additionally, the top wires exhibit capacitance directly to the vertical element. Together, each top wire contributes to the total capacitance. If the lines are of the correct length, they will add just enough capacitance to overcome the short vertical's capacitive reactance and bring it to resonance.

Some physical transmission-line variables, like the ground-conductor diameter, are not readily determined. And that in turn makes determining the line characteristics and top-wire lengths difficult. One solution is to use a modeling program, such as *EZNEC,* that takes into account the antenna environment. The ground conductivity and

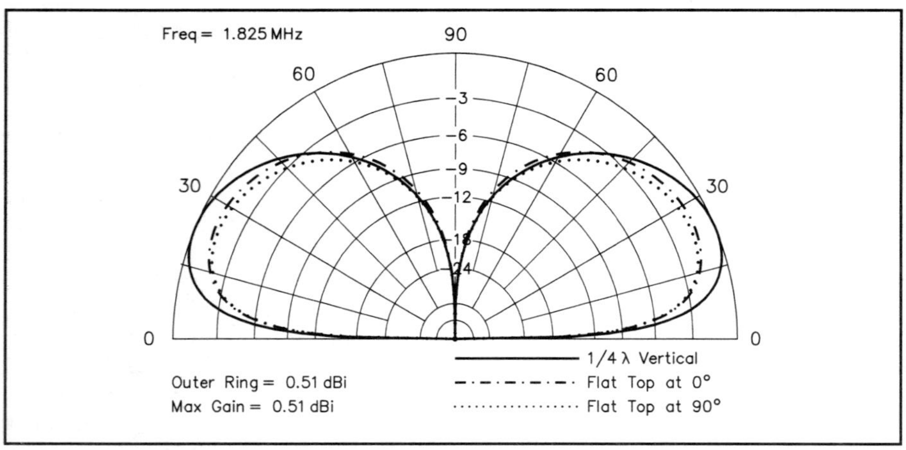

Fig 2—Elevation patterns for 1/4-λ and flattop verticals with 16 1/4-λ radials elevated 0.5 feet over real ground. The flattop pattern at 0° is for the plane that contains the flattop, and the pattern at 90° for the plain that is perpendicular to the flattop.

dielectric constant can be selected, and even two different ground media can be modeled.[5] Using this approach, and starting with a vertical slightly less than a full quarter-wavelength, the horizontal top-wire lengths can be adjusted until the antenna is resonant at the desired frequency.[6]

I used this method to establish the relationship between the flattop's vertical (V) and horizontal (H) lengths. Results from the modeling are shown in **Fig 3**, where the V and H relationship is expressed as the ratio V/H. For any vertical length (V), from 0.025 λ to 0.2375 λ, the corresponding top-wire length (H) can be determined. The vertical lengths are listed and entered as a decimal fraction of a wavelength on the horizontal axis. While the vertical axis gives the V/H ratio.

As earlier stated, conductor diameter is a factor in determining the transmission line impedance (Z_S); therefore, three different length-to-diameter (l/d) curves are given in Fig 3. I used ratios of 25,000:1, 1000:1 and 100:1 in the modeling. These nonlinear curves reveal why simply using V + H = ¼ λ is often inaccurate.

When relating a quarter-wavelength to a length in feet, as in Eq 1, the constant 234 provides the conversion factor that takes into account the speed at which radio waves travel in free space, and antenna end effects. Likewise, conversion constants are provided for each of the 3 flattop l/d ratios, but here they differ from Eq 1, in that they are being specified for a full wavelength:

For l/d = 25,000:1
λ (ft) = 961/f (MHz) (Eq 3)

For l/d = 1000:1
λ (ft) = 950/f (MHz) (Eq 4)

For l/d = 100:1
λ (ft) = 930/f (MHz) (Eq 5)

Since the flattop's V/H ratio will be a function of the selected vertical length, it was decided to use a full quarter wavelength as the l/d reference value. Just take a full-size quarter-wave vertical's length, the element diameter, and calculate the l/d ratio. Although a 25,000:1 l/d ratio may appear a bit high at first, a 160-meter flattop made with #14 AWG wire has this ratio. On the other end, a 10-meter flattop made from 1-inch tubing would have an l/d ratio of about 100:1. Therefore, the above l/d ratios pretty well cover most designs.

Due to the wide V/H range presented in Fig 3, resolving the ratio precisely can be difficult. Most installations will tend to use short vertical lengths, so expanding the scales for these lengths helps improve resolution. In **Fig 4**, the graphs have been expanded to better show the V/H details for vertical lengths of 0.025 λ to 0.125 λ.

Use of the V/H graphs is not difficult. Consider a flattop for 1.825 MHz. A vertical length must be decided upon: let's use the 44 foot length from Fig 1B. The antenna will be constructed of #14 AWG wire, which has a diameter of 0.064 inch. Plugging 1.825 MHz into Eq 1, and then dividing by 0.064 provides an l/d ratio of about 25,000:1. Next, using λ Eq 3, take 961 and divide by 1.825. One wavelength, for this antenna design, is then 526.6 feet. Now taking 44 feet, and dividing by 526.6 feet, yields 0.084—the vertical section's fraction of a wavelength (0.084 λ).

Using the expanded scale of Fig 4, enter the Vertical Element–Fraction of a Wavelength scale at 0.084 λ, and follow it up to where it intersects the 25,000:1 curve. Moving straight across to the V/H scale produces a ratio of about 0.66. This means the vertical (V) element has a length 0.66 times that of each horizontal (H) top wire. The 44 foot V length divided by 0.66 indicates a top-wire length of 66.7 feet. The total flattop span is twice this; about 133.4 feet—about the same length as shown in Fig 1B.

To test the V/H information of Figs 3 and

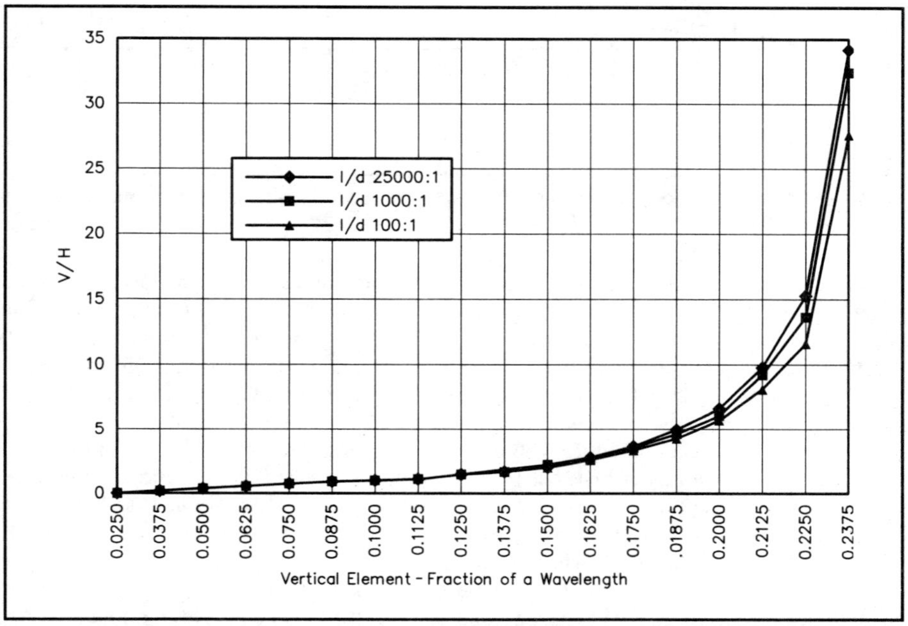

Fig 3—Ratio of vertical segment (V) to horizontal (H) for a single-wire flattop vertical.

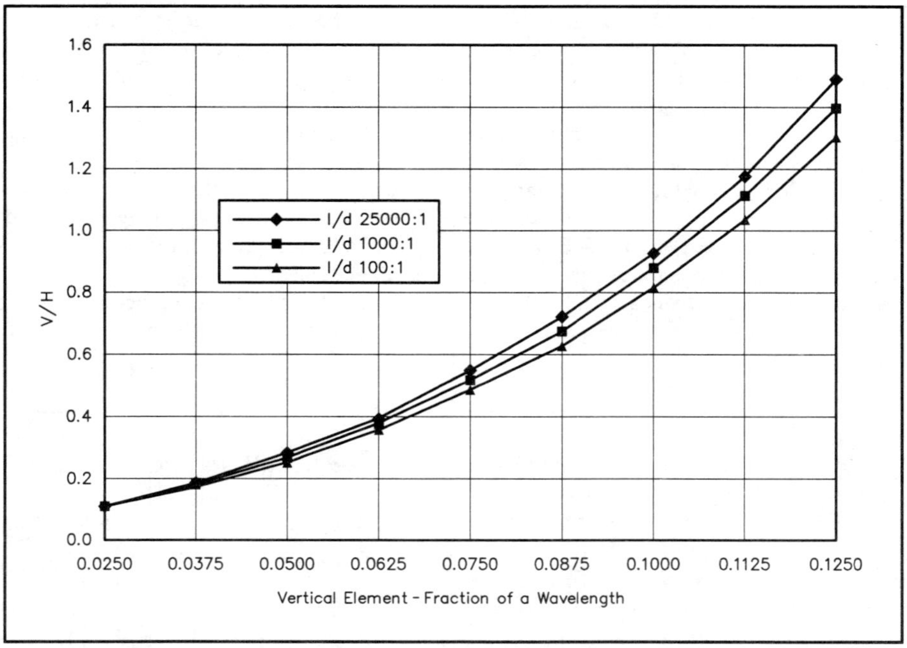

Fig 4—Expanded scale presentation of V/H ratio for the single-wire flattop vertical.

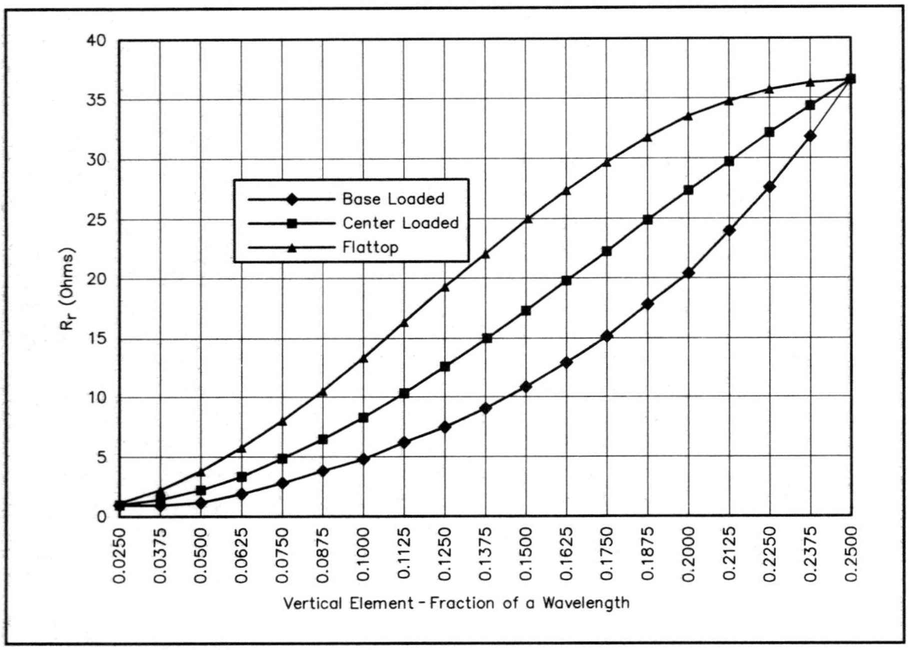

Fig 5—Comparison of radiation resistance (R_r) for the flattop and inductively loaded verticals.

4, I constructed a 6-foot tall 20-meter flattop from #12 wire. From Fig 4 the top-wire lengths (H) were derived and established at 8.5 feet per side. The test antenna was built using this length, but when all the pruning was done, the top wires were actually closer to 7.25 feet—about 15% shorter!

I was a bit concerned at first, but further analysis of the *EZNEC* model provides a clue why this happened. The V/H graphs were developed using R_g equal to 0 Ω. When a higher R_g was included in the model, the top-wire length decreased. The test antenna's ground return had been hastily constructed using four quarter-wave radials, suspended a few inches above landscaping pebbles. Certainly this is a high loss ground return at most locations. All this suggests that ground has a greater-than-expected effect on the antenna's *transmission line* characteristics.

To test the idea, 12 additional quarter-wave radials were added to the original 4. This would have the effect of substantially reducing R_g from its former value. SWR measurements made before and after the radial additions, indicated an increase in the resonant frequency by nearly 300 kHz! Additional top-wire length was required to reestablish resonance. In most practical applications where the ground return is not ideal, the top-wire lengths will be 5 to 15% shorter than called for in the ideal case, but it is easier to start out too long and then trim the wires shorter for resonance.

If a less-than-optimum ground return is employed, one with few radials, the return currents must flow through the lossy earth. The earth then plays a larger part in the transmission line characteristics; acting not only as one conductor, but also as part of the dielectric between the two conductors. As a result, the transmission line characteristics change considerably from the ideal case; not only in impedance (Z_S), but also by reducing the velocity factor (VF). The VF reduction causes the speed at which the electromagnetic wave propagates along the antenna length to slow. Therefore, for the wave to traverse the quarter-wavelength in the same allotted time, the length must be made shorter. This effect is also at work in coaxial cable, where the dielectric causes capacitance to increase and the velocity factor to decrease.

As I stated earlier, reducing the flattop's vertical length (V) results in progressively lower R_r, as illustrated in **Fig 5**. Following the flattop's R_r curve, it increases in a sine-like manner until it converges on 36 Ω maximum at a quarter-wavelength.

Curves for base and center-loaded verticals using lossless loading inductors are also included in Fig 5 for comparison. These produce exponential and *nearly* straight-line curves, respectively, which also converge on 36 Ω. At that point, the antenna is a quarter-wavelength vertical with no loading. Fig 5 shows that the flattop always provides a higher R_r than an equally tall base- or center-loaded antenna. For this reason, it will always operate with higher efficiency. And because the base and center-loaded antennas use loading coils, with their inherent losses, it further reduces their efficiencies.[7]

The flattop's advantage is even more dramatic when the vertical length is in the range of 0.125 λ to 0.2 λ, and a low R_g return is present. These longer lengths often use a tower as the vertical element on the low bands. An HF beam antenna atop the tower will provide capacitive loading in the same way as top wires.

Efforts must be made to minimize the ground return resistance, R_g, if maximum efficiency is to be attained with any ground-dependent antenna. This dictates that many buried ground radials, or alternately elevated radials, be employed at the base of the antenna.[8] Radio waves penetrate the earth more deeply at lower frequencies, and the earth becomes more lossy at higher frequencies.[9] An extensive ground-return system will lessen the antenna's sensitivity to the earth directly surrounding the antenna by making the antenna less dependent on the earth return.

One might assume that using large diameter conductors could provide an antenna with a substantially shorter horizontal span, but the curves of Figs 3 and 4 suggest the contrary. It is true that a top-hat with larger surface area will result in greater capacitance to ground, however, it is the current distribution along the vertical element that has a profound affect on the final dimensions. This can bring some surprising results.

Using the earlier 160-meter flattop example, **Table 1** gives the horizontal top-wire lengths required to bring a 44-foot

Table 1

Relationship of Element Diameter and Required Length for the 160-meter Flattop

Vert Element	*Vert*	*Horiz Element*	*Top Wire (per side)*	*Feed Point R_r*	*2:1 SWR BW*
#14 wire	44 ft	#14 wire	66.7 ft	9.8 Ω	38 kHz
1.5 in. tube	44 ft	#14 wire	78.8 ft	9.9 Ω	50 kHz
1.5 in. tube	44 ft	1.5 in. tube	68.5 ft	9.8 Ω	54 kHz
#14 wire	44 ft	1.5 in. tube	55.5 ft	9.7 Ω	40 kHz

vertical element into resonance when the vertical and horizontal sections use the same, or different, conductor diameters. Two element diameters have been selected for the example; 0.064 inch diameter wire (AWG #14) and 1.5 inch tube.

When the vertical and horizontal element diameters are the same, as in the first and third cases, the antenna dimensions are not considerably different, but note that the thinner diameter does result in a somewhat shorter top wire. Now look what happens when the element diameters are mixed; a difference of over 23 feet per side takes place between the second and fourth cases! The case of the thin vertical element, and the wide diameter horizontal element clearly provides the antenna with the most compact dimensions; while a thick vertical element and thin top-wire element produce the opposite effect—unfortunately!

A thin vertical element, while desirable from a size standpoint, may not be optimum from a bandwidth standpoint. As might be expected, the 2:1 SWR bandwidth is narrowest when the elements are made of thin wire, and broadest when the elements are made of tubing. These bandwidths will increase as greater ground loss (R_g) is introduced into a real antenna system. The bandwidth—narrow or wide—should not be of any consequence when a transmatch and low-loss feed line are employed at the station.

Multiple-Wire Flattop Antenna

The previous example shows that the horizontal top-wire length can be reduced significantly if its diameter is made large compared to the vertical element. In fact this may not be practicable. Conductors of large diameters can be heavy and difficult to support over a wide span, but the same benefit can be realized by using multiple parallel wires of smaller diameter. The more wires used, and the wider they are spaced, the greater the span reduction will be. The parallel-wire configuration changes the transmission line impedance (Z_S) such that it provides the required loading capacitance in a shorter length.

Fig 6A shows a simple, multiple-wire flattop using three parallel wires, each spaced 1 foot apart. For direct comparison, I've used a 44-foot vertical element tuned to 1.825 MHz. In this antenna the three parallel wires are connected together at the center and at each end. Each wire carries nearly identical current and maintains the same phase relationship with its adjacent wires.

This flattop arrangement produces an antenna with a total span of 78.8 feet, an R_r of 9.7 Ω, and a bandwidth of about 38 kHz. When compared to the single-wire equivalent, with a 134.4 foot span, the width is reduced by about 40% with almost no change in R_r or bandwidth! Such a design permits a 160 meter flattop in a space not much larger than a 40 meter dipole.

Fig 6B shows another three-wire flattop with one important difference; only the center wire connects to the vertical portion. In this configuration, the required horizontal span is reduced to 132.8 feet. This is not much difference in length, but it is quite different from a current distribution standpoint as you will see.

If the antenna of Fig 6B is momentarily opened at points "a" and "b" little change in resonant frequency will be noticed. (It may move up slightly.) That is because the current at these points is nearly zero and opening them has little affect. The majority of the current flow is in the center wire, and it divides as it reaches the end of its length. It then folds back in the outer wires flowing in the opposite direction. The outer wire currents are thus out of phase with that in the center wire, but drop off rapidly so their interaction is minimal. Performance of this design is virtually identical to that of Fig 6A.

Using the same method described for the single-wire flattop, I developed a similar V/H graph for a three-wire flattop, like that of Fig 6A. Since most applications will likely use the shorter vertical lengths, the graph shown in **Fig 7** is expanded like Fig 4. This graph represents a 25,000:1 l/d ratio, and a length-to-spacing (l/s) ratio of about 132:1. Once again, for purposes presented earlier, the spacing is also referenced to a full quarter wavelength.

A nice feature of the flattop is that for a given vertical element length the feed-point R_r is essentially independent of the top-wire design. Single and multiple-wire flat-tops, of most any design, will result in nearly identical R_r—provided the wires lie in the same plane or have the same slopes to them. This is shown in Table 1 for single-wire flattops when different top-wire lengths are employed. When the three-wire designs are used, they all have nearly equal R_r—about 9.8 Ω.

If the flattop spacing, shown in Fig 6A, is stretched and the ends are pulled in, so that a square is formed, a conventional top-hat structure is created. With appropriate sizing, the top-hat and flattop produce equivalent top loading and the same R_r. Note that the multiple-wire flattop can employ as many wires as can be practically accommodated resulting in even greater span reductions.

Coaxial Line Radiator Flattop

In a previous article, I suggested using an 80-meter dipole as a top-loaded vertical (flattop) on 160 meters.[10] This is accomplished by shorting the feed line at the end nearest the transmitter, and driving the feedline (the vertical) and dipole (the horizontal top wires) against a ground-return system. The coaxial line becomes the actual radiator. Typically, the coaxial line is whatever length happens to be in the installation and that length may not be optimum for this particular application. That may result in an inability to achieve resonance on the band, and the radiation pattern may differ from expectations. Therefore, give some thought to the antenna dimensions before attempting this approach.

Let's replace the 44-foot vertical wire with a 44 foot length of coax. The coax ends nearest to ground are shorted together, while the other ends connect across the dipole center insulator. It may not be obvious, but in operation the coax is now serving a dual role. First, the actual radiation takes place along the outside of the coax braid. Second, the inside of the outer braid and the outside of the inner conductor form a transmission line stub. This can best be visualized with the help of **Fig 8A**. Here a single wire rep-

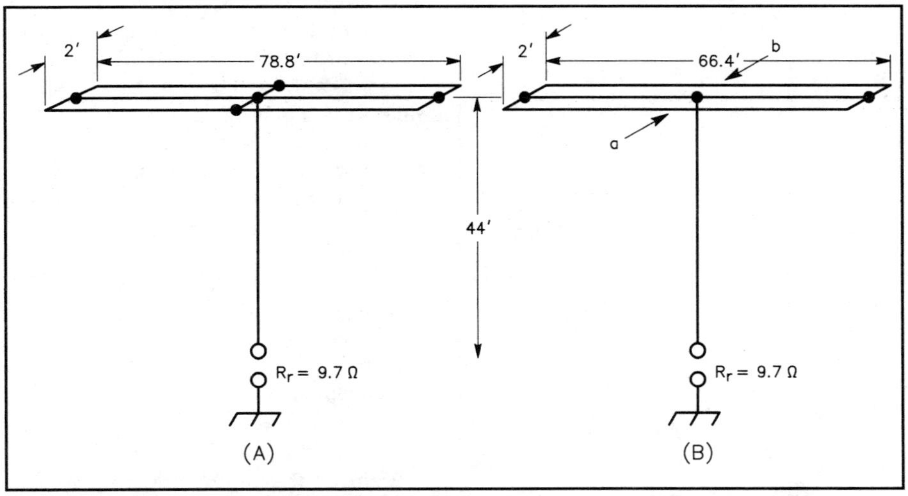

Fig 6—The three-wire flattop. At A, all horizontal wires are connected at the center and at each end. At B, only the the center wire connects to the vertical portion. All elements are constructed of #14 wire.

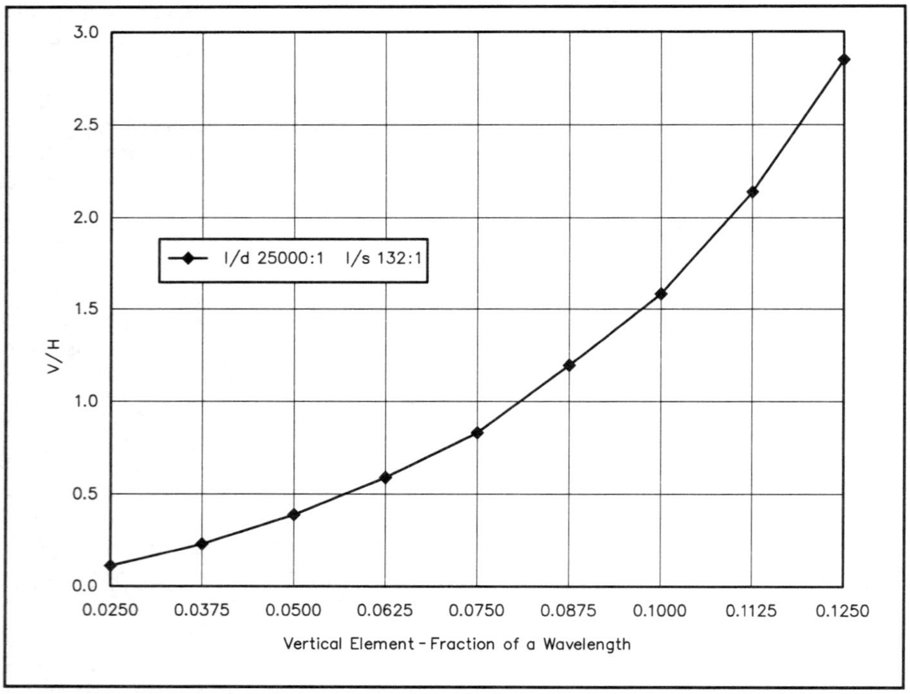

Fig 7—V/H ratio for the three-wire flattop.

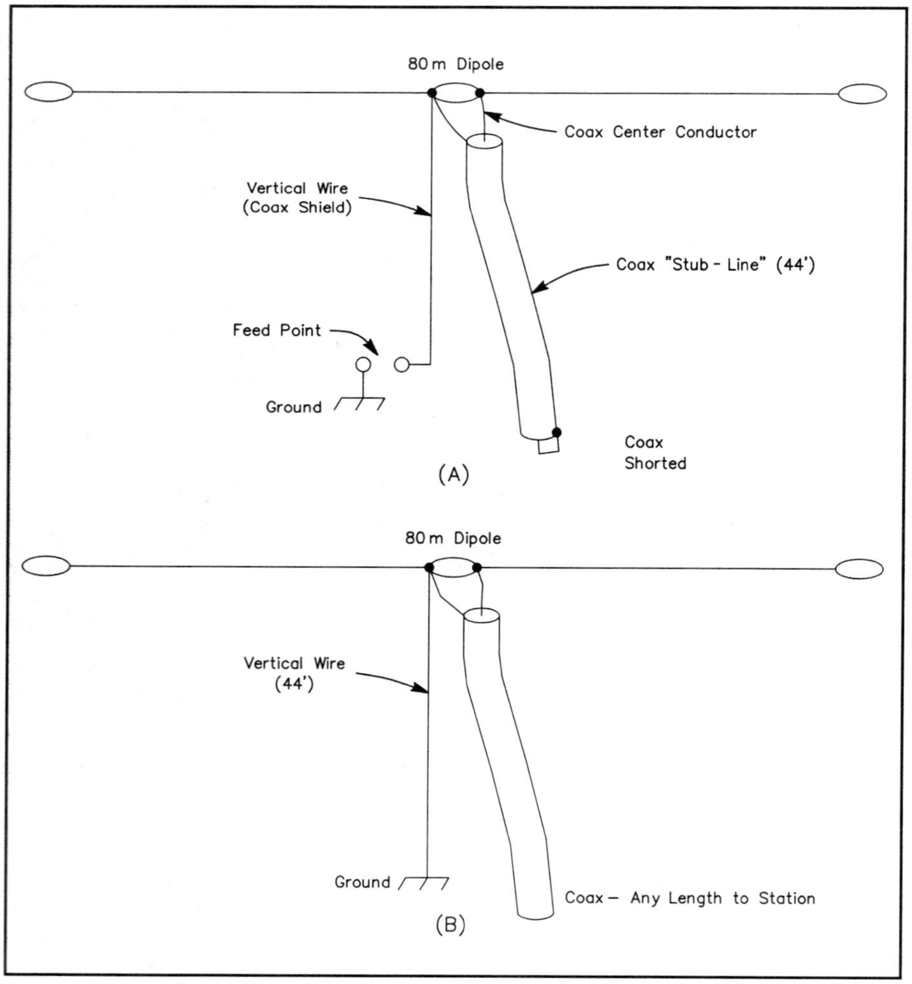

Fig 8—At A, *EZNEC* model of the 160/80-meter two-band antenna. At B, a design based on Eichenauer. See text.

resents the radiating coax braid, while a parallel coaxial line represents the stub. This is exactly the way it can be modeled in *EZNEC*.

A 44-foot coaxial stub, made from RG-8 or RG-58, will have an impedance of about 52 Ω, and have a velocity factor (VF) around 0.66. Divide 44 feet by the 0.66 VF, to find an electrical length of 66.7 feet—approximately 1/8 wavelength on 160 meters.

A 1/8 wavelength shorted stub will yield an inductive reactance (X_l) equal to the line's surge impedance (Z_s), or in this case; 52 Ω. Since the radiating braid connects to one side of the dipole, the stub lies in between it, and the other side of the dipole. Therefore, one side of the dipole has the equivalent of a 52 Ω inductor in series with it.

The addition of this inductor on one side unbalances the top-wire currents such that they are no longer equal and opposite. This effect results in some minor pattern distortion in the form of an increased horizontally polarized field component. Additionally, the inductor loads the antenna and causes it to resonate at a lower frequency, and reduces the antenna's bandwidth.

For this example, where the inductive reactance (X_l) is relatively low, the effects are small and the horizontal field increases by just a few decibels. Typically, it is down about 30 dB from the vertical field. But as the stub is made longer, X_l increases and the loading and pattern distortion effects become more pronounced. To compensate for the added loading, the top-wire lengths must be reduced to reestablish resonance on 160 meters. At the same time, the top wires are becoming too short for resonance on 80 meters.

Elimination of the stub's inductive effect can be achieved by placing a capacitor between the center conductor and braid at the feed point. The coax braid is then driven against ground. For the 44 foot, eighth-wavelength example, the capacitor should have a reactance (X_c) equal to the impedance of the transmission line, eg, 52 Ω. Connected in this fashion, the 52-Ω X_c will appear as a short (0 Ω) across the dipole insulator, completing the top-wire connection. Since high voltages can develop across the capacitor in this application, one should assess the voltage and current handling requirements to make certain the component can withstand the applied levels.

Use of a balanced feed line, such as open-wire or ladder line, eliminates the stub problem altogether. Ideally, the current divides equally between each side of the balanced line and flows to the corresponding top wire. Each side sees the same loading and no stub is formed.

Probably the biggest drawback to using this approach for a two-band antenna is that the ideal feed point R_r is significantly different for each band. For the previous ex-

Antenna control box used at AC7A. The box contains relays and a broadband torroidal matching transformer.

ample R_r will be about 10 Ω on 160 meters and 40 Ω on 80 meters. Even though ground loss resistance will raise the total antenna resistance on 160 meters, this suggests the use of an impedance matching network for 160, and then the ability to switch it out when using 80 meters. A better approach is one offered by Eichenauer in his 1983 *QST* article.[11] The basic layout is shown in Fig 8B. Although similar in design to the coaxial flattop, the feed line now only serves to feed the antenna. The antenna feed-point impedance is transformed to a value closer to 50 Ω on 160 meters, and is already near that on 80 meters. No complex matching or switching required!

Matching

A few words need to be said about matching the flattop's low feed point R_r. Fig 5 clearly shows that R_r will usually be much lower than the 52-Ω coax impedance often used in Amateur Radio applications. Even when R_g is summed with R_r, the total will likely result in a SWR that exceeds the 2:1 that most transceivers will tolerate. This can be handled in a number of ways: using a broadband multi-tap, autotransformer of the kind presented in the author's previous article, or a Sevick multi-impedance unun.[12,13] Similarly, a simple LC network such as an L network can be employed. Their design is covered in *The ARRL Antenna Book.*

Probably the easiest feed method to apply is to feed the antenna directly with a low-loss coaxial cable such as Belden 9913 or International 9089, and then rely on a transmatch to provide the 1:1 SWR. This would even relieve one from having to precisely tune the antenna for lowest SWR. Yes, the SWR may be high 3, 5, or even 10:1, but the losses will almost always be tolerable if the feed line length is kept reasonably short. This is easily realized at the lower frequencies where this type of antenna is most likely to be used. Operating with high SWR and high power levels can lead to high voltage levels being developed within the transmission line. But the voltage only increases by the square root of the SWR, over that of the 1:1 SWR condition; so in most cases the cable will be able to withstand the higher peak voltage.

Conclusions

The flattop antenna offers another possibility for an efficient, reduced-size antenna that should appeal to those who can't erect full-sized dipole or vertical antennas, but still want a low-angle radiator. Even though the text concentrated on the lower bands, these principles can be applied equally to antennas for the higher HF bands and beyond. Putting this in perspective, one can imagine an 11-foot, under-the-eve flattop for 40 meters, or a 2- or 3-foot 10-meter flattop barely visible above the roofline, or even a much improved replacement for your VHF hand-held's rubber ducky!

Acknowledgments

In particular, I would like to thank Ken, KO6RK, for his on-going e-mail discussions about various subjects presented in this article. And also Charlie, W4MEC, who raised the question about the difference in shorting, or not shorting the wires at the feed-point of the three-wire flattop.

A special "thank you" to my friend Jim, W7LM. He has always been very supportive of my writing efforts, and he comes to my rescue when I am having computer-related problems.

Notes

1W. J. Byron, "Short Vertical Antennas for the Low Bands: Part 1," *Ham Radio*, May 1983, p 36, Fig 1A.

2An impedance matched condition is assumed in each case.

3Richard C. Johnson and Henry Jasik, editors, *Antenna Engineering Handbook*, 2nd ed., (New York, McGraw-Hill, 1984), p 26-5.

4Thomas Kuehl, "Build Efficient, Short Vertical Antennas," *QST*, March 1998, p 39. A few readers suggested that there is more to the radiation efficiency than this simple equation involves. They pointed out that the imperfect ground surrounding the antenna, out many wavelengths, leads to what is referred to as "suck-in," a product of the Brewster effect. It results in signal cancellation at angles below the critical Brewster angle (see Note 9). If this effect is included in the equation, it would be equivalent to multiplying by a k factor (k < 1, at low angles). But there is little one can do to change the earth characteristics out tens or hundreds of wavelengths from the antenna, so one can realistically only affect R_a (by design) and R_g (by supplying the best possible return). Therefore, the equation is accurate considering those factors that we have some control over.

5Roy Lewallen, *EZNEC Users Manual*, Version 2.0, 1997, p 55.

6Modeling was accomplished using *EZNEC*, *MININEC* and high-accuracy grounds. When using the high-accuracy ground, 16 quarter-wave radials with tapered segments were employed. This ground type is recommended when dealing with horizontal wires at a height less than 0.1 λ. The *MININEC* ground type is an ideal ground, and allows the ideal radiation resistance (R_r) to be determined. In the model, I used a frequency of 1.825 MHz, an average ground conductivity, G=0.005 S/m, and an average dilelectric constant of 13.

7Kuehl, p 42, Table 1. Efficiency comparisons are made for base and center-loaded verticals. Refer to p 43, Table 2, example 1, for the flattop.

8Ibid, p 41.

9Capt Paul H. Lee, USN (Ret), *The Amateur Radio Vertical Antenna Handbook*, 2nd edition (Hicksville, New York, CQ Communications, 1996), Chapters 2 and 12. *Author note–this is an excellent source of information on vertical antennas.*

10Ibid, p 42.

11Carl Eichenauer, "A Top-Fed Vertical for 1.8 MHz–Plus 3," *QST*, September 1983, pp 25-27. For more information on feeding an antenna at the top-hat, see the article in note 4, pp 40 and 41.

12Thomas Kuehl, "An Efficient Multiband Vertical for 160 through 20 Meters," *QST*, October 1998, p 45. Also see, "Feedback," *QST*, November 1998, p 75, for a correction.

13Jerry Sevick, "A Multimatch Unun," *CQ*, April 1993, pp 28-30.

A Hanging 10-Meter Vertical

By Sam F. Kennedy, KT4QW
57 Huxley Pl
Newport News, VA 23606

For hams without a tower, but who do have at least one tree, this simple, low-cost, low-profile radiator may be the right choice!

This antenna was developed to encourage and support those members of the Peninsula Amateur Radio Club of Hampton Roads, Virginia who have only phone privileges on 10 meters. Since many new hams have limited radio resources, a good, easy-to-use 10-meter antenna is a handy thing to have. In our region of Virginia, most lots have tall pines and oaks. We designed an antenna that would cover the lower part of the 10-meter band and that can be installed by simply hanging it in a tree. This article describes the theory of operation, construction and tuning of this simple antenna. Several new hams have installed and used this antenna with good results.

The antenna has omnidirectional characteristics and suits the needs of 10-meter communication for both distant and local coverage. Although I have no facilities for measuring the antenna's gain, it consistently gives more than an S unit better received signal than my reference quarter-wave vertical. A hanging vertical is not the ultimate antenna, but you will be pleased with its performance! And it doesn't need radials. The design also lends itself well to a vacation antenna.

I have constructed more than 25 antennas using this same basic design. I've found this design effective for constructing antennas for 2 through 20 meters. This article pertains only to the 10-meter model. While construction is economical and fairly easy, it is a bit tricky to tune and is a time-consuming task without an antenna analyzer such as the MFJ-259. Tune-up and test procedures are included here.

To install this antenna, you need to get a line over a high tree limb. The 1998 Edition of *The ARRL Handbook* (page 20.7) has a good description of the Wrist Rocket slingshot method of hanging things from trees. You can usually place a line over the desired tree limb within three tries. It's not a tower, but it's the next best thing!

Theory of Operation

This antenna is electrically similar to several commercial models of fiberglass "stick" type antennas, but is packaged differently. This design provides a 50-Ω match over a reasonable bandwidth around the design frequency. SWR is typically 1.2:1 or better from 28.2 to 28.6 MHz. These antennas have been used at the 150-W level extensively, but should be able to handle 1.5 kW easily.

The antenna circuit is basically a half-wave radiator—similar to a half-wave dipole, but fed at the end rather than the center. Electrically, it is much the same as the classic J-pole antenna. The J-pole uses a quarter-wavelength shorted stub to accomplish impedance matching, whereas the hanging vertical uses an inductive link coupled to a parallel-resonant LC network. Since the significant part of the radiation from this antenna is from the half-wave section, it has no gain as over a dipole, and it gives an omnidirectional pattern.

The advantage of the hanging vertical over the J-pole is that the hanging vertical is one-third shorter than the J-pole, is much easier to weatherproof and is simpler to manage mechanically—especially as the length increases in the HF range. The hanging vertical's radiation characteristics and efficiency are approximately the same as those of a J-pole.

Fig 1 shows a schematic representation of the antenna, where C1 and L2 form a parallel-resonant circuit at the operating frequency. L1 inductively couples to the parallel-resonant circuit. C2 capacitively couples the half-wave radiating element to the parallel-resonant circuit. Capacitance and inductance values, experimentally determined, provide optimum coupling. The matching network provides a good match from the antenna's high impedance to 50 Ω.

The bottom end of the network is connected to the transmission line through a female UHF connector. Connector sealing is critical to long-term antenna reliability.

Construction

The construction sequence for the hanging vertical is as follows:

- Build the parts for the coil assembly
- Build the parts for the dual trombone tuning capacitor
- Prepare the parts for the PVC capsule
- Cut and fasten the radiator wire to the PVC capsule
- Assemble the parts
- Tune-up

The hanging vertical is constructed of widely available, low-cost materials. Although the total cost of material and parts for this project is low, the copper and PVC pipe/tubing must be purchased in minimum

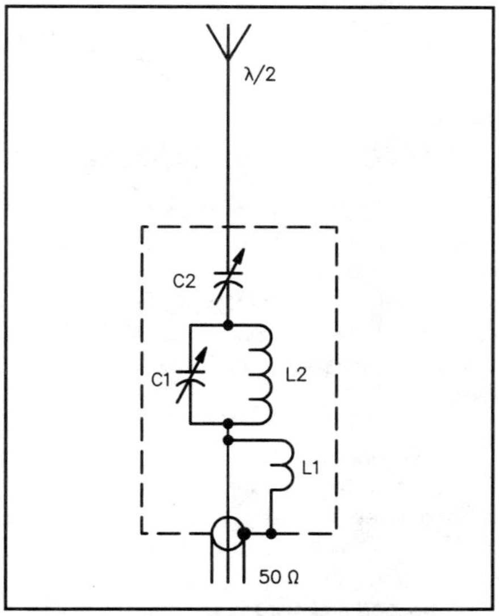

Fig 1—Schematic diagram of the hanging vertical. L1 couples energy from the antenna connector to the parallel-resonant circuit formed by C1 and L2. C2 couples RF to the radiating element. C1 and C2 are "trombone capacitors" that share a common plate, which is formed by a short section of copper pipe.

quantities, typically 5 or 10-foot sections. Unless you have access to suitable scrap materials, it makes sense to organize a club project and build several units.

Although it's not complicated, building this antenna does require basic mechanical skills as well as a general understanding of the electrical considerations. As with any project of this sort, many variations may be incorporated without degrading performance. **Fig 2** shows an overview of how the feed-point end of the antenna goes together. **Fig 3**, Fig 1 and Table 1 also help with construction.

Coil Assembly, Outer Coil

This is the larger coil, L2. Using #12 enameled copper wire, wind 17 close-spaced turns on a scrap PVC pipe as a temporary form. The outside diameter of the pipe I use is 0.625 inch. The wire measures 0.085 inch in diameter, including the enamel coating. The tuning capacitor has enough range to compensate for small variations in construction. This coil wound for the 10-meter band requires approximate 4 feet of wire; I cut 5 feet for convenience. See Fig 3 for details.

Coil Assembly, Inner Coil

This is the smaller coil, L1. Using #12 enameled copper wire, wind 7 close-spaced turns on PVC tube as a temporary form. The outside diameter of the form is 0.375 inch. Approximately 18 inches of wire is required. Cut a 24-inch length to make winding and handling easy. Note: Wind this coil in the opposite direction as the large coil (L1). Route the end back through the center of the windings. This requires a tight bend made with long-nose pliers. Wrap the pliers' gripping surfaces with a couple of layers of tape to reduce the possibility of scarring the enameled wire.

Connecting Link

To allow connection of the lower portion of L1/L2 to the tuning capacitor, fashion a piece of #12 wire in a 2.5-inch length, as shown in **Fig 4**.

Coil Spacer

Cut a piece of 0.020-inch polyethylene sheet (cut from a plastic milk jug) $1^1/_2 \times 2$ inches. This spacer is later installed between L1 and L2 to assure adequate and stable spacing between them.

Scrape and Tin

Remove the enamel from the end of the wires to allow soldering during assembly. Tin these parts of the wires to allow ease of soldering later. Using a sharp pocketknife or other scraping tool, carefully remove the enamel on the magnet wire without nicking the wire. All traces of enamel must be removed from the surface to permit proper soldering.

Bend to Shape

Bend the wire parts into shapes that will allow effective assembly. Use sturdy long-nose pliers for the bending operation. Wrap the pliers gripping surfaces with a couple layers of tape to prevent damage to the wire. This operation is not critical and it can be adjusted later if necessary.

Trombone Capacitor, Outer Section

Using a hacksaw, cut a 5-inch section of $^1/_2$-inch copper water pipe, lightweight type M. (Pipe is specified in inner diameter, so the $^1/_2$-inch pipe measures 0.636 inch OD and 0.575 inch ID.) Do not use a tubing cutter to cut the pipe, as it crimps the cut inward, causing a fit problem. Debur the inside surface of the unit using a pocket knife or deburring tool. Debur the outside of the unit with a file or sandpaper. Using a propane torch or heavy duty soldering iron, flow solder down one side of this unit to prepare for soldering on the coil unit later. It helps to use a small amount of rosin (not acid) flux.

Trombone Capacitor, Inner Sections

Using a tubing cutter, cut one 2.5-inch section and one 1.5-inch section of $^1/_2$-inch soft copper tubing. Make sure you get *tubing*, and not pipe, for these pieces. The tubing fits into the outer piece with just enough clearance to accommodate the plastic di-

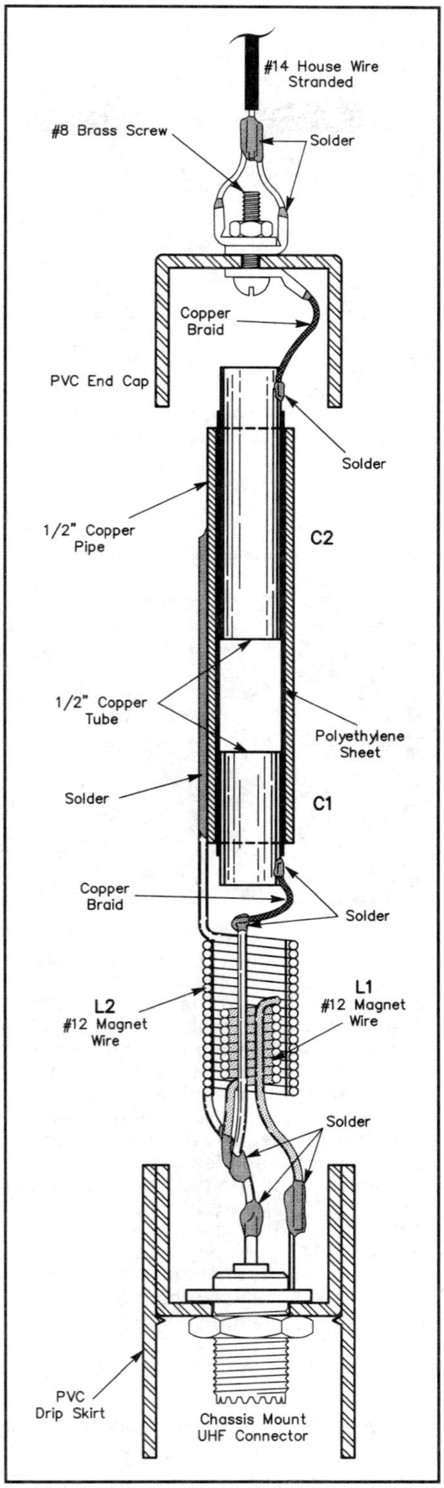

Fig 2—An overview of the feed-point assembly. The main body of the PVC pipe that encloses this assembly is omitted for clarity. L1 sits inside L2. L2 is soldered to the antenna connector at one end and to the copper pipe that makes up the common plate of C1 and C2 at the other. L1 is soldered to ground at one end and to the center conductor of the coax connector at the other. At the coax connector's center conductor, L1, L2 and a 2-inch wire are also soldered together; the 2-inch wire is routed through the center of L1 and L2 to the copper tubing that forms the inner plate of C1.

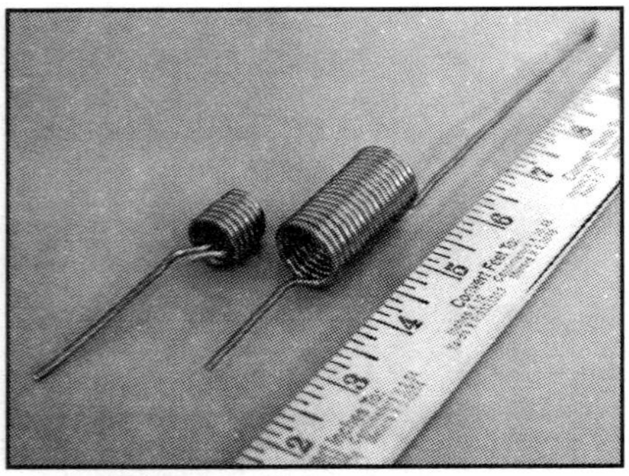

Fig 3—Coil-winding detail for the hanging 10-meter vertical. One lead of the smaller coil, L1 (at left), must be routed back through the coil body after winding is complete. *(Photos by the author.)*

Table 1

Materials

Component(s)	*Description*
L1	0.8 turns, #12 enameled magnet wire; coil is 0.6" outside diameter.
L2	17 turns, #12 enameled magnet wire; coil is 0.8" outside diameter
C1	1.5 inches × $^1/_2$-inch copper tubing (capacitor slide)
C2	2.5 inches × $^1/_2$-inch copper tubing (capacitor slide)
C1, C2	5.0 inches × $^1/_2$-inch copper water pipe (lightweight type M)
C1, C2	2 × 5 × 0.02-inch polyethylene dielectric sheet (cut from milk jug)
Separator	1.5 × 2, 0.02-inch polyethylene dielectric sheet (cut from milk jug)
End Caps	$^3/_4$-inch PVC pipe cap (2 used)
Drip Skirt	1-inch PVC pipe coupling
Barrel	$^3/_4$-inch × 10-inch schedule 20 PVC water pipe
Connector	Female UHF chassis connector, single-hole mount
Top Screw	#8-32 × $^3/_4$-inch brass round head screw
Nut	#8-32 brass nut
Radiator	18 feet, #14 stranded THHN wire (vinyl insulation)
Terminal Lug	Uninsulated crimp lug for #8 screw and #14 wire (2 used)

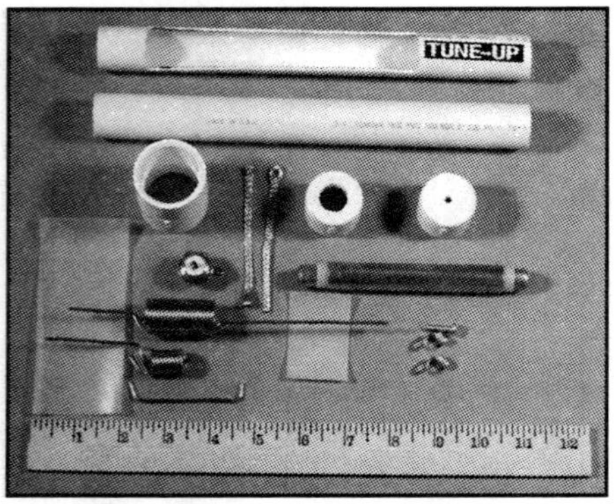

Fig 4–Part-preparation details for the 10-meter vertical. The plastic sheets, cut from a milk bottle, form the dielectric of the two-section trombone capacitor; the finished trombone-capacitor is shown in the right center of the photo. Also shown here are the coils; pre-bent coil-link wire; braid straps used for interconnection; bent solder lugs used to attach the antenna wire to the assembled matching section; and, at the top, the PVC pieces prepared for assembly. The upper section, marked TUNE UP, is used during tuning to facilitate adjusting the matching network. Then, the solid tube just below it replaces this section during final assembly.

electric. You may use a hacksaw for cutting these pieces; however, a tubing cutter works best here because it crimps the tubing inward and forms the ends to the proper shape for easy insertion into the outer piece. If you cut these pieces with a hacksaw, file or sand the outer edges to ease fitting them into the larger piece. Using a large-tip soldering iron, flow solder onto the outer surface at the end of each of these pieces to allow soldering the connecting braid to them. Tin about a $^1/_4$-inch spot.

Dielectric

Using sheet-metal shears, sewing shears, or a sharp knife, carefully cut out a $2^1/_2 \times 5^1/_2$-inch rectangle of 0.020-inch polyethylene. This will form the dielectric between the inner and outer capacitor plates (cylinders). This dimension allows a $^1/_4$-inch overlap, which will protect from arcing and will provide a firm fit between the inner pieces. This material is cut from the flat section of a clean, one-gallon plastic milk jug. To avoid melting the polyethylene dielectric, do not solder with it inserted.

Connecting Braid

Cut two 4-inch lengths of $^3/_{16}$-inch tinned copper braid. One piece of braid is used to connect the top slide and the other to connect the bottom slide.

Top Piece

Apply solder to one end of the braid—use just enough to keep the braid in place. Using a pointed object such as a small nail or a meter test probe, carefully separate the braid and make a hole in the braid about $^1/_4$ inch back from the soldered end. Enlarge the hole until a #8 machine screw can be easily passed through. Press the braid flat and sparingly flow solder over the entire hole area. Use diagonal cutters to round the end so that a neat lug is formed. This will later be used to connect the capacitor to the radiating element.

Bottom Piece

Save this piece for use during final assembly.

PVC Outer Enclosure, Bottom End Cap

Drill a carefully centered pilot hole through the end cap using a $^5/_{32}$-inch drill bit. Carefully hold the end cap in a vise or using suitable locking pliers. Drill a $^5/_8$-inch hole using a hole cutter or wood bit. This hole is for mounting the coaxial connector.

PVC Outer Enclosure, Top End Cap

Drill a hole through the center of the end cap, using a $^5/_{32}$-inch drill bit. This hole is for the #8 brass machine screw that connects loading capacitor C2 to the radiating element. This hole is slightly undersized to cause the screw to fit tightly.

Sleeve

Using a hacksaw or hand saw, cut two 10-inch sections of $^3/_4$-inch schedule 20 (thin-wall) PVC pipe. Keep cuts as straight as possible at 90°. One piece will be used temporarily for tune-up, and the other for permanent assembly. Using an appropriate tool, cut an opening into the side of the tune-up section to allow access to the tuning capacitor. If you are making more than one antenna, only one tune-up section will be necessary.

Connector

Check to be sure that the connector fits properly into the bottom end cap. Make certain that the connector can be locked into position. If it becomes loose after the final assembly, you will have a difficult problem to solve.

Drip Skirt

No modification is required to this 1-inch PVC coupling. It will be glued onto the bottom end cap after final tune-up.

Radiating Element

Measure an 18-foot length of #14 stranded-copper THHN wire to form the half-wavelength radiator. The electrical length of the radiator is approximately 16.47 feet at an operating frequency of 28.4 MHz (length [feet] = 468/f [MHz]). The electri-

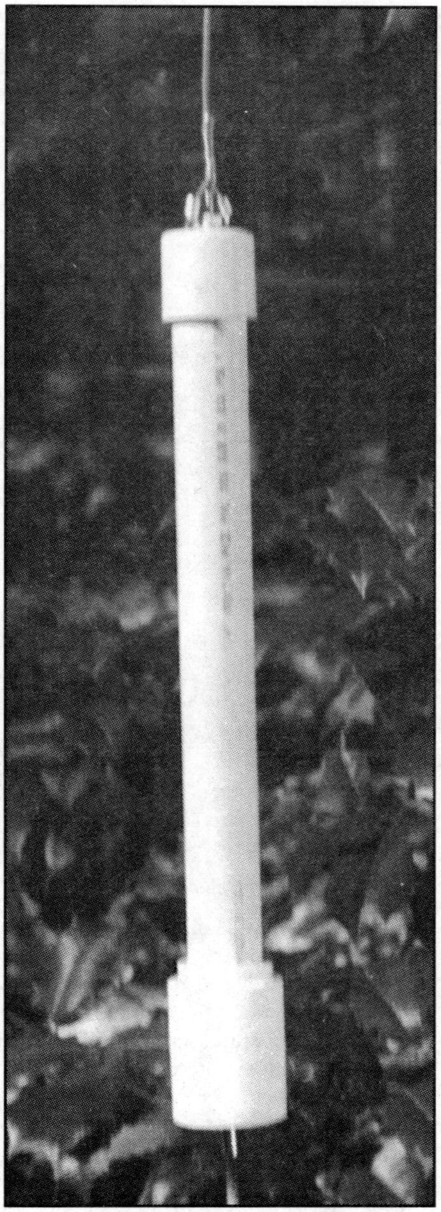

Fig 5—The finished feed-point assembly. The antenna wire is attached to the matching assembly by two solder lugs, which reduces the likelihood of flex-related failure in this connection. The large PVC pipe section, glued to the bottom of the main assembly cover pipe cap, protects the feed-line connector from rain and wind-driven moisture.

cal length is measured from the bottom of the trombone tuning capacitor to the end of the top fastening loop. Strip about 4 inches of this wire to connect it to the two terminals on the top end cap. Form a hanging loop at the top of approximately 6 to 12 inches. Twist the part of the wire that is doubled back. Note the length of the radiating element is adjusted by changing this loop. The doubled-back portion of the wire does not contribute to the electrical length.

Assembly

Slide the inner coil into the outer coil, along with the polyethylene spacer in between. Install the connecting link to the junction of the inner and outer coils at the lower end of the assembly. Using a scrap piece of fine copper wire to hold the parts in place, wrap the connection together tightly and solder it.

Using a heavy soldering iron or propane torch, solder the upper end of the larger coil to the outer part of the trombone capacitor. Make certain that the dielectric is not in place during this operation so that it won't melt.

Solder the braid sections to each of the capacitor inner slide units. The braid with the eyelet is used for the upper section, and the plain braid for the lower section.

Curl the dielectric sheet and insert it into the outer capacitor, leaving approximately $^1/_4$ inch exposed at each end. Slide the inner pieces into the outer pieces. Dress the lower braid for connection to the wire link. Trim and solder the lower braid.

Remove the upper tuning section from the trombone and pass the #8 brass machine screw through the eyelet. Screw it through the pilot hole (top PVC end cap). This forms a weathertight seal.

Remove the insulating plastic from the two lugs and bend the lugs as indicated in **Fig 5**. Mount them on the protruding screw using the #8 brass nut. Loop the stripped portion of the radiating wire through the two lugs, wrap, and solder. Using two lugs distributes the force over twice the area and prevents flexing that can easily break a single lug.

Solder the two lower leads from the coil assembly to the connector as indicated in Fig 1.

Tune-up

Install the coil assembly into the tune-up PVC enclosure, then install the top end cap with the radiating element attached. A press fit should be adequate to hold things together temporarily. Use tape at the pipe junctions to add strength, if necessary. Suspend the complete antenna vertically from a tree limb. Suspend the antenna so that the capsule end is at least $^1/_4$ λ away from any surrounding objects.

Insert the upper capacitor slide as far in as it will go without electrically shorting to the other lower section. Using an SWR analyzer, adjust the lower trombone capacitor slide until resonance is obtained, as indicated by a pronounced dip in SWR. (This is most easily done if the feed-line length is a multiple of $^1/_2$ λ.) If the SWR is greater than 1.2:1, adjust the two capacitors in small increments to improve the match. If not, try lengthening or shortening the radiator length by adjusting the connecting loop at the top. An SWR of 1.1:1 can usually be obtained. Note, however, that SWR changes with antenna height. If possible, check the SWR in the antenna's final position. When the antenna is tuned satisfactorily, carefully tape the connecting loop on the radiator to set its length.

Mix a small amount of epoxy and apply it to lock the lower copper slide in place. When the epoxy has hardened, carefully mark the upper capacitor slide's position so it can be reset to that point. Remove the tune-up sleeve and reinstall the coil assembly in the permanent sleeve. Glue the sleeve in the bottom end cap using PVC cement.

Install the upper copper capacitor slide, observing the depth marker applied during tune-up. Use a small amount of epoxy to stabilize this slide. (Reasonable amounts of clear epoxy do not affect antenna tuning.) Glue the top cap, with the radiating element attached, to the top of the sleeve using PVC cement.

Using a file or pocketknife, remove any protrusions from the side of the bottom end cap. Cement the 1-inch PVC pipe coupling to the bottom cap to form a drip skirt around the connector.

Your antenna is now ready for operation!

Antennas Here are Some Verticals on the Beach...

By R. Dean Straw, N6BV
Senior Assistant Technical Editor, ARRL
5328 Fulton Street
San Francisco, CA 94121

A DXpedition to Jamaica for the CQ World Wide Contest, "Field Day" Style

To me it was a contest adventure of a lifetime—I was to be a member of the team operating 6Y2A on the 1998 DXpedition for the CQ World Wide DX CW Contest. My 84-year-old father came along too. Although he's not a ham, for almost 40 years he's watched me doing radio and has helped me put up lots of antennas. On this trip, he had ample opportunity to put up and take down lots and lots of antennas!

As a result of a lot of planning, hard work and dedication, the 6Y2A team set a new world's record for the multi-operator, multi-transmitter classification. The world record was particularly remarkable since we accomplished it from the continent of North America, where contacts with the USA and Canada only count for two points, rather than the three points accorded QSOs made with other continents, such as Europe or Asia.

We operated from a rented villa on the north coast of Jamaica. This location had been carefully chosen for its clear, over-the-ocean shot towards Europe, the USA and Japan. While long-path directions toward the south were blocked by a nearby steep volcanic cliff, the spectacular short path more than made up for that deficiency! As hams are fond of saying, "We could practically *see* Europe from our front porch."

Located about an hour from Montego Bay, we were well removed from the usual tourist haunts. This meant that expenses were moderate rather than tourist-level. And most importantly for our mission, we didn't suffer from the electrical noise typical in a densely populated urban environment. We had rented two villas. One was for operating the contest; one was for sleeping. Since the villas were side-by-side we had a beach front of about 800 feet on which we could place antennas. The story of the planning and execution of those antennas is the real subject of this article.

Planning, Planning, Planning

Except for myself, the rest of the 6Y2A team had operated the CQWW CW contest as 6Y4A in 1997 from the same location. 6Y4A had come in second worldwide to 5V7A in Togo. The 5V7A team was intending to go all-out to defend their title in 1998, and the 6Y2A crew publicly announced that we were going to give them some heavy competition, even if only from a "two-point" country. Then we heard that there were going to be two other serious North American continental entries: the J6DX group was trying out the multi-multi category and so was a team headed by world-famous contester N6TJ, operating the superstation of TI2CF in Costa Rica. They were going to reactivate the TI1C call sign. I felt some irony in the situation, because the last time I had done a serious multi-multi from outside the USA was at TI1C in 1984, some fourteen years ago, together with fellow six-landers N6TJ, N6AA, N6ZZ and N6AW. I'll never forget the kindness and gracious hospitality of Carlos, TI2CF, on that trip. He had put together a really terrific station, with very big antennas.

The three teams from the North American continent were all gunning for a new North American record, but few among us hoped to seriously challenge the world's record, simply because we were operating from two-point land. The team at 5V7A was operating from the continent of Africa, where every European, USA or Japanese QSO was worth three points. Then, just before the contest we heard that there were some problems in Togo with the hotel from which 5V7A operated last year. This was really becoming an interesting horse race!

In 1997 the team at 6Y4A had engaged in what seemed then to be a daring experiment—they had used verticals near the saltwater, rather than horizontal Yagis on high towers. We intended to expand this approach in 1998.

What Elevation Angles Did We Need?

The first step planning an antenna system is to determine what elevation angles must be covered. Our operating credo could be summarized in the personal motto of Dave Patton, W9QA, the 40-meter operator at 6Y2A: "Be loud or be hosed!" We wanted to be loud in all areas around the world with large ham populations, especially on three-point continents. That meant that we needed to be loud, nay *very loud,* in Europe and Japan. Since the path to Japan from Jamaica is right through downtown Kansas City, we figured we might just as well be loud in the USA also.

Fig 1 shows an overlay of the enhanced elevation-angle statistical data from the 18th Edition of *The ARRL Antenna Book* on CD-ROM. This is for the 80-meter band from Jamaica to the continent of Europe. The bar graph shows elevation angles ranging from 1° to 27°. This statistical elevation-angle data was generated from thousands of computer runs using the *VOACAP* program. The data show the percentage of time a particular angle is open to Europe, for all months of the year and for all levels of solar activity. The biggest bar-graph spike occurs at an elevation angle of 12°, representing about 11.5% of all the times the 80-meter band is open from Jamaica to Europe. A secondary peak occurs at 7°, for almost 10.5% of the time when the band is open.

That these angles are quite *low* angles is not readily apparent from a casual glance at Fig 1, at least not until the other lines on the graph are viewed carefully. The line with the small black rectangles represents the elevation pattern response of a really large 80-meter antenna—a three-element Yagi placed 200 feet above flat ground. Its peak response is at a 20° elevation, and at 7° the response is down some 5 dB from the peak.

Next, look carefully at the line with the small, lighter-shaded ovals. This represents the response of a two-element vertically polarized parasitic array on the beach right next to the saltwater. This array, a two-element Yagi turned on its side to be vertical—if you will, a *Vagi*—is just a bit stronger at an elevation of 7° than its 200-foot high three-element horizontal cousin. At 3° the two-element vertical array is some 8 dB stronger than its lofty cousin and at 17° elevation it is down only about 4 dB. Note how "flat" the response of this simple vertical parasitic array is over the range from 1° to 30° elevation. This is actually one of its distinct advantages, since there are no *nulls* in the response, especially at higher elevation angles where signals often occur. We'll examine this important consideration some more later.

Let's examine the responses for some other types of 80-meter antennas. The line on Fig 1 with the small upward-pointing arrows represents the same kind of two-element vertical parasitic array, but this time located over "good ground," well removed from saltwater. This is typical for the kind of farmland found in parts of the US Midwest, with a conductivity of 5 mS/m and a dielectric constant of 13. This landlocked vertical array is down uniformly some 7 dB from its brother located at the edge of the ocean. The only difference between the two vertical arrays is the quality—or should I say, the *lack* of quality, so far as RF is concerned—of the ground. Sea water is hard stuff to beat when it comes to making loud signals with verticals.

Next, we have what most people would consider a darned good antenna, a dipole located 100 feet above flat ground. At 7° elevation, the dipole is down about 12 dB compared to the saltwater verticals and about 6 dB down at 12°. In other words, a darned good antenna, something most hams could only dream about, would be about 2 to 3 S-units down from the 80-meter antenna planned for 6Y2A for the range of elevation angles needed to cover Europe. (This assumes that a typical modern transceiver is calibrated for about 4 dB per S-unit.)

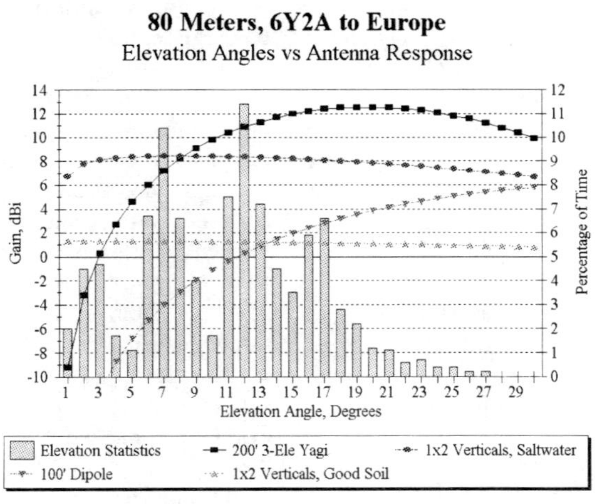

Fig 1—Statistical elevation angles needed for all months, all levels of solar activity, from Jamaica to Europe, compared with typical elevation-plane patterns for some 80-meter antennas. The array of two verticals (reflector/driver) located over salt water is superior even to a 200-foot-high three-element Yagi for angles less than 8°. Note that the same vertical array placed over soil with good conductivity is greatly inferior to the array over salt water, although the land-locked vertical would be superior to a 100-foot high flattop dipole until about 13° in elevation. Verticals do best, however, over salt water!

Fig 2 shows the another 80-meter plot, but this time for the path from Jamaica to Japan. The prevailing angles are even lower than for the path to Europe. The elevation response of a parasitic array on the beach would be even more advantageous on this path compared to even a 200-foot high three-element Yagi. By the way, this antenna is not just a figment of our fervid imaginations. Rumor had it that one of our competitors had just such an antenna ready to play during the contest. (It turned out that the antenna was tuned for phone and wasn't actually used during the CW contest.)

Fig 3 shows the elevation angles needed on 40 meters plotted together with the responses for two types of antennas. The one with black rectangles is a common type of contest antenna on 40 meters, a two-element horizontally polarized Yagi mounted 100 feet over flat ground. Compared to this is the elevation pattern of the antenna used at 6Y2A, a pair of "ZR" vertical dipole antennas by Force 12, with one dipole tuned as a parasitic reflector. This is a very interesting antenna in that it is only 15 feet high! I'll describe it in more detail later on.

Again, theory says that a two-element vertical parasitic array on the beach should be very competitive with a much higher horizontally polarized Yagi antenna. And like the situation on 80 meters, this setup would be even more favorable for the 40-meter path from Jamaica to Japan, where the prevailing angles are very low, peaking at 2° for some 22% of the time when the band is open. Theory says that the vertical array should be loud on this band, and it was loud.

Fig 4 shows the situation on the 20-meter band from Jamaica to Europe. For angles less than about 11°, a vertical parasitic array located at the seashore would have the upper hand over a four-element Yagi located 100 feet over flat ground. At 6Y2A, we ended up using the array labeled "2×2 Verticals" in Fig 4 on 20 meters for some very practical reasons, despite the fact that the array labeled "4 Halfwave Verts" exhibits about 2 dB more gain over the range of angles needed. I'll discuss the different types of vertical arrays shown in Fig 4 in more detail later on in this article.

Fig 5 illustrates the important 15-meter situation towards Europe. We all felt during the planning stages that 15 meters was going to be a very important band in 1998, where many QSOs were going to be made. Here again, vertical arrays on the beach should be very competitive with an installation using a four-element Yagi mounted 60 feet over flat ground. This is the sort of antenna that an ambitious DXpedition might install. We intended to operate six stations from 160 to 10 meters. If we didn't put up a lot of verticals, we would have had to install no fewer

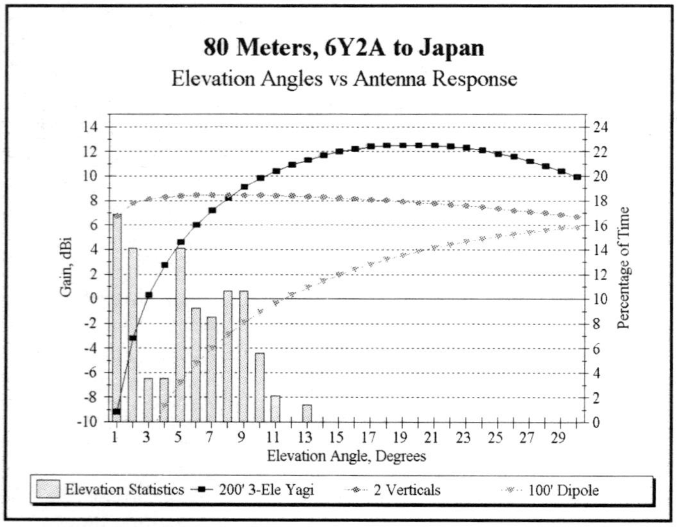

Fig 2—Statistical elevation angles for Jamaica to Japan on 80 meters, again for all months and levels of solar activity. The needed angles are much lower on this path than they are, statistically speaking, on the Jamaica-to-Europe path. Again, the two-element reflector/driver array over salt water has a decided edge over most of the elevation-angle range.

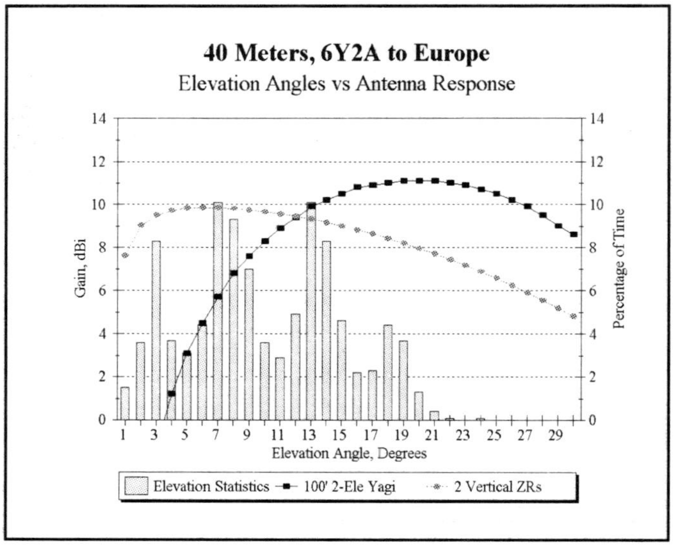

Fig 3—Comparison of needed elevation angles and typical antenna elevation-plane responses on the 40-meter path from Jamaica to Europe. Here, the antennas are a reflector/driver pair of 15-foot-high "ZR" verticals by Force 12 over salt water compared to a 100-foot high two-element Yagi. The vertical array over salt water is stronger for most of the range of necessary angles.

than three tall towers, probably even more to be really effective.

Note that the vertical array in Fig 5 should cover nicely all the angles needed from 1° up to as high as 26°, while the single 60-foot high Yagi on 15 meters would be at a considerable disadvantage at both ends of this wide range of angles. Although not shown, the situation towards Japan would involve even lower angles, with a peak elevation angle occurring at 1°. This would be the prevailing angle almost 24% of the time the 15-meter band is open from 6Y to JA. Statistically speaking, the highest possible angle to Japan would be 11°, which would occur only 1% of the time. Antennas with very low takeoff angles are vital to be really loud into Japan from Jamaica on 15 meters.

Fig 6 shows the 15-meter situation from Jamaica to all of the USA. Here we see that the range of elevation angles is surprisingly large, varying from 2° all the way up to 30°. Actually, this shouldn't be surprising, considering that the USA is a very large target area. The distance from Jamaica to the lower 48 states varies from a minimum of 575 miles to Florida, to a maximum of 3340 miles to Washington state. The vertical array labeled "2x2 Halfwave Verticals" covers this wide range of elevation angles much better than would a single Yagi at 60 feet, which would have degraded performance because of the null in its response from about 14° to about 28°.

This aspect of vertical performance—wide elevation coverage—is often overlooked in the Amateur literature. Avoiding such nulls is the reason why vertical stacks of horizontally polarized Yagis are becoming so prevalent among really serious contesters and DXers at their home stations. Putting up multiple stacks on a DXpedition would have required an effort beyond our enthusiasm for the sport of contesting.

Fig 7 shows the needed elevation angles and antenna patterns on 10 meters for the path to Europe. Once again, a vertical array on the seashore should be a potent performer over the whole range of angles needed—and it proved so in reality. The pileups calling us on 10 meters from Europe were truly phenomenal.

We also wanted to be loud on 160 meters into our targeted geographic areas. The elevation-angle statistics from the 18th Edition of *The ARRL Antenna Book* don't cover 160 meters, because *VOACAP* is not designed to operate reliably below 2 MHz. The team knew from experience in 1997 at 6Y4A, however, that a single 160-meter vertical on the beach could do very well. And we were prepared to up the ante for 1998!

The choice of what types of antennas to use at 6Y2A was set—it would mainly be verticals, backed up by a few rotatable horizontal Yagis at relatively low heights.

Why Verticals?—Some Other Parameters

Other than raw gain at low elevation angles, we had other important parameters to consider about the 6Y2A antennas. First, they had to be light and compact enough to qualify as regular carry-on baggage on the airplane. Remember, this operation was supposed to resemble a Field Day setup (minus generators, of course), not a typical major multi-multi contest station with huge towers and monoband Yagis.

We also had to take our transceivers, power amplifiers, computers, coax, and other peripheral equipment in our carry-on baggage. Luckily, the weather in Jamaica does not require much heavy clothing. Shorts and tee-shirts were perfect for our operation. I wore a set of blue jeans on the plane for "formal" occasions, and because the temperature in San Francisco when we left was about 50° F, just a bit different from 85° F in Jamaica.

Second, we didn't want to pay excess baggage fees, as we would have had to do if we were hauling hundreds of feet of tower, guy wires, rotators and large Yagis! Getting large antennas and towers through Customs has always been a challenge I've tried to avoid as much as possible—both arriving in a foreign country and returning to the USA.

Tom Schiller, N6BT, of Force 12 Antennas and Systems, made all of the antennas, some of them entirely custom-manufactured specifically for 6Y2A. Tom was one of our "secret" advantages. Not only is he a superb operator, but he managed to fit all 19 of the antennas we used into six fiberglass "golf-bag carriers." Each golf-bag carrier could transport individual pieces up to four feet in length, setting the size limit for each telescoping piece of aluminum tubing. Each item was carefully inventoried and packed so that each carrier weighed 69 pounds, just under the 70-pound limit for a carry-on item. One other advantage with golf-bag carriers—Airline personnel are very familiar with them, especially when your destination is a tourist spot like Jamaica. By the way, not only antenna parts went into the golf-bag carriers. They also held about 14 pounds of 6Y2A tee-shirts, all the tools and four large containers of Gatorade (another secret weapon in the tropics).

Bigger and Better

Experience from the 1997 6Y4A operation indicated that more gain would be desirable on 20 through 10 meters. On the higher-frequency bands, it is easier for others to erect reasonably competitive antenna systems. Remember, "Be loud or be hosed" was our operating credo. Fig 5 shows that a relatively modest station with a four-element Yagi at 60 feet would have a peak gain that is just a bit more than the two-by-two vertical array at 6Y2A. While such a horizontal Yagi at 60 feet would not have the same elevation response at very low elevation angles, it would still be a very competitive station for much of a typical opening to Europe. Fig 4 shows that more heroic efforts are required on 20 meters to be really competitive with our verticals on the beach, but nonetheless, a 100-foot high Yagi on 20 meters will be much more competitive than will an antenna at that height on 40 or 80 meters. The lower bands are where the verticals really shine in terms of dB per foot of height.

With the wide expanse of beach-front real estate available to us, we chose to use three sets of "two-by-two" vertical arrays for the all-important path into Europe on 20, 15 and 10 meters. The basic building block is a parasitic reflector and driver, spaced approximately 0.25 λ, tuned mainly for gain, with a compromise 9 dB front-to-back ratio. This first set of verticals (a "two-by-one") is fed in phase with another such set, spaced by about 0.6 λ. **Fig 8** is a bird's-eye view of the physical layout of this two-by-two array.

Each half of the phasing harness was made up of a 0.25-λ long RG-59 cable, in series with a 0.25-λ long RG-8X cable going to the feed point—enough to make up half the 0.6 λ physical spacing between the driven elements. The quarter-wave sections transformed 50 Ω at the end of the RG-8X sections to 100 Ω at the tee, where connecting two such impedances in parallel resulted in 50 Ω. From the tee to the shack we used RG-213 coaxes (or RG-8X on the lower bands). The coax run to the 160-meter vertical array (yes, we used a parasitic array of two verticals on 160 meters too) were rather long; about 500 feet long, in fact.

For 20 meters we chose to use vertical elements that were 1/4-wave high. For 15 and 10 meters half-wave long vertical dipoles were used. We did this because half-wave long dipoles compress the elevation-plane pattern down further towards the horizon than do quarter-wave elements, yielding additional gain. **Fig 9** shows elevation-plane patterns comparing four quarter-wave-long and four half-wave-long dipoles in the two-by-two array shown in Fig 8. The two curves cross over at about 25° elevation and the peak gain of the array using half-wave long elements is a significant 2 dB more than the array using quarter-wave verticals.

However, there is a physical problem associated with half-wave elements on 20 meters—they would have to be about 35 feet long, almost the same length as the shortened 80-meter vertical elements (which will be described in detail later on). For this reason, the 20-meter two-by-two arrays at 6Y2A used quarter-wave high vertical elements, with two elevated radials. Only two radials, you say? Yes, only two radials were used for each quarter-wave vertical element. Computations showed that there would be a negligible effect on the gain or the pattern. The system must have worked, because we had more than 4200 QSOs on 20 meters. You don't achieve these kinds of numbers without being really loud.

The half-wave vertical dipoles on 15 and 10 meters required no radials, a definite advantage when trying to walk through the forest of 28 verticals on the beach. **Fig 10** is a photograph looking down the beach showing some of these verticals. It was bad enough having to gingerly hop over the eight 20-meter radials, which were about 3 feet above the incredibly sharp volcanic rock and coral. It was, however, just a little weird to see that the 15-meter arrays were taller than the 20-meter arrays located nearby.

Besides compression in the vertical pattern, what else happens when a two-by-two array is used to increase the gain of a system? **Fig 11** shows the azimuth-plane response of a 15-meter half-wave dipole array at a 10° elevation angle, compared to the pattern of a single set of verticals arranged as a reflector/driver pair. The peak gain of the four-vertical array is increased by about 4 dB compared to the two-vertical array, coming at the expense of a narrower azi-

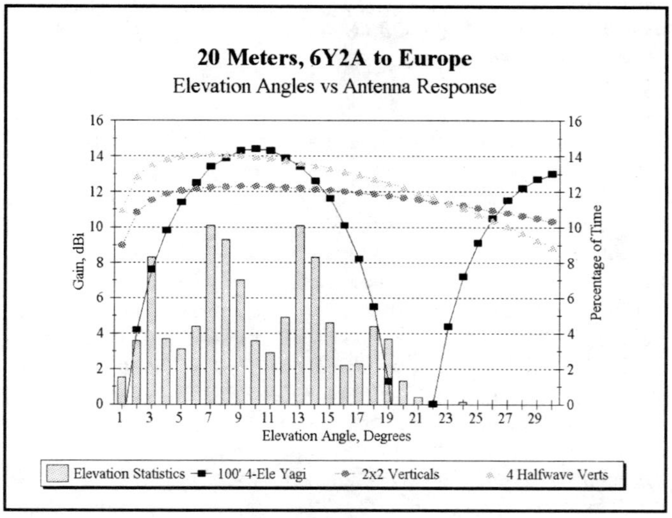

Fig 4—Comparison of needed elevation angles and typical antenna elevation-plane responses on the 20-meter path from Jamaica to Europe. Here, the antennas are a 100-foot-high four-element Yagi and two types of two-by-two vertical arrays over saltwater. The first type uses quarter-wave elements, while the second type uses half-wave long elements, compressing the elevation-plane response somewhat, but increasing the gain at lower angles.

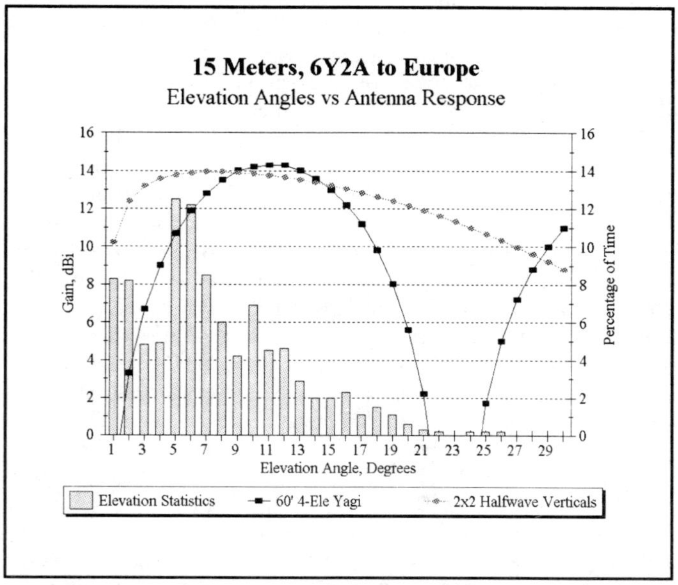

Fig 5—Comparison of needed elevation angles and typical antenna elevation-plane responses on the 15-meter path from Jamaica to Europe. At low angles the two-by-two array of half-wave verticals over salt water has the advantage, but a 60-foot- high Yagi is still quite competitive.

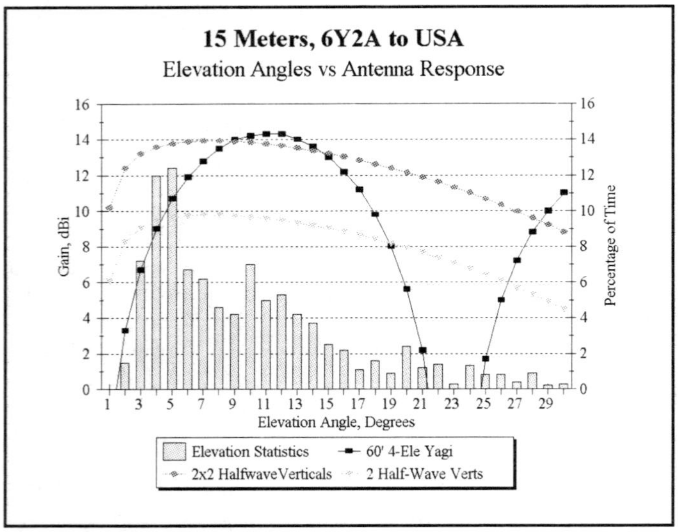

Fig 6—Comparison of needed elevation angles and typical antenna elevation-plane responses on the 15-meter path from Jamaica to the USA. Although it would have been nice to have a two-by-two vertical array on this path, a single pair of half-wave verticals over salt water still does a very credible job, and its high-angle response is better than a single 60-foot-high Yagi above about 20° in elevation, where signals occasionally come in on this band at high levels of solar activity.

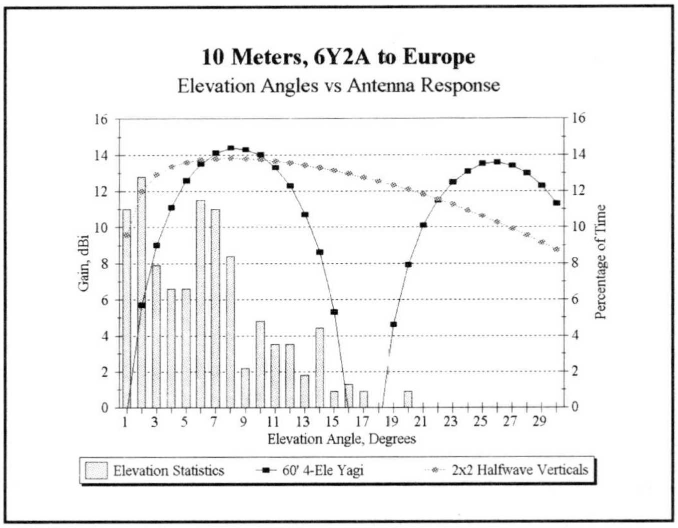

Fig 7—Comparison of needed elevation angles and typical antenna elevation-plane responses on the 10-meter path from Jamaica to Europe. While the 60-foot-high four-element Yagi is a serious competitor on this band, the two-by-two vertical array over salt water still holds its own, particularly at extremely low elevation angles.

muthal beamwidth, as you would expect. The higher-gain array also has much more distinct nulls in the azimuthal pattern.

In our case, the nulls in the higher-gain pattern were used to advantage. You will remember that we wished to improve our performance working three-point QSOs from 6Y2A. This means that not only did we need more gain into Europe but we actually needed to discriminate against North American stations, which by nature of their proximity would be much louder than the Europeans. The European antenna arrays at 6Y2A were carefully aimed to null out USA stations off the sides. That helped equalize the playing field for the Europeans. Our final percentage of Europeans (three points) versus USA stations (two points) was higher in 1998 than in 1997.

One disadvantage to narrow-beamwidth arrays aimed into Europe should be obvious. We needed separate arrays for working Japanese and USA stations. Thus we installed five separate two-element vertical arrays aimed at Japan/USA for 80 through 10 meters, in addition to the three two-by-two vertical arrays for Europe for 20, 15 and 10 meters and the two-element European vertical arrays for 160, 80 and 40 meters.

You Can Never Have Too Many Antennas

Another possible disadvantage associated with the use of narrow-beam two-by-two European arrays is that off-azimuth stations from Africa and South America would be tougher to work than in 1997, when only the wider beamwidth two-element arrays were employed. To cover these directions we erected specially designed, very lightweight two-element horizontal Yagis for 20, 15 and 10 meters. These were placed on portable masts that were a half-wave high. Now "half-wave high" sounds pretty impressive, until you realize that this means 15 feet on 10 meters! Rotating these Yagis was by the tried-and-true Armstrong method; that is, by hand. When pointing in the same direction as the two-by-twos, the Yagis were usually down 2 to 3 S-units, sometimes more. The 20-meter two-element Yagi shows up in the background of Fig 10.

Another "secret weapon" we employed at 6Y2A in 1998 was WXØB Array Solution's *StackMatch* boxes.[3] These allowed our 20, 15 and 10 meter operators to selectively use one, two or all three of the antennas they had on those bands. Fred Cady, KE7X, one of the 15-meter operators, built special control boxes for the StackMatches tailored to our antenna configurations. (Fred and Dennis, AF7Y, also custom-built our operating tables from scratch!) For example, the 15-meter operator could simultaneously distribute his transmitter power equally among the European two-by-two verticals, the JA/USA one-by-two verticals and the rotatable 2-element Yagi. When a multiplier called in from, say, Africa, he could flip a switch to select the horizontal Yagi, which was often aimed in that direction (when it wasn't aimed into the Pacific).

We proved the old ham adage "you can never have too many antennas" many times over during the contest. We only wish we had some antennas that were high enough to peek over the 200+ foot high cliffs located about a quarter mile behind the villa. I'm sure we

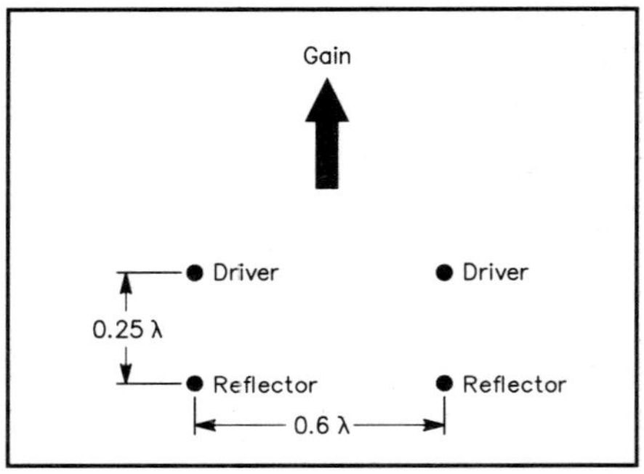

Fig 8—Bird's-eye view of two-by-two vertical array. The spacing between reflector and driver is approximately 0.25 λ, and the side-by-side spacing is 0.6 λ.

missed some multipliers because of the cliffs, but you can't have everything, can you? **Fig 12** shows the propagation prediction for 15 meters for 6Y2A for the month of November 1998, with a High level of solar activity (a Smoothed Sunspot Number of about 100, equating to a solar flux of 140). This is one of the 70,000 pages of customized predictions that are included on *The ARRL Antenna Book*, 18th Edition CD-ROM. These predictions cover a full spectrum of amateur QTHs around the world, for all months of the year and all levels of solar activity.

The 40 CQ Zones are listed in the left-hand column and the UTC hours are listed in the top row. The table is calibrated in S units, assuming that S9 is 50 µV on 50 Ω at the target receiver's input and that the receiver is calibrated at 4 dB per S unit. The transmitting and receiving antennas for this 15-meter table are assumed to be four-element Yagis

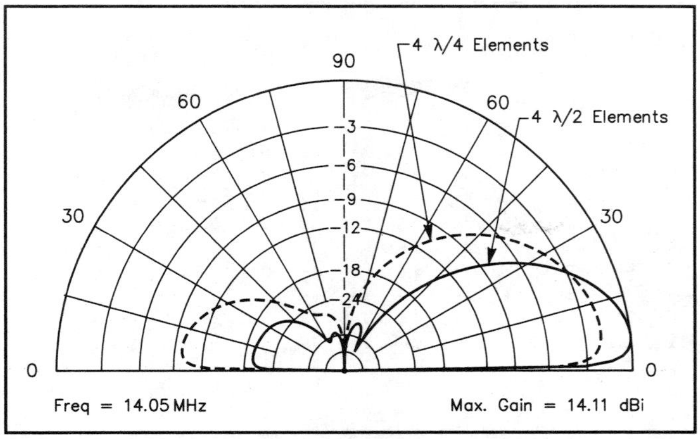

Fig 9—Elevation-plane pattern for a vertical array of four half-wave dipoles compared to four quarter-wave vertical monopoles with radial systems. The half-wave dipole system has about 2 dB more gain, at the expense of a somewhat compressed elevation lobe.

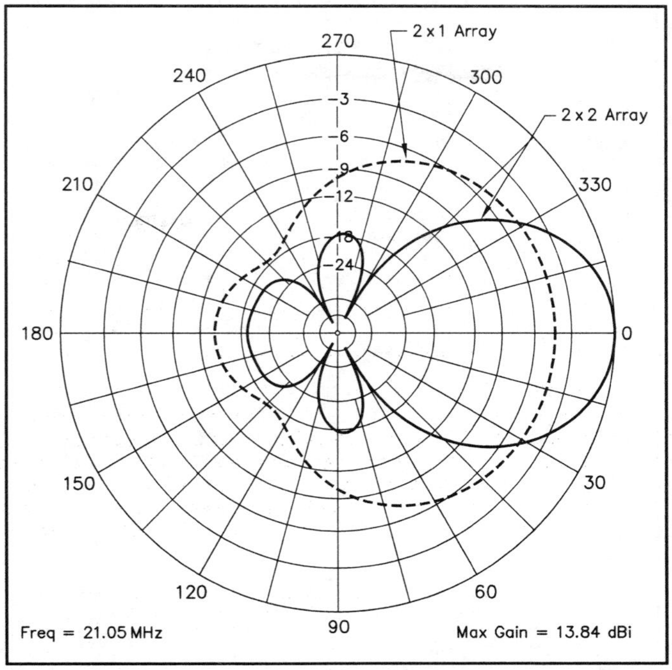

Fig 11—Azimuth-plane patterns for two-by-two (four element) 15-meter vertical array over salt water compared to an array with a single pair of verticals. Each element in this comparison is a half-wave-long vertical dipole. The two-by-two significantly compresses the pattern in the azimuthal plane, achieving about 4 dB of extra gain in the process, but making aiming of the main lobe more critical.

Fig 10—Photo looking along the beachfront of the 1998 6Y2A operation from Jamaica. In the foreground is the two-by-two vertical half-wave array for 15 meters into Europe. The 10-meter two-by-two array is behind this, and the two-element horizontal 20-meter Yagi at 30 feet shows up in the background.

that are 60 feet above flat ground, and the transmitter power is assumed to be 1500 W. Look at the table entry for 10 UTC in Zone 14, located in England. The band opens from Jamaica at 10 UTC, with a fairly weak signal of S2. By 11 UTC, the band has opened better, with an S8 signal in London from Jamaica. From 12 to 14 UTC, the signal peaks at S9+, which means that it is at least 10 dB above S9. The band closes at 20 UTC, for a total predicted opening of 11 hours at this level of solar activity in November.

Now, look at the entries for JA1, in Zone 25. Note that there are some signal predictions with asterisks (*). For example, see the entries for 03 and 04 UTC and for the times between 11 and 14 UTC. The asterisks indicate that the dominant signals are by means of the long path. Although they won't be very strong (peaking only S4 at 11 UTC), they will be there. In comparison to signals coming from Europe at S9+, however, it's unlikely that the long-path JA signals will be very useful. And besides, JA signals are predicted to hit an S9 level in the afternoon opening starting at 21 UTC. We concluded that the long path blocked by the cliff behind us would not be a major factor to us. It would have been nice to have a good shot in all directions, but we figured that we should find the multipliers we needed on short-path beam headings over the ocean.

SOME PRACTICAL TIPS

The first practical tip for someone who goes to operate from Jamaica is "keep moving." Now why would I say *that*? We arrived at the villa a full week before the contest was scheduled to start, so that we could put up all the antennas and the six stations. As enumerated previously, we ended up with 28 vertical elements, not to mention the three horizontal Yagis and the receiving Beverages, the EWE and the low dipole! Every day out on the rocky coral beach we would be joined by a crew of very large, very ugly turkey buzzards, "John Crows" as the locals call them. These things spent all their time gliding in lazy circles high overhead, watching us expectantly, or so it seemed. We made sure that we looked very much alive! One morning after a big storm blew by overnight, about 20 of these disagreeable-looking birds noisily congregated on the rocks overlooking the 80-meter arrays. They kept busy fighting with each other and scanning the seashore. I figured that their fondest hope must be that a dead whale would wash up on the beach.

Beware the High Voltage

The second practical tip for those who would install verticals on the beach is to have a very healthy respect for the levels of voltage that exist on the ends of radials and wherever high impedances exist. Salt spray is *guaranteed* to provide a ready path for arcing to occur when you transmit at full power. The ends of radials are particularly vulnerable points if you don't pay a lot of attention at installation.

Fig 13 shows what I call the "turkey-buz-

15 Meters: Nov., Jamaica (Kingston), for SSN = High, Sigs in S-Units. By N6BV, ARRL.

UTC -->

Zone		00	01	02	03	04	05	06	07	08	09	10	11	12	13	14	15	16	17	18	19	20	21	22	23
KL7 =	01	9	9	3	-	-	-	-	-	-	-	-	-	-	-	-	-	4	8	9	9	9+	9+	9+	9+
VO2 =	02	9	7	3	1	1	1	1	-	-	-	-	2	6	9+	9+	9+	9+	9+	9+	9+	9+	9+	9+	6
W6 =	03	4	9	6	1	-	-	-	-	-	-	-	-	-	-	5	9+	9+	9+	9+	9+	9+	9+	9+	9+
W0 =	04	9+	9+	9	4	-	-	-	-	1	-	-	-	9	9+	9+	9+	9+	9+	9+	9+	9+	9+	9+	9+
W3 =	05	9+	8	2	-	-	-	-	-	-	-	-	-	9+	2	8	9	9	8	8	8	8	6	-	9
XE1 =	06	4	9+	9+	9+	8	6	2	2	1	-	-	-	5	-	9+	9+	9+	9	8	8	8	8	7	2
TI =	07	8	3	-	-	-	-	-	-	-	-	-	-	-	9	9+	9+	9+	9+	9+	9+	9+	9+	9+	9
VP2 =	08	5	2	1	-	-	-	-	-	-	-	-	-	7	9+	9+	9+	9+	9+	9	9	9	9	7	8
P4 =	09	9	6	4	1	-	-	-	-	-	1	-	-	2	9+	9+	9+	9+	9+	9	9+	9+	9	8	9+
HC =	10	9+	9+	9+	9+	9+	6	-	2	-	-	-	1	9+	9+	9+	9+	9+	9+	9+	9+	9+	9+	9+	9+
PY1 =	11	9+	9+	9+	9+	9+	9+	7	1	-	-	1	9	9+	9+	9	9	9	9	9	9+	9+	9+	9+	9+
CE =	12	9+	9+	9	9+	9+	9+	9+	9	1	-	-	6	9+	9+	9+	9	9	9	9	9	9+	9+	9+	9+
LU =	13	9+	9+	9+	9+	9+	9+	9	7	-	-	-	9	9+	9+	9	9	9	9	9	9	9+	9+	9+	9+
G =	14	-	-	-	-	-	-	-	-	-	-	2	8	9+	9+	9+	9+	9+	9	9	8	2	-	-	-
I =	15	-	-	-	-	-	-	-	-	-	-	3	9	9	9	9	9	9	9	9	4	-	-	-	-
UA3 =	16	-	-	-	-	-	-	-	-	-	-	-	6	9+	9	9	9	9	4	-	-	-	-	-	-
UN =	17	-	-	1*	3*	1*	1	-	-	-	-	-	1	9+	9	6	-	-	-	-	-	-	-	-	-
UA9 =	18	-	4*	2*	2*	1*	-	1	-	-	-	-	-	2	-	-	-	-	-	-	-	-	-	-	-
UA0 =	19	9+	8	4	1*	-	-	-	-	-	-	-	1*	-	-	-	-	-	-	-	-	-	5	9+	9+
4X =	20	1	2	1	-	1*	1*	6	-	-	-	5	9	9	8	8	8	9	9	9	7	6	-	1	-
HZ =	21	1	-	-	1*	2	5	3	-	-	-	3	8	8	8	8	8	8	9	5	5	2	1*	1*	3
VU =	22	1	2	2*	3*	1*	-	-	-	-	-	-	4	9	9	8	3*	1*	-	-	-	-	-	-	1
JT =	23	4	8	2	2*	1*	-	-	-	-	-	-	-	5*	1*	-	-	-	-	-	-	-	-	-	-
VS6 =	24	9	6	2	1*	2*	-	-	-	-	-	-	1*	6	5	2*	1*	-	-	-	-	-	-	-	9+
JA1 =	25	9	8	5	1*	1*	-	-	-	-	2	-	4*	3*	3*	1*	-	-	-	-	-	1	9	9	9
HS =	26	9	5	4*	3*	1*	-	-	-	-	-	-	-	9+	9+	9	6	4	1	1	1	-	-	-	1
DU =	27	9	8	5*	1*	-	-	-	-	-	-	-	3*	4*	9	5	1*	1*	1*	-	-	-	-	9	9
YB =	28	9	7	2*	2*	2*	-	-	-	-	-	-	1*	5	9+	9	9	8	8	8	9	5	-	1	9+
VK6 =	29	8	9	8	8	8	8	8	9	8	5	1	4	8	8*	5*	4*	2*	1*	-	-	-	-	4	8
VK3 =	30	7	8	8	8	9	9+	9	8	7	2	-	2*	3	9	7	-	-	-	-	2	7	7	6	6
KH6 =	31	9	9	8	8	6	3	-	2	2	1	-	-	-	-	-	-	9	9	8	8	8	8	9	9
KH8 =	32	8	9	9+	9+	9+	9	8	5	2	-	-	3*	4*	9+	9	4	9	8	8	7	8	8	8	8
CN =	33	-	-	-	-	1	-	-	-	-	-	5	9	9+	9+	9	9	9+	9+	9+	1	9	8	-	-
SU =	34	4	7	6	-	1*	1*	6	2	-	-	5	9	9	8	8	8	9	9	9	9	8	7	5	2
6W =	35	9	6	5	2	2	1	-	-	-	-	8	9+	9+	9	9	9	9	9+	9+	9+	9+	9+	9+	9
D2 =	36	9+	9+	9	9	9	8	7	1	-	-	6	9	8	8	8	6	8	9	9	9+	9+	9+	9+	9+
5Z =	37	9+	8	8	6	7	8	6	-	-	-	5	8	6	6	6	7	8	9	9	9+	9+	9+	9+	9+
ZS6 =	38	9	7	2	-	5	6	3	-	-	-	4	7	6	6	6	5	6	8	9	9	9+	9+	9+	9+
FR =	39	7	3	-	8	8	7	2	-	-	-	1	6	5	5	6*	5	6	8	9	9	9+	9+	9	8
FJL =	40	2	2	1	-	-	-	-	-	-	-	-	-	1	3	3	4	5	5	6	4	2	-	-	3
Zone		00	01	02	03	04	05	06	07	08	09	10	11	12	13	14	15	16	17	18	19	20	21	22	23

UTC -->

Expected signal levels using 1500 W and 4-element Yagis at 60 feet at each station.

Fig 12—Page from *The ARRL Antenna Book* CD-ROM, 18th Edition, showing propagation prediction from Jamaica to 40 Zones around the world for the month of November, at a High level of solar activity. Note signals levels followed by an asterisk (*), which indicates long-path predictions.

zard-wing" (TBW) radial system, in honor of our omnipresent feathered companions. (Most folks would probably call them "gull-wing" designs, but I'm fond of the term TBW. Curiously, we didn't see many gulls on Jamaica. Maybe the turkey buzzards intimidated them.) We supported the ends of each radial with pieces of driftwood, most often bamboo, that washed up on the shore. The relatively short radials used on 20 meters are pretty much self-supporting in the arrangement shown in Fig 13. However, the longer radials used on 80 and 160 meters required intermediate supports, since the radials started out typically about 8 to 10 feet above the ground at the vertical element. Sag in any long span of unsupported wire would cause them to droop down onto the coral and volcanic rock, which was washed constantly by salt water.

One evening before the contest K7CO/6Y5 was operating on 75-meter phone when I took a stroll out in the moonlight. I walked past the 80-meter array and noticed a peculiar flashing near the end of one of the reflector radials. This radial had been propped up near its end with a piece of semi-wet driftwood. Close (but careful) examination showed that the wood was arcing and actually smoking each time there was a transmission! I asked the operator to stop transmitting and performed an emergency repair, consisting of putting lots of black vinyl tape at the point where the wire ran over the driftwood. By the way, all the driftwood pieces used for such supports were either held up by placing them in natural holes in the volcanic rock or by placing a pile of coral and rock around the base. Not all the beaches on Caribbean islands are made of pure white sand like the ads show.

Another example of how high the RF voltage can get was vividly demonstrated on the half-wave dipoles used on 15 and 10 meters. Vertical dipoles have a high-voltage point at their top and bottom ends. The 10-meter arrays were located within 10 feet of the ocean and were regularly washed by salt spray when the wind picked up, as it often did. The bottom of the aluminum tubing used to make each half-wave dipole was perhaps a foot off the rock. In broad daylight I observed arcing at the bottom insulator of the 10-meter JA/USA driver element. The insulator itself was about six inches long, made of solid fiberglass rod. The arcing created a permanent carbon track that had to be carved off the

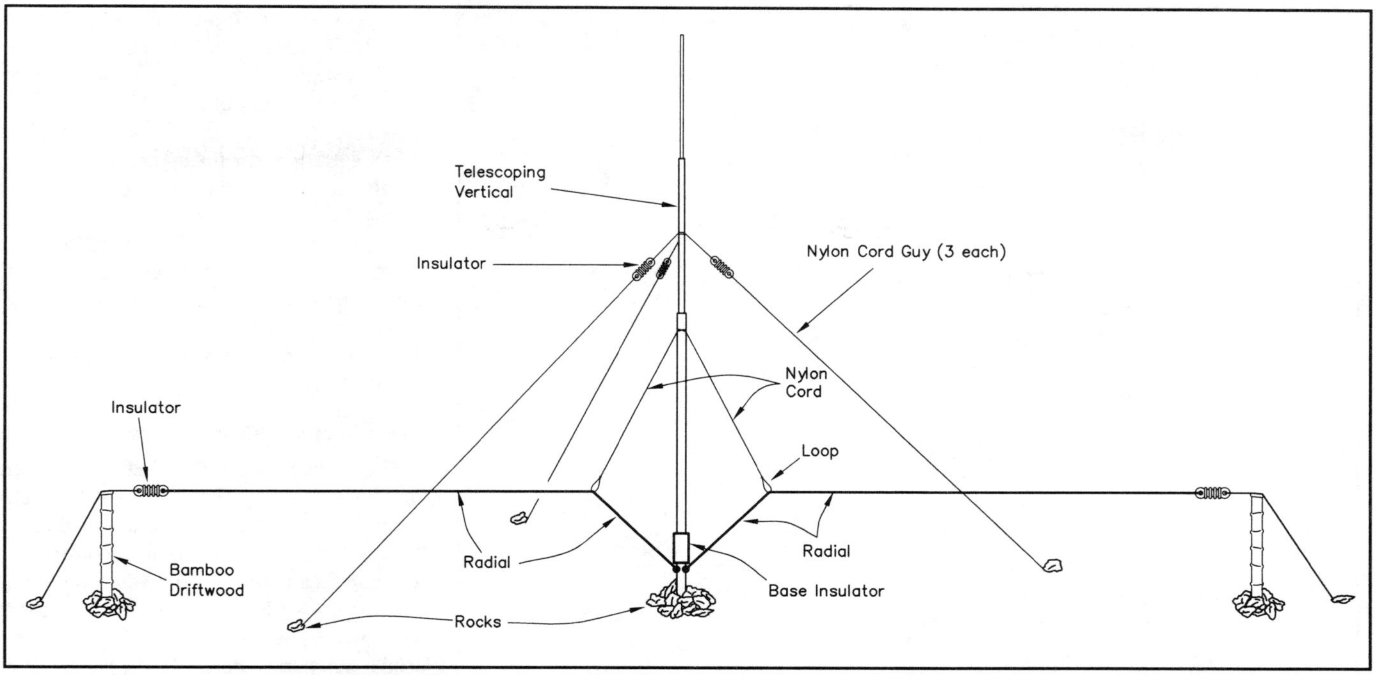

Fig 13—Drawing of 20-meter quarter-wave high vertical system, including the two "gull-wing" shaped radials used to fine-tune the resonant frequency. At 6Y2A, rocks were piled up at the insulated base of the verticals to help hold them vertical, along with nylon "parachute cord" used as guys.

insulator with a knife and the insulator washed with fresh water before operation could resume. The reason I was out investigating such a phenomenon was that the arcing in the 10-meter array created a broadband noise that interfered with both the 20 and 15-meter radios during the contest.

Another place where the RF voltage can rise to surprising levels is in the loading system used to tune the 80 and 160-meter verticals. See **Fig 14**. I refer to this scheme as "Deltoid Loading," rather than "linear loading" since the shape of the wires resembles the Greek uppercase Delta symbol (Δ), rather than the usual form of linear loading made with two parallel wires. The Deltoid Loading wires are at a high-impedance point, although not as high an impedance as the end of radial wires. Every morning we would walk the "vertical forest" to see what damage the wind, salt and RF had wrought overnight, especially while we were operating on 80 and 160 meters, where the radials were long and the Deltoid Loading was necessary.

For several days before the contest, we struggled with arcing at the Deltoid insulators due to salt-spray accumulation. Then we finally resorted to a home-grown solution—we used *Red Stripe* beer bottles as insulators! Tie-wraps, overlaid with black vinyl tape to make sure they didn't drift, were used around each end of these improvised insulators and they worked just fine. *Red Stripe* bottles also served as excellent coil forms for the hairpin-match inductors at the 80 and 160-meter feed points.

Incidentally, we standardized on the *Red Stripe* bottles after a smaller bottle we found on the beach (it looked like some sort of medicine bottle) broke at one of the Deltoid loading points. Apparently, this occurred as a result of RF heating of the Deltoid wire, which in turn cracked the glass. The broken medicine bottle was discovered when the 160-meter array began putting broadband noise into the other radios. Although it was dangerous to do so, N6BT walked out at night onto the jagged coral and found the 160-meter vertical was acting like a flame-thrower!

Another word of caution: You don't want to fall down on coral, day or night. Not only do you get instant nasty cuts and bruises, but you usually get a nasty infection later.

Tuning the 20, 15 and 10-Meter Arrays

As alluded to previously, the vertical arrays were all designed by computer. Actually implementing the antennas required a disciplined approach in the field, or rather, on the beach. The following procedure describes how we tuned the 20-meter two-by-two array (a reflector and a driver in each "cell" of the two cells). In essence, each reflector is tuned without the presence of the driver, which is physically installed, but the feedpoint is left open. Once a reflector is tuned, its vertical element and radials are shorted together to make it into a parasitic reflector and then the associated driver is resonated. Finally, each driver/reflector cell is impedance-matched at the desired frequency for the array. (With a 0.6-λ spacing between the two parasitic reflector/driver systems, the mutual coupling between each reflector/driver cell can be ignored—each system is tuned independently and then the phasing harness is connected.)

All of the antennas consisted of telescoping aluminum tubing sections that were pop-riveted with aluminum rivets. This made for secure electrical connections that could survive the rigors of the salt-spray environment. We used WD-40 to "grease" the sections so that they'd go together and come apart easily; WD-40 is marvelous stuff. (And yes, we had to drill out the rivets when we disassembled the antennas after the contest was over. A battery-operated drill was essential to the 6Y2A operation.)

Computer modeling shows that a vertical with radials can be tuned by varying the length of the vertical radiator and/or the length of the radials. What I didn't realize (until N6BT showed me) was that a variable TBW radial design (Turkey-Buzzard-Wing, or "gull-wing radial" if you insist) can be used to fine-tune the radials and hence the system. See Fig 13 again. The radial wires were supported from the vertical element by nylon cords ending in a loop made with a plastic tie-wrap. By sliding the loop along the radial wire, the effective height of the radial could be varied slightly, fine-tuning the radial/vertical system.

We coarse-tuned the radials by connecting them as a dipole (a rather low dipole, to be sure) and tuning them to resonance using an MFJ-259B *SWR Analyzer* connected through a multi-bead Force 12 choke balun. The closer this pseudo dipole was brought to the very lossy volcanic rock, the broader

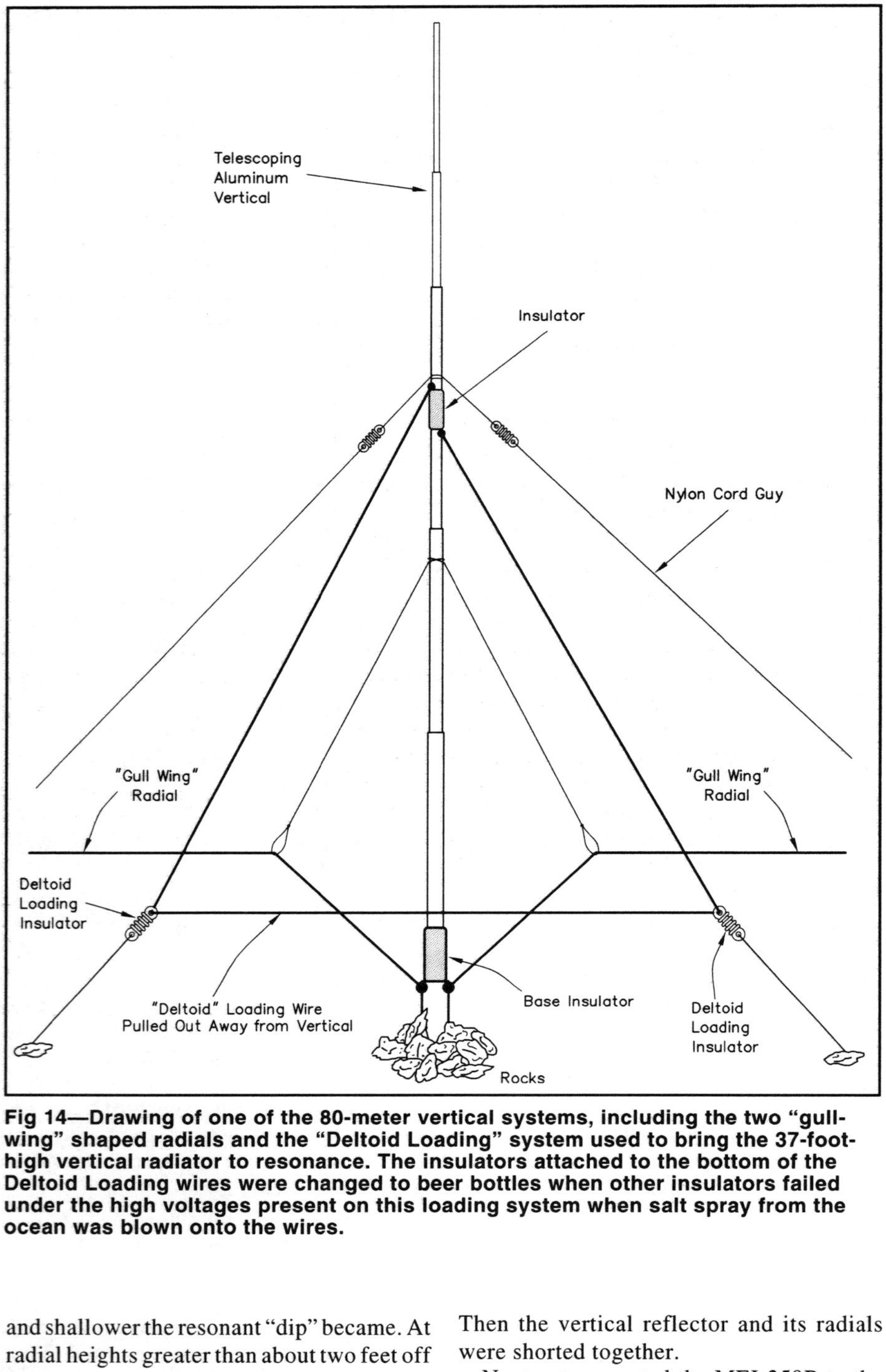

Fig 14—Drawing of one of the 80-meter vertical systems, including the two "gull-wing" shaped radials and the "Deltoid Loading" system used to bring the 37-foot-high vertical radiator to resonance. The insulators attached to the bottom of the Deltoid Loading wires were changed to beer bottles when other insulators failed under the high voltages present on this loading system when salt spray from the ocean was blown onto the wires.

and shallower the resonant "dip" became. At radial heights greater than about two feet off the rock on 20 meters the dip was sharp and distinct and the feed-point impedance was close to the theoretical value for that height.

For the reflector, the frequency was set to the resonant frequency determined in the computer model when the driver was removed from the model. This was to eliminate the effect of the strong mutual coupling that occurs between the driver and the reflector, which are spaced apart by about 0.25 λ. Once a reflector radial pair was coarse-tuned, the two radials were shorted together at the base of the vertical and the resonant frequency of the vertical/radial system was fine-tuned, using the gull-wing height, to the desired reflector frequency. Then the vertical reflector and its radials were shorted together.

Next we connected the MFJ-259B to the driver radials and resonated them to the desired array operating frequency. For example, on the 20-meter arrays this was at 14.05 MHz. Finally, we connected the driver radials to the base of the driver vertical and fine-tuned this system (which now included mutual coupling to the parasitic reflector) using the gull-wing radial height again. Impedance matching to bring the feed-point impedance to 50 Ω was done using a small "hairpin" coil at the drive point.

Then the second driver/reflector cell was tuned. Finally, the phasing harness was connected and the SWR at the feed-point tee was verified to make sure the whole two-by-two was properly configured. This procedure sounds complicated, but we managed to build, install and fine-tune a 20-meter two-by-two array in about six hours from start to finish, using a crew of two people most of the time, supplemented by a few others during the actual installation phase.

Installation and tuning of the half-wave dipole arrays on 10 and 15 meters took less time, since they didn't have radials with which to contend. Imagine, however, how much time it would have taken to install three or four 60 to 100-foot towers and rotatable Yagis.

The 160 and 80-Meter Arrays

The verticals for 160 and 80 meters were custom designs by Force 12. The 80-meter verticals were 37 feet long and weighed a mere 9 pounds. These were made up of pop-riveted, telescoping aluminum tubing that was strong and supple, yet light enough to be hoisted to the vertical position by a single person. We used nylon parachute cord for guys, attached about halfway up the vertical element with small egg insulators and tie-wraps.

The same tuning procedure outlined above was used for each 80-meter element in each two-by-one array, except that the vertical element was tuned (after its two resonant radials had been separately resonated) with the Deltoid loading system.

The 160-meter array was truly a thing of beauty. Tom Schiller had TIG-welded aluminum tubing and rod to create what I like to call "Rohn 12 and a half" mini-tower sections that fit inside the golf-bag carriers and yet connected together to form a sturdy 20-foot high base for each 57-foot long vertical radiator. The face of each triangular mini-tower section was about 6 inches wide and each element weighed just 25 pounds. A pair of Deltoid loading systems were used with each vertical radiator, along with the now-standard pair of gull-wing radials. Six men were required to install each vertical—three on the top guys, two at the base and one pushing upward along the mast as it rose to a vertical position. Each vertical went up with surprising ease thanks to such a coordinated effort. **Fig 15** is a photograph looking up from the base of one of the 160-meter "mini-towers."

We installed a single 160-meter vertical and tested it out for the first few nights we were there on the island. The reports from all around the USA indicated that we were *very* strong. One W7 in Arizona said that we were louder than most of the W6s that he heard calling us! So we were confident that when the reflector was added to steer the beam North we would have a really outstanding signal on Top Band. The numbers tell the story: During the contest 1161 QSOs were made on 160 meters, with 22 Zones and 86 countries. Not bad for a couple of short verticals on the beach, don't you think?

The 40-Meter ZR Arrays

As promised, I'll try to describe the 40-meter antennas used at 6Y2A, which are the same arrays used so successfully in 1997 at 6Y4A. "ZR" stands for "Z-Axis Radiator," Tom Schiller's unique name for what amounts to a vertically positioned, loaded half-wave dipole. To me, the ZR looks like a rather large box kite, without paper on the ribs, of course. It is 15 feet high and 8 feet wide, and is fed in the center through a current-mode ferrite-bead balun. At the top and the bottom, both ends of the short dipole are wound into a sort of spiral. Although they superficially resemble a classic "top hat," the top and bottom is really a continuous conductor extending the electrical length of the dipole.

We set up two ZRs pointed into Europe as a reflector/driver pair, with another pair pointed into Japan/USA. They were placed at the edge of a small embankment about 8 feet above sea level and about 20 feet from the ocean. Did they work? You bet they did—W9QA had 3976 QSOs, with 36 Zones and 134 Countries during the contest. After the contest we pulled Dave's chain, telling him that in both 1997 and 1998 he had definitively proved that it's impossible to work 4,000 QSOs on 40-meter CW...

And If We Do It AGain?

It's hard to imagine topping the exhilaration of making 19,100 QSOs (18,000 after dupes) in a weekend and ending up with a raw (unchecked) score that is about 20% higher than the existing world's record. Perhaps we'll all simply rest on our laurels and sit out the next competition, but then again maybe not. . .

What would we do differently if we were to make a new assault on the world's record? One obvious thought is to seek a three-point location; one that is easy to get to and that has a lot of ocean-front property in the right directions. We're convinced that vertical arrays on-the-beach are the right way to go.

We're also convinced that there are a *lot* of reliability problems associated with any antenna located where salt spray can get directly to it. The amount of corrosion we found on the antennas after taking them down was pretty astounding—and these antennas had only been up in the air for about a week. It would be useful to know exactly how far from the ocean's edge you can move (to get away from the direct spray) without losing the advantage of the over-the-saltwater shot. Preliminary modeling indicates that it may be possible to move back from the shoreline some, but not too far. The main effect appears that the wide "window" of elevation angles illuminated by a vertical array from the horizon up to about 30° in elevation is narrowed the further you move from the ocean's edge. The surprising part is that the model shows that the higher elevation angles are the most affected, not the lower angles as intuition would indicate.

Fig 15—A view looking up from the base of one of the 160-meter "mini-towers" constructed by N6BT especially for this DXpedition. The overall height of each radiator is 57 feet and the tower face is about 6 inches across. All the welds were made with a TIG-welding system.

Another interesting aspect of the modeling for verticals located back from the ocean is that the actual front-to-back ratio is higher than if the array were located directly over saltwater in all directions. The difference is due to the poor ground characteristics in the back of the antenna compared to the saltwater in the foreground. We experienced a sort of "double whammy" due to this effect during the contest. The cliffs toward the south have been mentioned previously. Add in an increased front-to-back ratio due to the poorer ground in the rearwards direction and that means that South American signals were weaker than a glance at Fig 9 or Fig 11 would seem to indicate.

Groundwave measurements done in the salt flats of San Francisco Bay a couple of years ago don't seem to follow part of what the computer model is showing, however. We're not sure whether to believe the models in this regard, although the models certainly seem to mirror reality for how well verticals can do right at the shoreline. To make a very bad pun, stay tuned. More experimental work needs to be done in this area.

While the antennas and the location were major factors in the success of the 6Y2A operation in the 1998 CQ World Wide CW DX contest, other factors were equally important. It's hard to imagine putting eleven "Type-A" personalities together in the same room without friction and bickering. You know the old saw about "Put 5 contesters in the same room and there'll be 6 opinions," don't you? But this team worked and played together exceedingly well.

The leadership and planning were superb. Kenny Silverman, K2KW, is an incredibly detail-oriented manager who kept things on track for the entire year it took to plan and execute this operation. Not incidentally, having a year to plan everything down to the last detail gave us a material advantage over other teams we knew we were going to be competing against—detailed planning for 6Y2A started as soon as the 1997 6Y4A operation ended.

Having team members who were versed in many different aspects of contesting was another crucial element in the success too. If something broke down, there were many willing hands capable of fixing it.

And what did my Dad think of all this? He had a *great* time and would do it again at the drop of a hat! The raw numbers for the contest from 6Y2A were:

Band	*QSOs*	*Zones*	*Countries*	*Band Captain*	*Antennas*
160 m	1161	22	86	N6BT	2×1 Vertical, 57′ high
80 m	1913	29	107	N6BV	2×1 Europe, 2×1 JA/USA, 37′ high
40 m	3976	36	134	W9QA	2×1 Europe, 2×1 JA/USA, 15′ high ZR arrays
20 m	4213	38	151	N6TV, W4SO	2×2 Europe, 2×1 JA/USA, λ/4 high, 2-ele. Yagi @30′
15 m	3508	39	151	AG9A, KE7X	2×2 Europe, 2×1 JA/USA, λ/2 high, 2-ele. Yagi @25′
10 m	3228	33	126	K2KW	2×2 Europe, 2×1 JA/USA, λ/2 high, 2-ele. Yagi @15′

Broadbanding the Elevated, Inverse-Fed Ground Plane Antenna

By Sam Leslie, W4PK
1038 Lone Pine Terrace
Goode, VA 24556-2545

N4KG's elevated ground-plane has become a popular antenna for the lower frequencies. W4PK offers some ideas on making it cover both CW and SSB portions of the 80-meter band.

Constructing an antenna that will operate with reasonable SWR in the CW DX segment at the bottom end of 80 meters *and* the SSB DX segment at 3.8 MHz has always been a challenge. I discovered the approach I'll describe here quite by accident while evaluating Tom (N4KG) Russell's inverse-fed, elevated ground-plane concept (described in June 1994 *QST*) on my tower.

To make a long story short, what I found was that two sets of radials at 90° with respect to each other could each be cut to achieve a low SWR at their respective parts of the band, producing two dips in the SWR curve. As ON4UN states in his ARRL book, *Low Band DXing,* two in-line radials provide good ground-plane performance. Thus, one pair is dedicated for the CW portion of the band and the other pair is set for the SSB portion of the band. This, of course, assumes that the radials are sufficiently elevated to avoid coupling losses with the ground.

The challenging part is to find the right spot on the tower to make this technique work. Basically, this is where the electrical length of the tower, as loaded with antennas, is equivalent to 90° (λ/4) above the feed point. N4KG's article describes a procedure you can use to estimate this, both in terms of an equation derived from ON4UN's book, and with a table describing the effective equivalent loading effect of several types of Yagi antennas.

A step-by-step approach to constructing this antenna is as follows:

1. Approximate the electrical height (in degrees) of your tower from the equation in N4KG's article as follows:

$L = 0.38F[H+(2S)^{1/2}]$

where F is in MHz, H is in feet, and S is the equivalent area of the Yagi(s) in square feet of wind area. This number must be greater than 90° ($^1/_4$ λ) for this concept to work. An electrical length of 120° or more will assure that the radials are sufficiently elevated above ground for reasonable performance.

2. Next subtract 90° from the number obtained above to determine the electrical height (in degrees) above ground that represents the target "sweet spot."
3. Now convert degrees into feet or meters for the frequency halfway between your two target SWR minima (ie, 3.65 MHz). For feet, the relationship is simply:

$$\text{Feet} = \frac{(\text{Electrical length, degrees}) \times 492}{180° \times F_{MHz}}$$

In meters, the relationship is:

$$\text{Meters} = \frac{(\text{Electrical length, degrees}) \times 150}{180° \times F_{MHz}}$$

This will be the starting point for finding the "sweet spot" above ground level for your tower.

4. Cut one pair of $^1/_4$-wavelength radials near the bottom end of the 80-meter band. I used 69 feet each, or a total of 138 feet from radial tip to radial tip for 3.550 MHz.
5. Cut the second pair of radials for the phone portion of the band (65 feet each for 3.8 MHz)
6. Join all four radials together at an insulator temporarily hooked to the calculated "sweet spot" in step 1 above. Make sure that the CW radial pair is at right angles to the SSB radial pair. Also, pull the radials out to temporary posts such that the ends are at least 8-10 feet off the ground, and that the sag is minimum (less than 3 feet or so).
7. Connect the center lead of a short piece of coax cable (less than 2 feet) to the cross point where the radials are joined together, and temporarily connect the braid of the coax to the tower. Since the impedance of this antenna will be quite low, connecting a 50:25-ohm unun transformer at the end of this short piece of

coax may make measuring SWR easier. Connect the 25-Ω port of the unun to the short piece of coax, and connect the SWR meter to the 50-Ω port.

8. Measure the SWR, and determine whether the SWR minima are low or high with respect to the target. NOTE: *Do not attempt to adjust the length of the radials to move the SWR minima!* You will likely not be successful if you are too far from the actual "sweet spot" on the tower. The goal is to move the connection point on the tower to where a reasonable SWR match is obtained for both portions of the band.

If the SWR minima are low in frequency, then you need to raise the connection point on the tower. Likewise, if the SWR minima are higher than desired, you will need to lower the tower connection point. An easy way to do this is to climb the tower with your handy battery-powered SWR box and have a helper loosen and then retighten the radials as you work your way up or down the tower.

The hard way (my way) is to climb down each time you make an adjustment and check the SWR. If you check the SWR at the bottom of the tower, connect a length of coax from the 50-Ω port on the unun, and run this coax down *inside* the tower to the base. I made my measurements at the tower base primarily to remove any temptation to drop the SWR analyzer while I was on the tower. With either method, you will find that the final adjustment will be a matter of inches to balance the SWR minimum at each target frequency.

I used Amidon's multiple-tap unun (part number W2MFI-1.78:1-HMMUSD) since I was expecting a fairly low impedance (below 18 Ω or so). However, I found that drooping the radials at an angle of 35-45° from horizontal raised the impedance above 20 Ω. Be careful about getting the ends of the radials too close to the ground. I had observed some unpredictable effects when the ends were less than 6 feet or so from ground. Eight to 10 feet or higher should be okay, and would be preferred anyway if you have large mammals roaming the area. The 50:22.16-Ω tap on the Amidon unun proved to be the best match in my case. Any 50:25-Ω unun rated at your preferred power level should work just as well.

A quarter-wave matching network cut for 3.65 MHz and made from 75-Ω coax as described in N4KG's article may work just as well. It might be worth a try if you wish to save the cost of an unun transformer and minimize complexity.

Note that the above approach assumes that your guy wires are insulated from the tower and that the guys are either broken into nonresonant lengths or made of insulating material.

An Example

Fig 1 represents my specific application. I have a rather large multiband beam on a 24-foot boom that covers 40 through 10 meters. In addition, I have a dual-band VHF/UHF gain antenna on top of this tower (more on this later). Applying the formula in step 1 above (and ignoring the VHF/UHF gain antenna on top), I calculated an electrical height of 144°.

Subtracting 90° from this number gives a point on the tower where a quarter wavelength at 3.65 MHz remains above the target "sweet spot." This point is thus 54° above ground. The initial starting point is therefore (54°× 492)/(180° × 3.65 MHz), or 40.4 feet above ground. My initial try resulted in two SWR minima, with one at around 3.2 MHz and the other at about 3.5 MHz. This indicated that I was too low on the tower, and that the VHF/UHF gain antenna was contributing, as expected, to the tower's electrical length.

I then raised the attachment point until I found the best compromise for operation in the desired band segments. This point was 46 feet above ground. Again, do not succumb to the temptation to trim the radial lengths to move the points where the SWR minima occur. I found that it's better to move the attachment point on the tower in increments as small as six inches, if necessary, to achieve the desired balance. The key point here is that you should try to find the point where an equivalent antenna length of 90° above the connection point is halfway between your two desired band segments. With the 50:22.16 transformer in place, the SWR curve that I ended up with for this arrangement is shown in **Fig 2**.

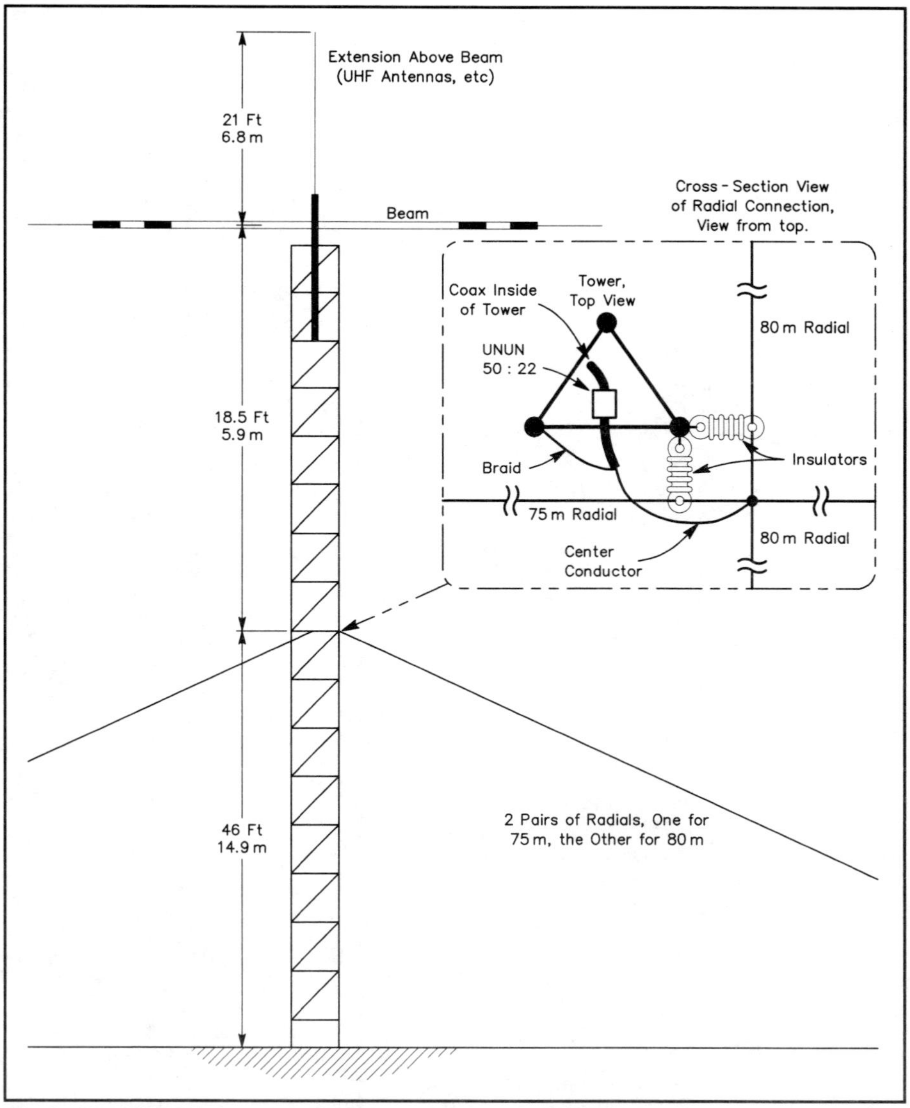

Fig 1—Details of the reverse-fed ground-plane at W4PK. The inset shows the feed-point details. Keep the two sets of radials as close to 90° from each other as possible. Don't try to resonate the antenna by adjusting radial lengths.

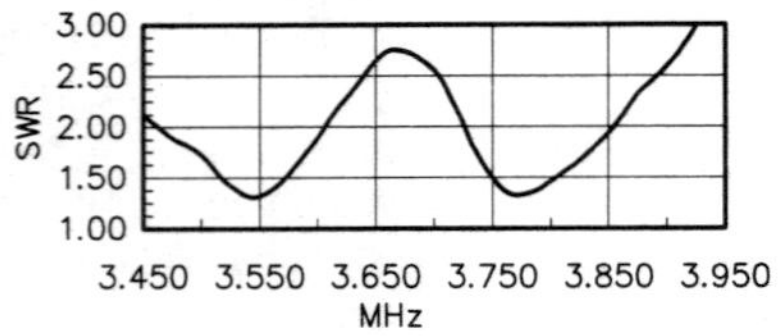

Fig 2—SWR curve of the reverse-fed ground-plane antenna installed on W4PK's tower. The antenna is placed on the tower at a height that places the SWR minima at 3.55 and 3.8 MHz. This equates to 90° above the feed point at 3.65 MHz. The antenna is fed with a 50:25-Ω unun transformer or quarter-wave coaxial matching section.

Note that the VHF/UHF antenna on my tower contributes as a part of the radiator for the elevated ground plane. If you have a similar arrangement, be sure to check the 3.5-MHz RF signal level at the input to your VHF/UHF equipment. Likewise, check the level of your VHF/UHF transmitted signal at the input to your HF transceiver. In my case, I routed the coax for the VHF/UHF antenna down the inside of the tower to reduce coupling from the 3.5 MHz signal onto the coax shield. In addition, I employ two VHF/UHF lightning protectors (one at the tower base, the other at the entry point to the house) rated for 30 MHz and above. They provide some degree of attenuation at 3.5 MHz, and the signal at the input to my VHF/UHF equipment is at a very safe –40 dBm level with a kilowatt applied to the 80-meter antenna.

You may need to employ a high-pass or band-pass filter if the level that you measure is significantly higher. I also found that the signal level from the VHF/UHF equipment at the HF transceiver input to be negligible. Note that the VHF/UHF antenna is not necessary for the successful application of this approach.

Results

I have been very pleased with this antenna's performance. I installed it as a temporary arrangement while I was constructing a full-size 4-square antenna for 75/80 meters. I had the opportunity to compare its performance with the first vertical of the 4-square, under which I had 72 buried radials of nominally 0.31 λ each. Basically, I found that the difference in signal levels with a local ham about two miles away was less than 1 dB. On-the-air listening tests on low-angle signals from Europe further confirmed that there was no discernible difference in received signal levels between the full-size, ground-based vertical and the elevated, top-loaded groundplane. For reference, I have since completed the 4-square array, and as expected, the 4-square does show a solid S-unit (4-5 dB) improvement in its favored direction over the elevated ground-plane.

Dual-Mode Elevated Verticals

By Al Christman, K3LC
Grove City College
100 Campus Drive
Grove City, PA 16127-2104

Making an antenna work in the DX windows on both ends of the 80/75-meter band is not always easy to do. K3LC describes a nifty solution to this problem for a vertical antenna with elevated radials.

The 80-meter band is so broad in terms of frequency that it is difficult to build an antenna that can cover both the phone and CW DX portions of the band with satisfactory SWR. This paper discusses a method for switching an elevated vertical between two different band segments using one SPST relay.

Basic Concept

This antenna system utilizes a technique described in ON4UN's well-known book on low-band DXing.[1] There, John illustrates an elevated four-square 80-meter phased vertical array, composed of wire elements suspended from a central tower (ON4UN's quarter-wave transmit vertical for 160 meters). The feed points are 17 feet above ground. Interestingly, each element has only a single horizontal radial, rather than two to four as are often found with elevated verticals. The monopoles are 64 feet long, while each radial is just 61.35 feet in length, yielding resonance at 3.75 MHz. When operating on CW, a shorted stub is added to the feed end of every radial. Each stub consists of two parallel #9 AWG copper wires 7.4 feet long, spaced 8 inches apart. At each vertical monopole, a set of relay contacts at the upper end of its stub is closed for SSB, and opened to work CW, with resonance at 3505 kHz.

Design Procedure

John's clever idea clearly shows (and computer analysis confirms) that the resonant (zero-reactance) frequency of an elevated vertical antenna may be shifted by altering the length of *either* the monopole *or* the elevated radial(s). Since the radials are situated much closer to the ground, it is often easier (and safer) to adjust the length of the radials, as opposed to changing the overall height of the vertical element.

With this in mind, I created a computer model of a single elevated vertical radiator with four elevated horizontal radials, using the *EZNEC/4* computer program.[2] All conductors were #12 AWG copper, and the base of the antenna was placed 12 feet above average ground (soil conductivity = 0.005 Siemens/meter and dielectric constant = 13).

I assumed that the center of the DX phone band was at 3790 kHz and selected 3510 kHz on CW, for a median value of 3650 kHz. Then I adjusted the length of all five wires in my computer model to produce resonance at 3650 kHz. The required length for each wire was 67.25 feet, and the resulting feed-point impedance was 33.45 – j 0.06 Ω. *EZNEC/4* showed a peak gain of 0.33 dBi at a take-off angle of 22°, and the azimuthal pattern was almost perfectly circular, with the gain falling slightly to 0.31 dBi at compass angles midway between the radials.

SSB Mode

Initially, I wanted to make my reference antenna resonant at a frequency midway between the desired band segments, because I wasn't sure how the length of the vertical element would impact the SWR readings once the dual-mode design was completed (more on this later). After the antenna had been resonated at 3650 kHz, the next task was to offset the radials slightly from their original center point, which had been directly beneath the vertical radiator. I did this to make room for the shorted stub (see **Fig 1** for a detailed drawing).

I decided to add the stub-wires to my computer model, even though I wouldn't activate them until the antenna was switched to the CW mode. Since the base of the vertical was 12 feet above ground, I made the stub wires 8 feet long, and spaced them six inches apart. When operating on SSB the stub would be shorted at its upper end, so I simulated this short-circuit with a six-inch jumper wire between the base of the vertical element and the common point where the inner ends of all four radials are joined together. I didn't bother to include the shorting bar on the stub, just the two parallel wires.

Surprisingly, these small changes at the base of the antenna necessitated a drastic shortening of the radials—to only 53.25 feet—in order to shift the resonant frequency upward from 3650 to 3790 kHz. Once I did this, an input impedance of 39.06 – j 0.05 Ω was achieved, and the maximum gain in SSB mode was now 0.20 dBi at a 22° take-off angle. With a six-inch offset be-

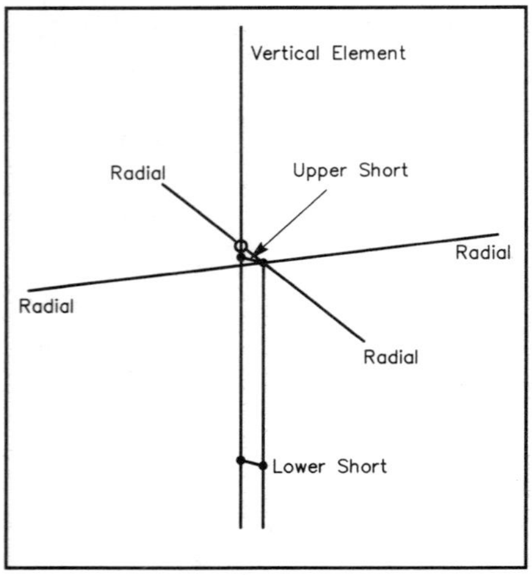

Fig 1—Close-up of the area near the feed point of a single elevated vertical antenna, showing the inner ends of the four elevated radials. These are offset slightly from the base of the vertical radiator to allow for placement of the shorted stub. The lower short (on the stub) is always present and its position sets the CW resonant frequency. The upper short consists of a closed pair of relay contacts and is present only in SSB mode (relay contacts are opened for CW).

tween the base of the monopole and the common point of the radials, the non-circularity of the azimuthal-plane radiation pattern is only 0.12 dB, which seems tolerable. The highest gain is in the direction of the offset, with the minimum (0.08 dBi) occurring in the opposite quadrant.

A frequency-sweep was carried out with *EZNEC/4* to determine the bandwidth of the antenna in SSB mode, and the results are listed in **Table 1**. Notice that the SWR (referred to 50 Ω) remains below 1.5:1 for about 100 kHz.

CW Mode

The resonance of the antenna system is moved down into the CW band by activating the stub (removing the short at its upper end) and adjusting the position of the shorting bar to achieve zero reactance at the desired frequency. With the shorting bar located 5.635 feet below the top of the stub, *EZNEC/4* predicts an input impedance of 31.28 + *j* 0.02 Ω at 3510 kHz. The peak gain now is 0.25 dBi (still at an elevation angle of 22°), with a non-circularity of just 0.03 dB, a negligible value. The frequency-sweep data is displayed in **Table 2**; the SWR is quite a bit higher in CW mode than on phone, which is disappointing. However, all of the numbers are well below 2:1.

The SWR on both CW and SSB can be improved by adding an "un-un" (unbalanced-to-unbalanced) transformer at the feed point of the antenna. Since the input resistances at resonance are roughly 39 and 31 Ω for the two modes, an un-un with a 35:50-Ω ratio would be a good compromise. **Table 3** shows the results, which are certainly worth the cost of the un-un transformer. Now the SWR is below 1.12:1 at the center frequencies for both modes.

Table 1

SWR versus Frequency for the Elevated Vertical Antenna in SSB Mode

Frequency (kHz)	*SWR (50-Ω reference)*
3750	1.451
3760	1.387
3770	1.335
3780	1.299
3790	1.280
3800	1.281
3810	1.301
3820	1.338
3830	1.387
3840	1.448
3850	1.516

Table 2

SWR versus Frequency for the Elevated Vertical Antenna in CW Mode

Frequency (kHz)	*SWR (50-Ω reference)*
3500	1.619
3510	1.599
3520	1.596
3530	1.611
3540	1.641
3550	1.690

Table 3

SWR versus Frequency for the Elevated Vertical Antenna in both CW and SSB Modes, using a 35:50-Ω un-un

Frequency (kHz)	*SWR (50-Ω reference)*
3500	1.162
3510	1.119
3520	1.146
3530	1.220
3540	1.312
3550	1.424
—	—
3750	1.366
3760	1.274
3770	1.195
3780	1.136
3790	1.116
3800	1.148
3810	1.212
3820	1.291
3830	1.380
3840	1.479
3850	1.582

Variations

On my first attempt to model this antenna with *EZNEC/4*, I used a twelve-inch spacing between the two parallel stub-wires, but this produced a non circularity in the SSB-mode azimuthal-plane radiation pattern of fully half a decibel, which I felt was excessive. Because of this, I decreased the spacing to six inches, which yielded a pattern shape that is much more circular, as mentioned above.

Since the SWR of the dual-mode antenna (without un-un) is higher on CW than on phone, I decided to increase the length of the vertical monopole to see what effect this would have. Therefore, I added 1.75 feet to the radiator, making it 69 feet tall instead of 67.25 feet. This changed the SSB resonant frequency of the antenna, so then I had to shorten the length of the four radials to 46.65 feet to achieve resonance once again at 3790 kHz.

Now the input impedance was 43.04 + *j* 0.027 Ω, with a maximum gain of 0.12 dBi at a takeoff angle of 22°. The azimuthal-plane radiation pattern is very close to perfectly circular, with a minimum gain (at a 22° elevation angle) of 0.04 dBi, in the direction away from the offset.

With the stub energized, the shorting bar was then adjusted to achieve CW resonance at 3510 kHz. An active stub length of 5.80 feet yielded an input impedance of 34.35 + *j* 0.045 Ω, with a peak gain of 0.18 dBi at a takeoff angle of 22°. Maximum non-circularity in the azimuthal-plane radiation pattern was only 0.03 dBi. The SWR values for both modes of operation are given in **Table 4**. Note that the SWR has improved on both phone and CW, but the CW values are still a bit high, around the 1.5:1 mark. As before, an un-un (this time with a 38:50-Ω impedance ratio) at the feed point can be used in an attempt to equalize the SWR readings on the two modes, as shown in **Table 5**. Now the SWR on phone is actually slightly worse than on CW; choosing an un-un impedance-transformation ratio of 39:50 would make the phone values lower and the CW numbers higher.

A Dual-Mode Four-Square Array

The classic four-square phased-vertical array was introduced to low-band operators more than 20 years ago[3] and it continues to be widely used by DXers and contesters around the world. A simplified version of the four-square, incorporating only two elevated radials per vertical element, was

Table 4

SWR versus Frequency for the Elevated Vertical Antenna with 69-foot Radiator, in both CW and SSB Modes

Frequency (kHz)	*SWR (50-Ω reference)*
3500	1.476
3510	1.456
3520	1.456
3530	1.474
3540	1.510
3550	1.561
—	—
3750	1.352
3760	1.284
3770	1.226
3780	1.183
3790	1.162
3800	1.168
3810	1.198
3820	1.246
3830	1.305
3840	1.373
3850	1.446

Table 5

SWR versus Frequency for the Elevated Vertical Antenna with 69-foot Radiator, in both CW and SSB Modes, using a 38:50-Ω un-un.

Frequency (kHz)	*SWR (50-Ω reference)*
3500	1.146
3510	1.106
3520	1.135
3530	1.204
3540	1.292
3550	1.391
—	—
3750	1.342
3760	1.262
3770	1.193
3780	1.146
3790	1.133
3800	1.161
3810	1.216
3820	1.287
3830	1.367
3840	1.454
3850	1.549

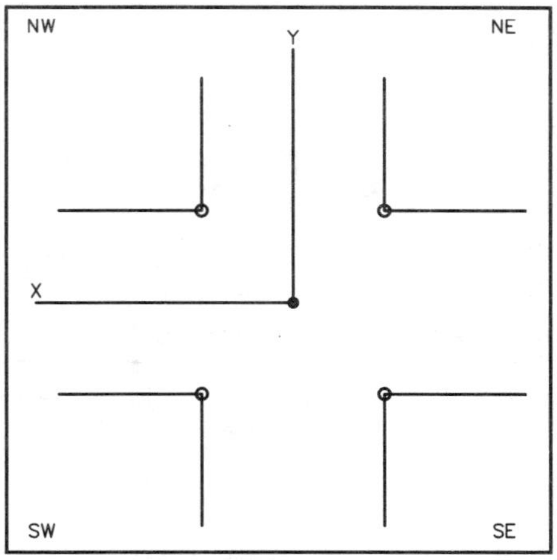

Fig 2—Plan view of the elevated dual-mode four-square array, which utilizes only two radials with each vertical element. Four shorted stubs are present but are too small to be discernible in this drawing. Each pair of radials is offset in the direction of the associated forward lobe (ie, the radials for the northeast element are offset toward the northeast from the base of the vertical radiator).

modeled using *EZNEC/4*. A plan view of this array positioned at a height of 15 feet above the ground is illustrated in **Fig 2**.

The first step in the design process would normally be to generate a computer model of a two-radial antenna resonant at 3650 kHz, midway between the center frequencies of the CW and SSB sub-bands (3510 and 3790 kHz respectively). However, many hams use the ComTek Systems hybrid unit[4] with their four-squares: Jim Miller, K4SQR, suggests a different procedure. He recommends that each full-sized element of the array be resonated at a frequency that is 100 kHz *lower* than the desired center frequency, which would be 3550 kHz for this example. The length of all three wires turned out to be 68.82 feet, producing an input impedance of 37.92 + *j* 0.009 Ω at 3550 kHz.

Next, I inserted a 10-foot-long stub, made from two #12 AWG copper wires spaced 6 inches apart, at the inner ends of the radial pair. Initially, the shorting bar was omitted from the *EZNEC/4* model, and the upper end of the stub was short-circuited (using a six-inch wire jumper) for SSB operation. I wanted to resonate the antenna at 3790 – 100 = 3690 kHz so the ComTek hybrid would work correctly. When each of the radials is shortened to a length of 59.8 feet, *EZNEC/4* predicts a feed point of 41.25 + *j* 0.024 Ω at 3690 kHz.

For CW operation, the short-circuit at the top of the stub is removed, and the adjustable shorting bar is moved downward along the stub until resonance is achieved at 3510 – 100 = 3410 kHz. I found that an active stub length of 7.39 feet produces an input impedance of 33.47 + *j* 0.031 Ω at 3410 kHz.

Don't try to improve the SWR by installing an un-un or other impedance-matching circuit at the base of the vertical element. If only a *single* antenna is contemplated, then it may be worthwhile to match the antenna impedance to 50 Ω to improve the SWR bandwidth. However, with an array like a four-square, efforts to improve the impedance match of each individual element may produce undesired consquences when these verticals are connected to the phasing/matching system, possibly generating incorrect drive-point currents.

Now at last we can create three more identical copies of our two-radial dual-mode vertical antenna to finish the construction of the entire four-square array, shown previously as Fig 2. Generally speaking, it is not a good idea to build into the model more than one vertical antenna at the start, unless *all* of the quarter-wave wires making up the other antennas are electrically isolated from one another. (If just two quarter-wave elevated radials are joined together at their inner ends, then this floating half-wave wire can interact strongly with the active antenna; simply open-circuiting the base of the vertical element is probably not sufficient.) For best accuracy, all pruning and tuning in the model should be carried out on a single,

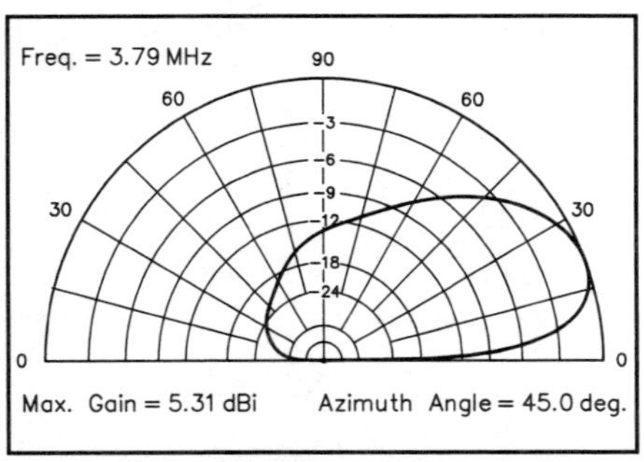

Fig 3—Elevation-plane radiation pattern for the elevated dual-mode four-square phased vertical array, in SSB mode.

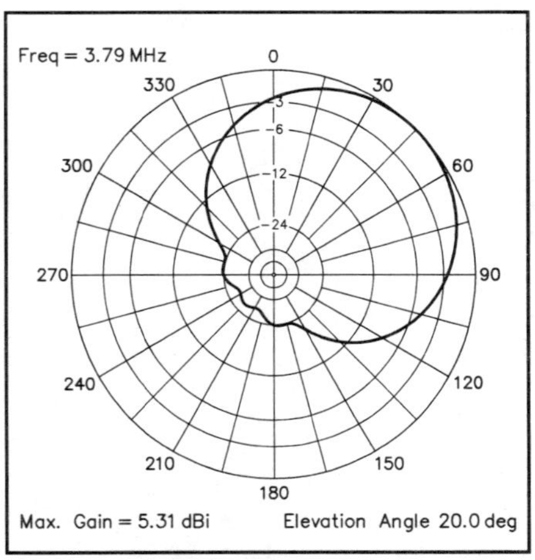

Fig 4—Azimuthal-plane radiation pattern for the elevated dual-mode four-square phased vertical array, in SSB mode.

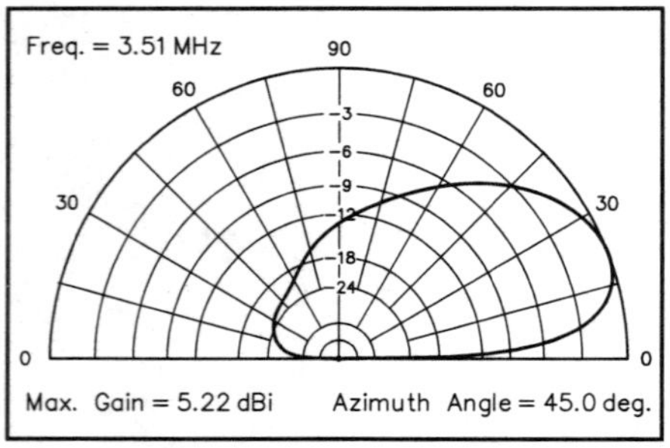

Fig 5—Elevation-plane radiation pattern for the elevated dual-mode four-square phased vertical array, in CW mode.

isolated vertical antenna and the other three clones should not be built into the model until all of these preliminary adjustments have been completed. This also holds true for the actual array.

When driven at 3790 kHz in the center of the 75-meter DX phone band, the elevation and azimuthal-plane radiation patterns that result are displayed in **Figs 3** and **4**. Forward gain is 5.31 dBi at a 20° elevation angle, and the front-to-back ratio at this takeoff angle is 27.37 dB. On 3510 kHz in the CW subband, the gain is slightly lower and the take-off angle a bit higher (5.22 dBi at 21°), since the physical height of the array is fixed at 15 feet. The front-to-back ratio is also not quite as good, at 24.48 dB for the 21° take-off angle. The principal-plane CW-mode radiation patterns are shown in **Figs 5** and **6**.

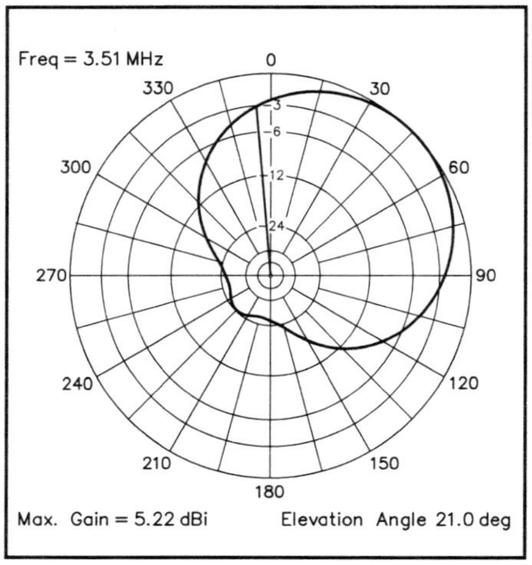

Fig 6—Azimuthal-plane radiation pattern for the elevated dual-mode four-square phased vertical array, in CW mode.

Please be aware that the radiation patterns realized in practice may not be as good as those portrayed in the figures. It isn't possible to simulate the ComTek hybrid with *EZNEC/4*, so idealized feed-point currents are utilized in the computer model. The perfect driving-point currents assumed for the software simulation may not exist in an actual array, especially if a single set of quarter-wave 75-Ω phasing lines is used with the ComTek hybrid. Generally, these phasing lines are cut to be a quarter-wave long at the 3650 kHz center frequency, which is a compromise for both phone and CW operation. If this arrangement isn't satisfactory, it might be better to cut the phasing lines for quarter-wave resonance at 3790 kHz (SSB), and then use relays to add short extender cables to increase their electrical length to a quarter-wave at 3510 kHz (CW).

Conclusion

I've shown in this paper how to switch an elevated vertical antenna from one frequency subband to another, using a single shorted stub to effectively increase the length of several elevated radials simultaneously. Computer modeling indicates that the design can be used with an isolated vertical or with multi-element phased arrays. An actual antenna system has not been constructed by the author, although a similar elevated four square is being used successfully by John Devoldere, ON4UN, who has previously described this basic technique in the literature.

References

[1]John Devoldere, ON4UN, *Antennas and Techniques for Low-Band DXing*, second edition (Newington: ARRL, 1994).

[2]Several versions of *EZNEC* antenna-modeling software are available from Roy Lewallen, W7EL, PO Box 6658, Beaverton, OR 97007.

[3]D. Atchley, W1CF, H. Stinehelfer, ex-W2ZRS, and J. White, *"360°-Steerable Vertical Phased Arrays,"* *QST*, Apr 1976, pp 27-30.

[4]ComTek Systems, 11600 Hilda Court, Charlotte, NC 28226-3825.

Elevated Vertical Antennas Over Sloping Ground

By Al Christman, K3LC
Grove City College
100 Campus Drive
Grove City, PA 16127-2104

One important part of antenna analysis that's frequently neglected is ground slope. Here's a look at how to configure a single elevated vertical, a pair in a cardioid array, and a four-square array to meet your needs under sloping-ground conditions.

In mid-1994, I received a letter from some hams in Sweden who were preparing to construct an elevated four-square vertical array. The ground at their site was not level, but sloping, and they wanted to know the best way to orient the antennas and the radials. I gave them an initial answer, based on some analyses that I had performed using *ELNEC*.[1] Since then, I've obtained better software, and this article presents the results of that updated study.

Computer Simulation

All of the computer modeling shown here was carried out using *EZNEC pro*.[2] This is an improved version of Roy (W7EL) Lewallen's widely used *ELNEC* software, but it incorporates *NEC-4*[3] as its computing engine rather than *MININEC*.[4] Because of this upgrade, *EZNEC pro* can model antennas that touch or even penetrate the ground, and it also handles real ground much more accurately than *ELNEC*. My 1988 *QST* article on elevated vertical antennas[5] was derived from computer analysis I had completed using *NEC-GSD*, a special version of *NEC-3* that was configured to efficiently model vertical antennas with large numbers of symmetrically disposed radials and/or top-hat wires. *NEC-4* is, itself, a descendant of *NEC-3*, and is similar to it in many ways.

Fig 1 shows a single vertical monopole antenna with four horizontal radials; each conductor has an electrical length of 0.25 λ at a frequency of 3.8 MHz. All wires are #12 AWG (radius = 1 mm) and the base of the antenna is 15 feet above the ground. The monopole and the outer ends of the radials are supported by metallic masts, which are anchored to the earth via 5-foot ground rods. The height of the radial-support masts is 15 feet, and they are separated laterally from the tips of the radials by a distance of 6 inches. The central mast supports the monopole, but is only 14.5 feet high, so it is *not* electrically connected to the vertical radiator. This arrangement, which I called "isolated feed" in the earlier *QST* article, reduces the amount of current that flows on the center mast.

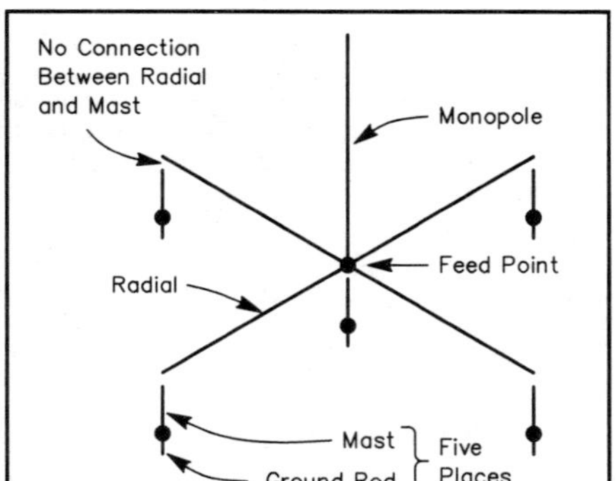

Fig 1—Physical layout of an elevated-vertical antenna with four elevated, horizontal radials. The coax shield is connected only to the four radials, and not to the central support mast (so-called isolated feed). Height=15 feet.

At this point I ran into a minor problem caused by a limitation in *EZNEC*. My earlier *NEC-GSD* model had used an aluminum monopole and copper radials, plus steel masts and ground rods, while *EZNEC* requires that all metallic conductors be composed of the same material. To overcome this difficulty I used an *EZNEC pro* model in which all the conductors were made of aluminum. The conductivity of aluminum is intermediate between those of copper and steel, so I felt that selecting aluminum for the entire antenna would be a good compromise.

Finally, I decided to change the conductivity of the soil for the *EZNEC pro* computer models. Back in 1988, I had used a conductivity value of 0.003 Siemens/meter, in conjunction with a dielectric constant of 13. Since W7EL includes several standard sets of ground constants, I chose to follow his lead and assume a soil conductivity of 0.005 S/m, which allows me to keep the dielectric constant fixed at the previous value of 13.

Sloping-Ground Strategy

The next area of concern is the technique for modeling sloping ground. With *NEC* it is possible to construct a ground composed of two regions, each with its own conductivity and dielectric constant. However, the authors of the code didn't anticipate the need to model a large expanse of smooth but *sloping* ground. The easiest way to do this is simply to tilt one or more portions of the antenna so that the geometrical orientation of the tilted antenna over flat ground is identical to that of a normal upright antenna over sloping ground. This concept is illustrated in **Fig 2**; rotating the drawing at (a) by α in a clockwise direction produces the drawing at (b). Of course, *EZNEC* doesn't know that you have done this, so all of the elevation-plane radiation patterns that it produces will be correct for flat ground, and these must also be rotated clockwise by α to make them correct for sloping ground.

Throughout this article I have used a slope angle of 8°, which means that the ground is fairly steep. I chose such a large angle for two reasons. First, I am sure that some amateurs have antenna farms situated on terrain similar to this, or even worse. In the mid-1970s I worked in the coal mines of southern West Virginia, and I had almost two hours less daylight at my QTH "down in the hollow" than the fellows who lived 50 miles away in the flatlands. Second, I wanted to make sure that any slope-induced changes in the radiation patterns were big enough to be easily discernible on the plots, and the best way to accomplish this was to use a large angle.

Evaluation Approach

To appropriately answer the original question, namely, what's the best way to erect a four-square over sloping ground, it's necessary to take some intermediate steps to get to the core issues. In this article, I'll first look at a few configurations for a single vertical with elevated radials, then a pair in a cardioid array (90° spacing and phasing), and finally the four-square. Each analysis is based on four radials per element, except in certain specific situations where the antenna geometry allows combining radials; in these situations, I'll state how the models are configured.

Once this basic analysis is done, I'll draw some conclusions about the best approach to take for the single element, cardioid, and four-square array, depending on your particular location and goals. You can draw additional conclusions from this material as well—such as the trade-offs and advantages of building vertical antennas and arrays using sloping radials over flat ground.

Single Elevated Vertical Antenna

There are a number of different ways to configure a single elevated vertical-monopole antenna over sloping ground, since either the radials and/or the monopole can be tilted with respect to the local earth surface. I decided immediately to always orient the monopole so it was perfectly vertical ("straight up and down"), because I don't think anyone would deliberately try to support a large quarter-wave radiator in an inclined position.

If we specify that all four of the elevated radials must lie in a single plane, then there are only two ways to orient this plane, as illustrated in **Fig 3**. Antenna A in the figure shows the radial-plane located parallel to the local (sloping) ground. Here, each of the four radial wires is always at the same height above the ground along its entire length. (Notice that the plane containing the radials is *not* perpendicular to the vertical monopole.) The other option is illustrated as antenna B, where the radial plane *is* perpendicular to the vertical monopole, but no longer parallel to the local ground-level. Now the radials form a plane that is perfectly horizontal, but their height above lo-

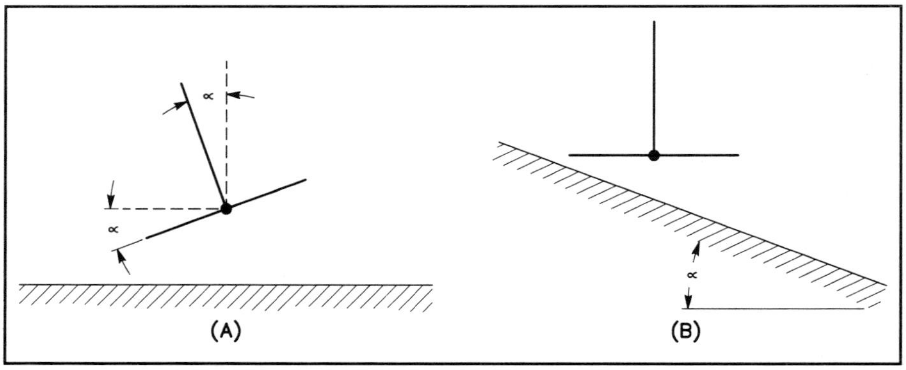

Fig 2–A tilted antenna over flat ground looks like a normal antenna over sloping ground. The tilt angle, a, is the same as the ground-slope angle.

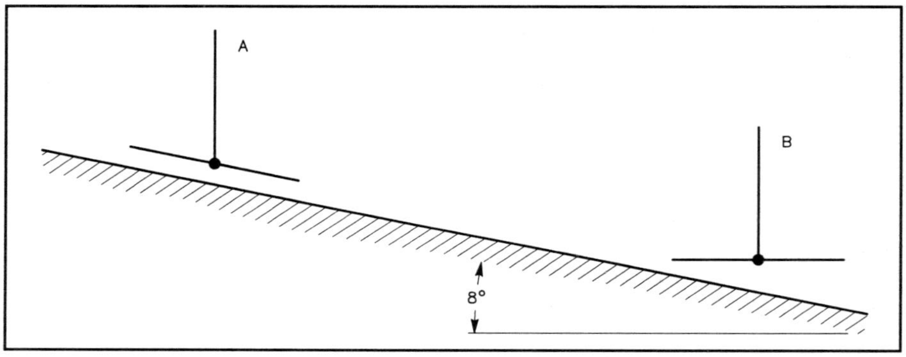

Fig 3–If all four radials must be coplanar, only two placement options exist: At A, the radial plane is parallel to local ground; at B, the radial plane is perpendicular to the vertical monopole.

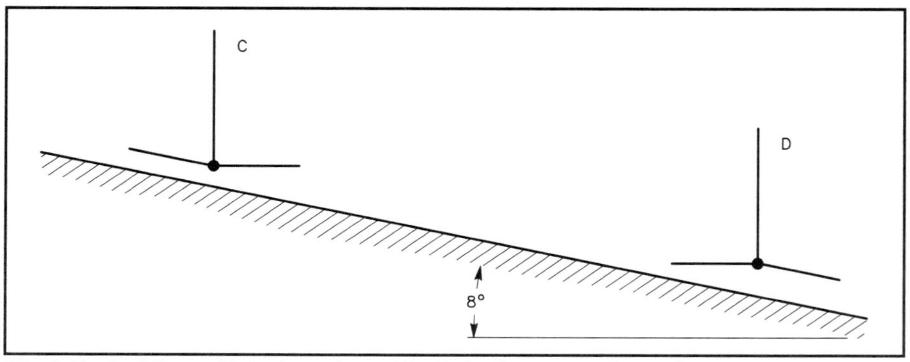

Fig 4–If all of the radials do *not* need to lie in the same plane, then two more options appear: At C, the uphill radials are parallel to local ground while the downhill radials are perpendicular to the vertical monopole; at D, the uphill radials are perpendicular to the vertical monopole and the downhill radials are parallel to the local ground.

cal ground may vary from place to place.

In contrast, if we do not force all four of the elevated radials to lie in the same plane, then two more variations in their positioning become possible, as shown in **Fig 4**. Antenna C of this figure has the uphill radial(s) drawn parallel to the local (sloping) ground, with the downhill radial(s) depicted as perpendicular to the vertical monopole. For antenna D the downhill radial(s) are parallel to the sloping ground, with the uphill radial(s) drawn perpendicular to the vertical monopole.

After the elevation-plane orientation of the radials has been selected, the next task is to determine the azimuth positioning of the four radials with respect to the sloping ground. There are many possibilities, but **Fig 5** shows the only two variations that will be discussed in this paper. In part (i), two of the radials lie parallel to the "line of steepest descent," while the remaining two radials are perpendicular to this line. For example, if the ground slopes directly downhill along a line that runs from north to south (toward 180° azimuth), then the four radials would point to the four cardinal points of the compass—north, south, east and west. Part (ii) illustrates the second alternative, wherein the radials all are turned at 45° with respect to the line of steepest descent. Thus the four radials point to the northeast, southeast, southwest and northwest, respectively.

The four elevation choices (A, B, C and D of Fig 3 and Fig 4) in combination with the two azimuth selections (i and ii in Fig 5) yield a total of eight different possible ways to construct a single elevated-radial vertical-monopole antenna. **Table 1** summarizes the computer-predicted performance of each configuration, showing the maximum gain in various directions (downhill, uphill, either side of the hill) as well as the amount of non-circularity in the azimuth-plane radiation pattern.

Table 1A

Power Gain versus Direction, Noncircularity, Front-to-Back Ratio and Front-to-Side Ratio for Eight Configurations of a Single Isolated-Feed Vertical Antenna with Four Elevated Radials with a Ground Slope Angle of 8°

	Power gain (dBi)					
Configuration	*Downhill*	*Uphill*	*Sides*	*Noncircularity (dB)*	*Front/Back Ratio (dB)*	*Front/Side Ratio (dB)*
A(i)	0.24	0.12	0.22	0.12	0.12	0.02
B(i)	0.77	–0.42	0.08	1.19	1.19	0.69
C(i)	–0.02	–0.77	0.44	1.21	0.75	–0.46
D(i)	0.78	0.07	–0.40	1.18	0.71	1.18
A(ii)	0.21	0.16	0.18	0.05	0.05	0.03
B(ii)	0.72	–0.60	0.16	1.32	1.32	0.56
C(ii)	0.47	–0.15	0.18	0.62	0.62	0.29
D(ii)	0.56	–0.21	0.21	0.77	0.77	0.35

Table 1B

Power Gain versus Direction, Noncircularity, Front-to-Back Ratio and Front-to-Side Ratio for Eight Configurations of a Single Isolated-Feed Vertical Antenna with Four Elevated Radials with a Ground Slope Angle of 4°

	Power gain (dBi)					
Configuration	*Downhill*	*Uphill*	*Sides*	*Noncircularity (dB)*	*Front/Back Ratio (dB)*	*Front/Side Ratio (dB)*
A(i)	0.25	0.19	0.23	0.06	0.06	0.02
B(i)	0.53	–0.09	0.19	0.62	0.62	0.34
C(i)	0.14	–0.22	0.39	0.61	0.36	–0.25
D(i)	0.56	0.22	–0.04	0.60	0.34	0.60
A(ii)	0.22	0.19	0.20	0.03	0.03	0.02
B(ii)	0.51	–0.16	0.20	0.67	0.67	0.31
C(ii)	0.36	0.03	0.20	0.33	0.33	0.16
D(ii)	0.39	0.02	0.21	0.37	0.37	0.18

Table 1C

Power Gain versus Direction, Noncircularity, Front-to-Back Ratio and Front-to-Side Ratio for Eight Configurations of a Single Isolated-Feed Vertical Antenna with Four Elevated Radials with a Ground Slope Angle of 12°

	Power gain (dBi)					
Configuration	*Downhill*	*Uphill*	*Sides*	*Noncircularity (dB)*	*Front/Back Ratio (dB)*	*Front/Side Ratio (dB)*
A(i)	0.20	0.02	0.21	0.19	0.18	–0.01
B(i)	0.94	–0.75	–0.15	1.69	1.69	1.09
C(i)	–0.21	–1.42	0.41	1.83	1.21	–0.62
D(i)	0.89	–0.22	–0.79	1.68	1.11	1.68
A(ii)	0.20	0.12	0.14	0.08	0.08	0.06
B(ii)	0.85	–1.11	0.10	1.96	1.96	0.75
C(ii)	0.54	–0.31	0.14	0.85	0.85	0.40
D(ii)	0.70	–0.51	0.20	1.21	1.21	0.50

This table contains several items of interest. First, if true omnidirectional coverage is desired, configuration A(ii) produces a radiation pattern that is almost perfectly circular, despite being constructed over sloping ground. This antenna has its four radials oriented at 45° to the line of steepest descent, in a plane that is parallel to the local ground.

All eight configurations have more gain in the downhill direction than uphill, so I have chosen to equate downhill with the front of the radiation pattern. Antenna D(i) has the most gain toward the front, and is especially notable for having the least amount of radiation to the sides, as well as the largest front-to-side ratio. The honors for highest front-to-back ratio belong to antenna B(ii), which also holds third place in both forward-gain and front-to-side ratio. Antenna B(i) is the runner-up in no less than three categories: front-to-back ratio, front-to-side ratio, and forward gain, where it loses to antenna D(i) by only 0.01 dB. The uphill-downhill elevation-plane radiation patterns for all three of these antennas are shown together in **Fig 6**; remember that (here) the ground slopes downhill toward the right edge of the page. **Fig 7** illustrates the side-to-side elevation-plane radiation patterns for the same three antenna configurations.

None of these antennas produces maximum gain in the uphill direction, but C(i) is very unusual because its peak gain occurs to the *sides* rather than uphill or downhill. As a result, its front-to-side ratio is actually negative, if we consider downhill to be the front. This configuration produces more signal to the sides, and less signal going uphill or downhill, than any of the others.

Rus Healy, K2UA, suggested that it would be helpful to examine what happens to the radiation pattern of a single elevated vertical antenna if the ground was sloped at some angle greater or less than 8°. Therefore, information for a slope angle of 4° is listed in **Table 1B**, and **Table 1C** includes data for a 12° slope. For both of these additional slope angles, all eight configurations *still* have the most gain in the downhill direction. Increasing the steepness of the ground causes the front-to-back and front-to-side ratios to increase proportionally, while the reverse is true when the ground becomes flatter. As before, configuration A(ii) always yields the most circular azimuth-plane pattern, while C(i) continues to produce more gain alongside the hill than it does downhill.

So, by adjusting the manner in which its four radials are positioned with respect to local ground, the enterprising ham can build a single elevated vertical-monopole antenna with either an omnidirectional radiation pattern, or a very modest beam effect. None of these antennas have any side-to-side directivity; that is, each radiates the same amount of signal on both sides of the line of steepest descent.

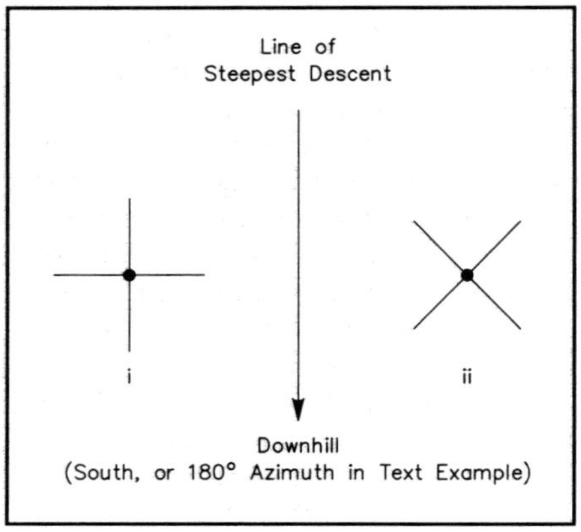

Fig 5–Top view of an elevated vertical with four elevated radials, showing two ways to position the radials (in azimuth) with respect to the sloping ground: At (i), the radials are parallel to the line of steepest descent; and at (ii), the radials are oriented at 45° to the line of steepest descent.

Two-Element Vertical Arrays

Normally, a ham constructing a two-element phased vertical array wants the antenna system to be electrically symmetrical. In other words, when building a cardioid array with 90° spacing and phasing, the operator would like the gain, front-to-back ratio and relative driving-point impedances to remain unchanged (or nearly so) as the array is switched from one direction to the other. Thus, the antenna should "look the same" and perform the same (electrically) in both directions of fire. Similarly, for bi-directional two-element arrays (such as those with 180° spacing and phasing) the goal would be to have two equal-sized lobes of radiation for a given directional setting.

However, there are two potential strategies that may be pursued by the amateur who has sloping ground at his disposal: One, construct an array designed to counteract the (undesired) directive effects of sloping ground, so that the resulting antenna system performs in a more or less symmetrical fashion, despite the sloping ground; or two, build an array that deliberately combines the normal directivity of the phased radiators with the previously undesired effects of the sloping ground to produce *enhanced* gain toward a particular compass heading. This array would, no doubt, be highly asymmetrical, and would probably be designed to beam in only *one direction* (downhill). We have already seen that an isolated vertical

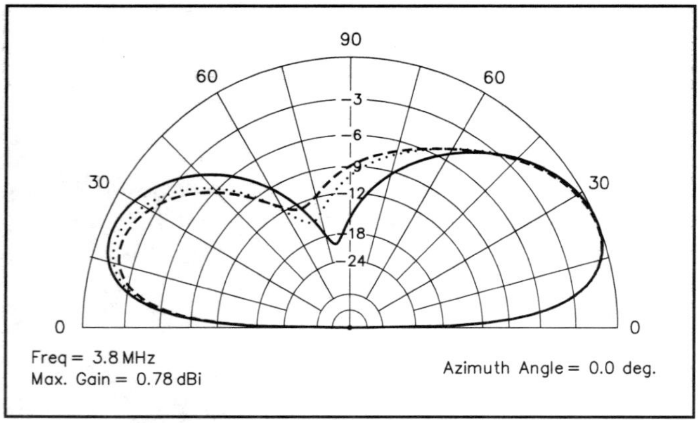

Fig 6–Principal elevation-plane radiation patterns for several elevated-vertical antennas. Downhill is toward the right edge of the drawing. The outer ring of the plot represents 0.78 dBi.

Solid line = configuration D(i)
Dashed line = configuration B(ii)
Dotted line = configuration B(i)

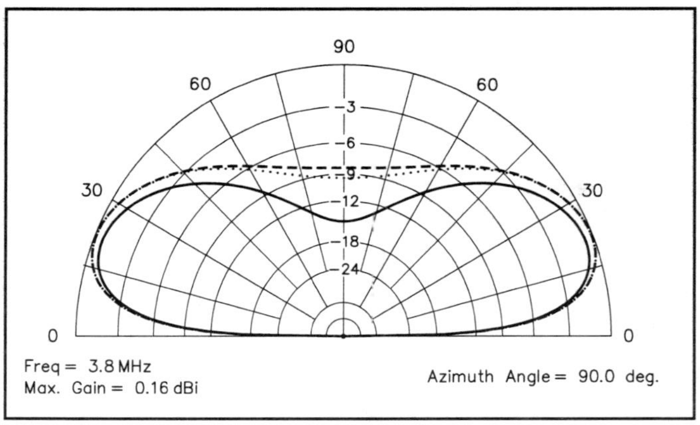

Fig 7–Elevation-plane radiation patterns at right angles to the uphill/downhill line shown in Fig 6. (Downhill is perpendicular to the paper.) The outer ring of the plot represents 0.16 dBi.

Solid line = configuration D(i)
Dashed line = configuration B(ii)
Dotted line = configuration B(i)

monopole, unless carefully oriented, tends to radiate mostly downhill, so it might be possible to use this to our advantage.

Fig 8 illustrates two ways to position a two-element array over sloping ground. At (A) the antennas are placed in an uphill-downhill fashion along the line of steepest descent, while in (B) the two vertical monopoles are located side-by-side at the same elevation. These are the two limiting cases, and intermediate orientations have not been considered, for the sake of simplicity. This figure shows configuration (ii) for the radials (all are oriented at 45° to the line of steepest descent), but this was done simply for convenience, and is purely arbitrary; configuration (i) could have been drawn just as easily. Both will be modeled, and are discussed below.

First I will analyze case (A), where one antenna is higher than the other in terms of absolute elevation. As before, eight permutations will be modeled, using the four possible choices (A, B, C or D) from Figs 3 and 4 in combination with the two different radial selections (i or ii) from Fig 5. Since this is an endfire array, the main lobe of radiation will be directed either uphill or downhill, when the antennas are positioned as shown in **Fig 8A**.

The results are summarized in **Table 2**, which lists the forward gain, front-to-back ratio, and input impedances for all eight possible antenna systems, firing both uphill and downhill. Several configurations have performance characteristics that are worthy of mention.

Array C(ii) has excellent gain in *both* directions, with a variation of only 0.03 dB between uphill and downhill. This system would be the logical choice for those who want an array that performs as if it were mounted on level ground. However, there is a noticeable change in the driving-point impedances when reversing the direction of fire, so it might be advisable to build *two* matching networks (one for each direction) to realize the full potential of this system. **Fig 9** shows both of the principal elevation-plane radiation patterns.

Array C(ii) would also be preferred if you only wanted to fire uphill, since it has the best uphill gain. On the other hand, if you were planning to just shoot downhill, array D(ii) has the highest gain in that direction, along with a decent front-to-back ratio. See **Fig 10** for the elevation-plane radiation patterns in this scenario.

Array D(i) is the optimal selection if you need an antenna that is easy to get working. Its driving-point impedances are very similar in both directions of fire, so a single matching network will do a good job, if properly designed. However, it has better gain and front-to-back ratio when aiming downhill than up, as illustrated by the radiation patterns shown in **Fig 11**.

Now let's review the case shown in Fig 8B, where the two antennas are mounted side-by-side at the same absolute elevation. Once again, eight permutations will be modeled, using the four possible choices (A, B, C or D) from Figs 3 and 4, in combination with the two radial configurations (i or ii) from Fig 5.

The results are summarized in **Table 3**, which lists the forward gain and front-to-back ratio for all eight different antenna systems. Since the two verticals are now installed side-by-side, rather than uphill-downhill, both directions of fire are *perpendicular* to the line of steepest descent. Because of this, the array looks the same, physically and electrically, in both directions. Thus, there is no change in the gain, pattern shape, front-to-back ratio, or relative driving-point impedances within the array, when the directivity is reversed. This is in marked contrast to the situation discussed earlier (as highlighted in Table 2) wherein most of the electrical parameters were highly dependent upon the direction of fire.

What *does* occur in the side-by-side case, however, is a certain amount of skewing in the azimuth-plane radiation patterns of these arrays. Normally we would expect the peak gain to occur in the plane that contains the two verticals, but this is no longer the case. As shown in Table 3, the nose or peak of the radiation pattern is actually shifted slightly downhill in seven of the eight mounting configurations. The column la-

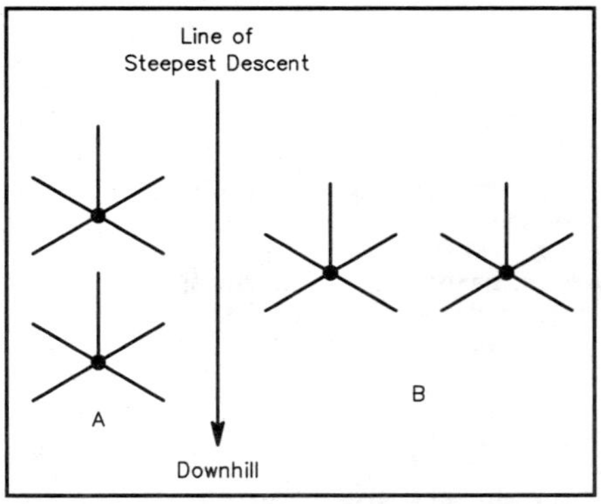

Fig 8—Two possible arrangements for a two-element cardioid phased vertical array (90° spacing and phasing) mounted over sloping ground: (A) uphill-downhill; (B) side-by-side. The orientation of the radials as shown in the figure is arbitrary; either configuration (i) or (ii) from Fig 5 may be used. Both are discussed (with their results) in the text.

Table 2

Power Gain, Front-to-Back Ratio and Input Impedances for Eight Configurations of a Two-element Cardioid Array of Vertical Antennas with Elevated Radials (90° Spacing and Phasing)

The elements are oriented in an uphill-downhill fashion as shown in Fig 8A.

Configuration	*Gain(dBi)*	*F/B (dB)*	*Input Impedances (Ω)*
A(i) Downhill	3.65	16.40	18.0 –*j* 12.0, 50.6 +*j* 15.2
Uphill	3.26	14.40	19.8 –*j* 10.5, 48.8 +*j* 13.6
B(i) Downhill	3.74	14.96	20.6 –*j* 11.2, 47.8 +*j* 16.1
Uphill	3.02	12.41	18.2 –*j* 10.4, 50.1 +*j* 15.2
C(i) Downhill	3.69	15.20	19.3 –*j* 11.2, 47.5 +*j* 15.8
Uphill	3.27	14.71	18.4 –*j* 9.2, 48.3 +*j* 13.7
D(i) Downhill	3.70	15.10	19.4 –*j* 12.0, 50.9 +*j* 15.5
Uphill	3.01	12.01	19.6 –*j* 11.7, 50.6 +*j* 15.1
A(ii) Downhill	3.70	13.06	20.8 –*j* 15.3, 59.6 +*j* 26.2
Uphill	3.19	12.40	27.1 –*j* 9.7, 53.3 +*j* 20.8
B(ii) Downhill	3.69	13.56	21.8 –*j* 15.3, 48.0 +*j* 19.3
Uphill	3.11	13.45	16.5 –*j* 13.5, 53.1 +*j* 17.5
C(ii) Downhill	3.63	11.81	20.2 –*j* 18.1, 49.8 +*j* 18.5
Uphill	3.66	14.76	17.7 –*j* 15.1, 52.3 +*j* 15.8
D(ii) Downhill	3.77	14.41	22.3 –*j* 13.6, 57.4 +*j* 25.1
Uphill	3.53	15.29	18.8 –*j* 11.7, 60.4 +*j* 23.0

Notes

Since the two antennas are spaced only 0.25 λ apart, and the radials are 0.25 λ in length, all the arrays which use arrangement (i) for their radials can use a common radial between them. Thus, these systems have a total of only seven radials rather than eight; the extra radial is omitted.

Two sets of radials in arrangement A(ii) cross at right angles, and could touch each other if suspended at the same height. Insulated wires would prevent this occurrence; I solved it in my computer model by offsetting the height of each antenna by 1 inch (the uphill antenna is 15 feet, 1 inch above ground, and the downhill antenna is at 14 feet, 11 inches).

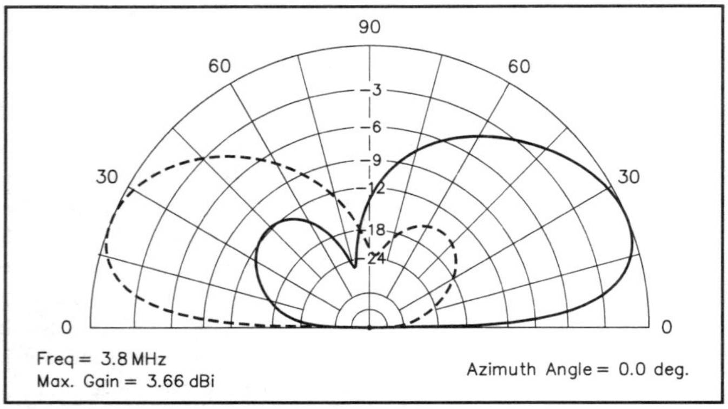

Fig 9—Principal elevation-plane radiation patterns for a two-element cardioid phased vertical array (90° spacing and phasing) positioned in an uphill/downhill fashion along the line of steepest descent, using radial configuration C(ii). The outer ring of the plot represents 3.66 dBi.

Solid line = firing downhill
Dashed line = firing uphill

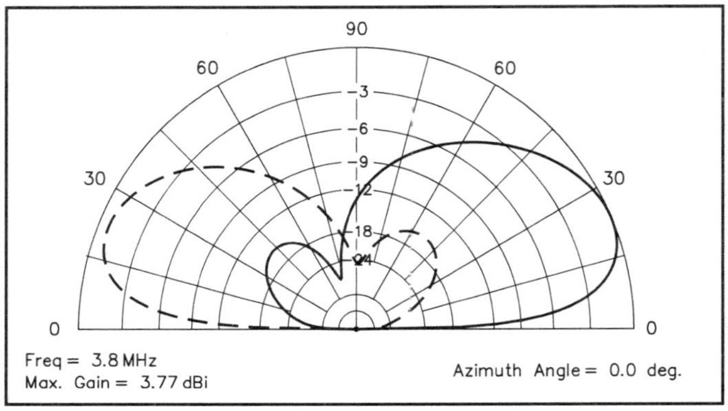

Fig 10—Principal elevation-plane radiation patterns for a two-element cardioid phased vertical array (90° spacing and phasing) positioned in an uphill/downhill fashion along the line of steepest descent, using radial configuration D(ii). The outer ring of the plot represents 3.77 dBi.

Solid line = firing downhill
Dashed line = firing uphill

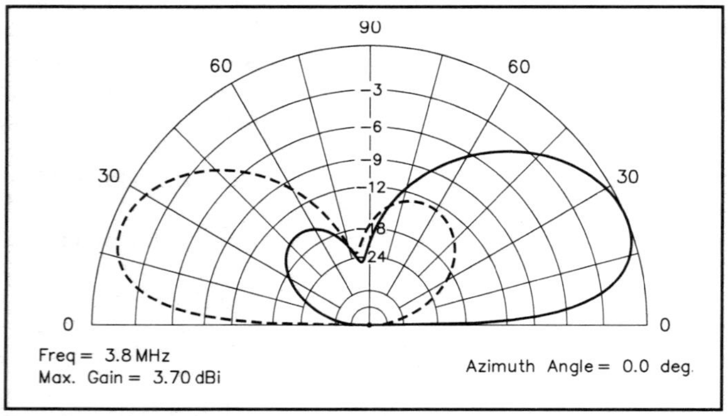

Fig 11—Principal elevation-plane radiation patterns for a two-element cardioid phased vertical array (90° spacing and phasing) positioned in an uphill/downhill fashion along the line of steepest descent, using radial configuration D(i). The outer ring of the plot represents 3.70 dBi.

Solid line = firing downhill
Dashed line = firing uphill

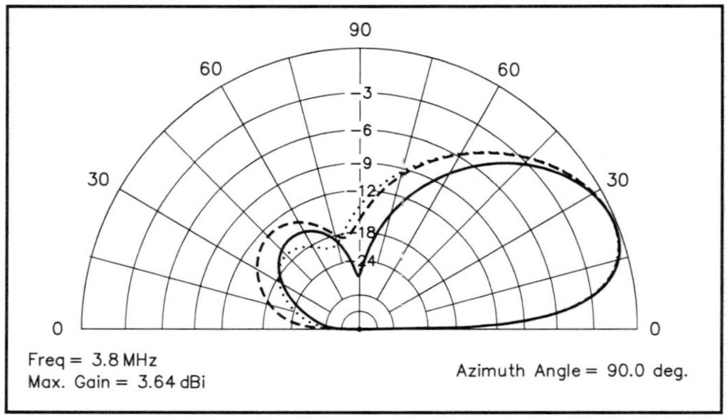

Fig 12—Elevation-plane radiation patterns for several two-element cardioid arrays (90° spacing and phasing) where the antennas are mounted side-by-side, perpendicular to the line of steepest descent. Downhill is perpendicular to the paper toward the reader. The outer ring of the plot represents 3.64 dBi.

Solid line = configuration A(i)
Dashed line = configuration A(ii)
Dotted line = configuration D(ii)

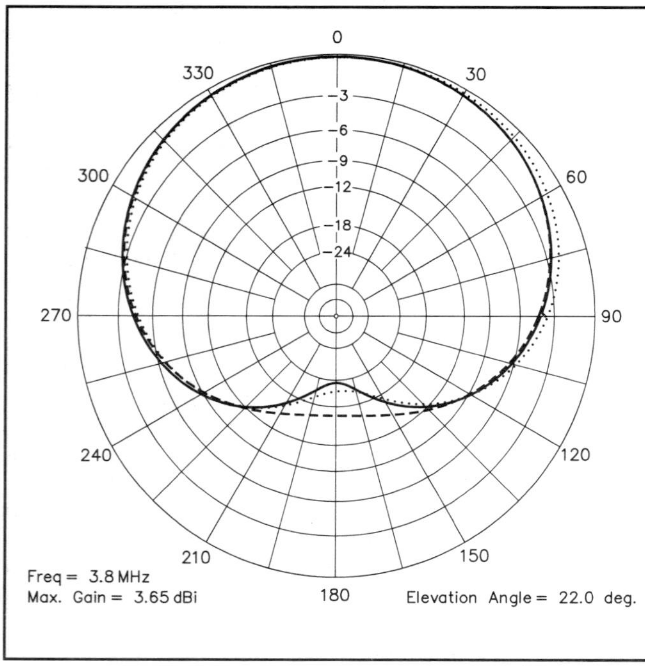

Fig 13—Azimuth-plane radiation patterns for several two-element cardioid arrays (90° spacing and phasing) where the antennas are mounted side-by-side, perpendicular to the line of steepest descent. Downhill is toward the right edge of the drawing. The outer ring of the plot repesents 3.65 dBi.

Solid line = configuration A(i)
Dashed line = configuration A(ii)
Dotted line = configuration D(ii)

Table 3

Power Gain, Elevation-Plane Front-to-Back Ratio and Azimuth-Plane Pattern Skewing for Eight Configurations of a Two-element Cardioid Array of Vertical Antennas with Elevated Radials (90° Spacing and Phasing)

The elements are oriented side-by-side as shown in Fig 8B.

Configuration	*Gain(dBi)*	*F/B (dB)*	*Azimuth Pattern Skewing*
A(i)	3.48	15.10	None
B(i)	3.41	15*	19° downhill (gain = 3.53 dBi)
C(i)	3.34	12.99	9° downhill (gain = 3.36 dBi)
D(i)	3.47	17.14	15° downhill (gain = 3.52 dBi)
A(ii)	3.46	12.88	1° downhill (gain = 3.46 dBi)
B(ii)	3.53	13.51	10° downhill (gain = 3.55 dBi)
C(ii)	3.39	11.92	5° downhill (gain = 3.40 dBi)
D(ii)	3.64	14.47	6° downhill (gain = 3.65 dBi)

Notes

Since the two antennas are spaced only 0.25 λ apart, and the radials are 0.25 λ in length, all the arrays that use arrangement (i) for their radials can share a common radial between them. Thus, these systems have a total of only seven radials rather than eight; the extra radial is omitted.

Two sets of radials in all of the arrays that use arrangement (ii) cross at right angles, and could touch each other if they were suspended at precisely the same height above the ground. In the real world, the use of insulated wires would prevent this occurrence; I solved it in my computer model by changing the height of each antenna by 1 inch (one antenna is 15 feet, 1 inch above ground, and the other is at 14 feet, 11 inches).

*The front-to-back ratio given for configuration B(i) is only an estimate, since there is no distinct rear lobe in the elevation-plane radiation pattern developed by this array.

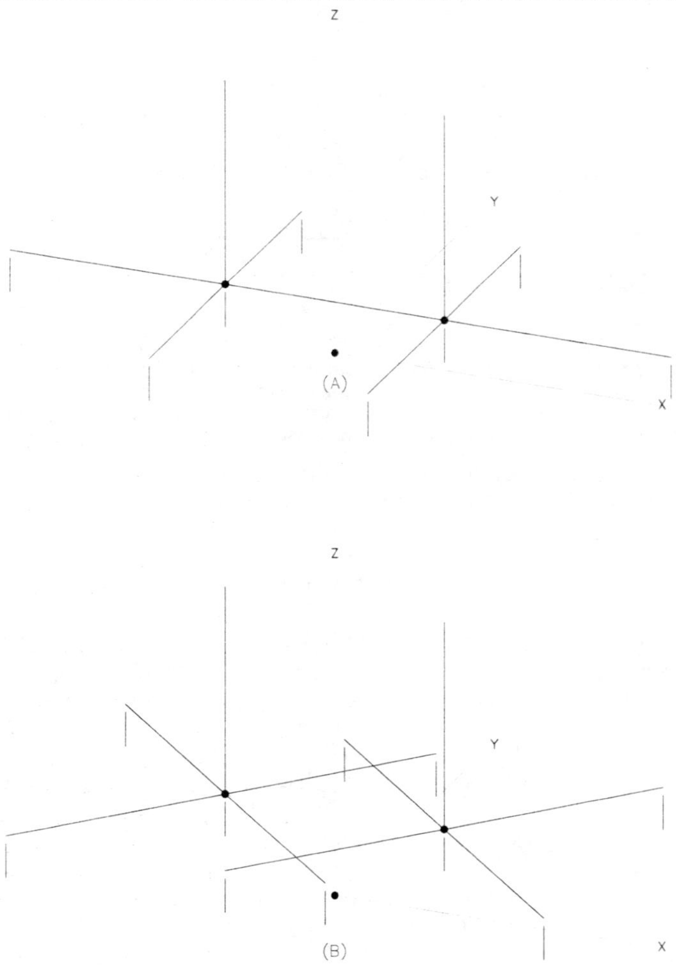

Fig 14—Two ways to configure a two-element cardioid array over flat, smooth ground: At A, radials are parallel and perpendicular to the array axis; and at B, radials are all at 45° to the array axis (the x-axis).

beled "Gain" in this table lists the maximum power gain along the theoretical direction of fire, which is perpendicular to the line of steepest descent. The right-most column shows how far the peak of the radiation pattern is skewed from the expected azimuth angle, and the maximum gain at this other compass heading.

Examining Table 3, we can see immediately that only configuration A(i) has *no* pattern skewing. The other seven versions exhibit varying amounts of pattern skewing, ranging from as much as 19° with antenna B(i) to as little as 1° for configuration A(ii). Version A(ii) has slightly less gain and front-to-back ratio than A(i). Antenna D(ii) has the highest forward gain of all eight configurations, and although it has 6° of pattern skewing, the difference in gain at the two azimuth angles is only 0.01 dB. At low takeoff angles, the front-to-back ratio of the D(ii) array is intermediate to that of configurations A(i) and A(ii). **Fig 12** illustrates the elevation-plane radiation patterns for these three antennas; the azimuth-plane patterns are shown in **Fig 13**.

By this time you may be thinking, "Fine, but what about the performance of the cardioid array when it is installed in a normal (flat-land) setting?" **Fig 14** shows two different ways to orient the radials with respect to the array axis. In Fig 14A we see that the radials are either parallel or perpendicular to the array axis. Only seven radials are needed, since one can be shared by both vertical monopoles. Eight support masts are required—two for the radiators themselves, and six more to hold up the ends of the radials.

Part B shows the alternative configuration, where each vertical has its own set of four radials, with no common connections. Now we need eight radials and ten support masts. Notice that if all the radials were actually at the same height above ground (15 feet) then there would be two places where pairs of radials would touch. To prevent this, I raised one antenna an extra inch (to 15 feet, 1 inch) and lowered the other by one inch (to 14 feet, 11 inches).

Fig 15 shows the principal elevation-plane radiation patterns for the two versions, and **Fig 16** does the same for the azimuth-plane patterns. Array (A) has slightly more forward gain (3.52 dBi versus 3.50) than antenna (B), but the take-off angle is the same (22°) in both cases. The front-to-back ratio of configuration (A) is clearly superior to that of (B), whether viewed in the elevation or azimuth planes. When combined with the lesser number of masts and radials required to build the antenna, system (A) is the clear winner.

A review of Tables 2 and 3 reveals that, as we might have suspected, several of the sloping-ground cardioid-array configurations

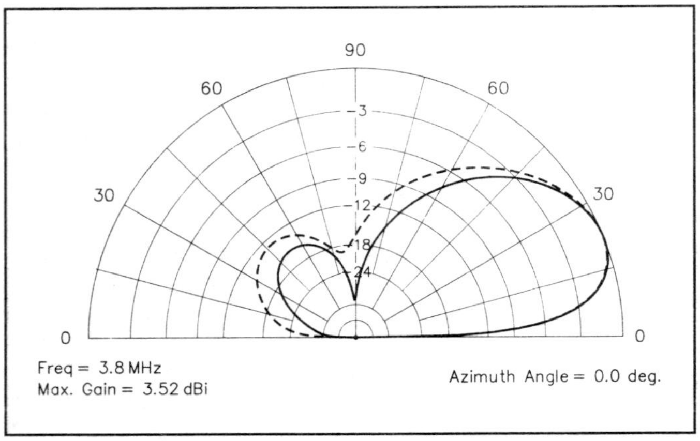

Fig 15—Principal elevation-plane radiation patterns for the two radial configurations illustrated in Fig 14. The outer ring of the plot represents 3.52 dBi.

Solid line = configuration A
Dashed line = configuration B

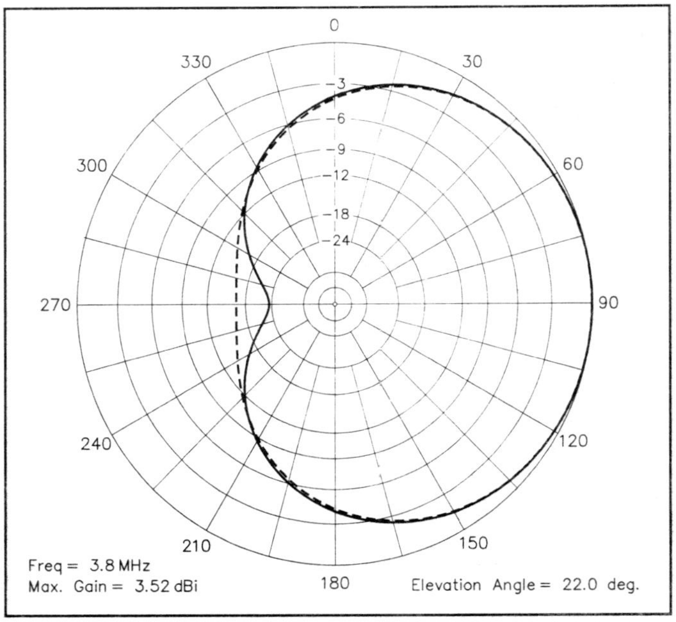

Fig 16—Principal azimuth-plane radiation patterns for the two radial configurations illustrated in Fig 14. The outer ring of the plot represents 3.52 dBi.

Solid line = configuration A
Dashed line = configuration B

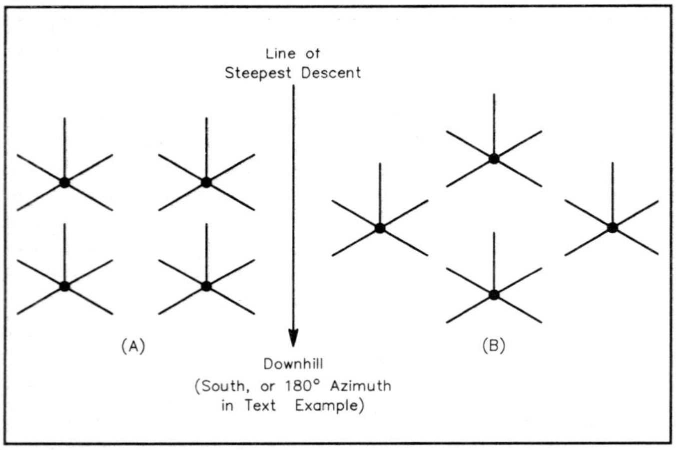

Fig 17—Two possible physical orientations for the four-square phased vertical array, when installed over sloping ground: At (A), a square; at (B), a diamond. The orientation of the radials shown in this figure is purely arbitrary; either configuration (i) or (ii) from Fig 5 may be used. Both are discussed (with their results) in the text.

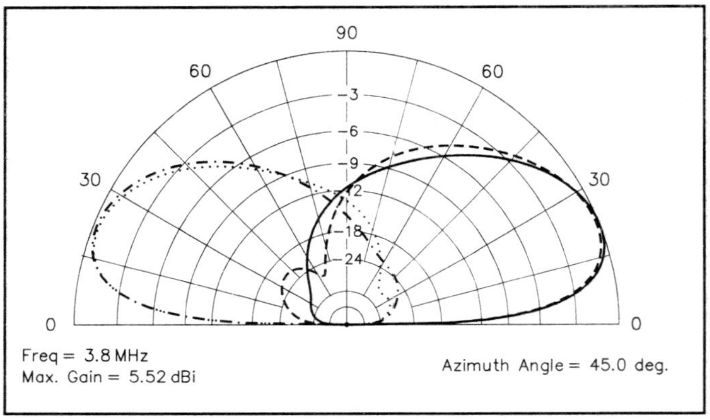

Fig 18—Elevation-plane radiation patterns for two of the four-square configurations discussed in the text. The monopoles are arranged in a square fashion. Downhill is toward the right, while uphill is toward the left edge of the drawing. The outer ring of the plot represents 5.52 dBi.

Solid & dotted lines = configuration D(i)
Dashed & dot-dashed traces = configuration D(ii)

produce more forward gain than can be achieved from the same antenna system when it is constructed over level ground.

Four-Square Vertical Arrays

The classic four-square array, using quarter-wave spacing and progressive 90° current phasing, is popular with low-band DXers. There are several ways to orient this antenna on sloping ground, and two are illustrated in **Fig 17**.

Version (A) shows a square which has two of its sides running uphill-downhill (parallel to the line of steepest descent) while the other two sides are perpendicular to this line, and extend alongside the hill. Since the antenna system actually transmits along a diagonal line connecting opposite corners, the four directions in which this configuration radiates are *all* oriented at an angle of 45° to the line of steepest descent. For example, if the line of steepest descent pointed directly toward the south, then the array would fire to the northeast, northwest, southwest, or southeast.

The second method for siting the four monopoles, drawn as version (B), looks somewhat like a diamond, and this array fires either straight uphill, straight downhill, or in either direction alongside the hill (perpendicular to the line of steepest descent). If downhill is due south, then this system radiates to the north, east, south, or west. I have chosen to call these two configurations *square* and *diamond,* respectively, even though both of them are actually square in their geometry, to distinguish one from the other.

Initially, I will analyze case (A), where the array resembles a square. Eight different versions will be modeled, using the four possible choices (A, B, C or D) from Figs 3 and 4, in combination with the two different radial selections (i or ii) from Fig 5. Recall that, with these eight permutations, the main lobe of radiation will be directed either partially uphill (the two lines of fire will be 45° to either side of the line of steepest ascent)

Table 4

Power Gain and Front-to-Back Ratio Data for Eight Configurations of the Four-Square Array of Vertical Monopoles with Elevated Radials

The elements are oriented in a square configuration, as shown in Fig 17A, and fire in four directions oriented either partially uphill or partially downhill.

Configuration	*Forward Gain(dBi)*	*Change in Gain (dB)*	*Front/Back Ratio (dB) Elevation*	*Azimuth*
A(i) Downhill	5.47		30*	29.55
Uphill	5.23	0.24	30*	33.13
B(i) Downhill	5.45		25*	27.60
Uphill	5.07	0.38	25.44	26.39
C(i) Downhill	5.39		30*	29.59
Uphill	5.13	0.26	29.82	31.60
D(i) Downhill	5.52		30*	34.03
Uphill	5.16	0.36	29.22	30.16
A(ii) Downhill	5.22		23.44	24.65
Uphill	4.86	0.36	22.56	23.60
B(ii) Downhill	5.12		20.58	23.89
Uphill	5.33	0.21	20*	25.30
C(ii) Downhill	5.04		22.00	23.60
Uphill	5.15	0.11	24.09	25.67
D(ii) Downhill	5.29		21.54	23.98
Uphill	5.43	0.14	23.36	25.41

Notes

Since all four sides of the Square are 0.25 λ, and the length of the radials is 0.25 λ, those arrays with arrangement (i) for their radials can share several common radials. Thus, these systems have a total of only 12 radials rather than 16, with the four redundant radials being omitted.

All of the arrays using radial arrangement (ii) have a quarter-wavelength radial extending from the base of each monopole directly toward the exact center of the array. Since these radials partially overlap each other, these four interior radials were deleted and replaced by two longer radials that span the diagonal distance across the center of the array. Thus, these systems have a total of 14 radials; 12 are 0.25 λ and the other two are 0.3535 λ.

Several sets of radials in the arrays that use configuration (ii) cross at right angles, and could touch each other if suspended at the same height. Insulated wire would prevent this occurrence; I solved it in my computer models by raising two of the radials by 1 inch (to 15 feet, 1 inch above ground) and lowering the other two to 14 feet, 11 inches.

*The actual front-to-back ratio for those configurations indicated by an asterisk is uncertain, since there is no distinct rear lobe in the radiation patterns of these arrays. The value given in the table is therefore an estimate.

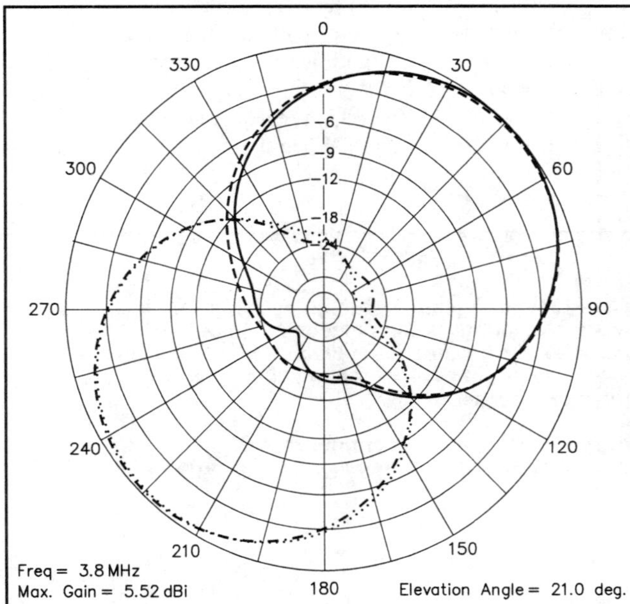

Fig 19—Azimuth-plane radiation patterns for two of the four-square configurations discussed in the text. The monopoles are arranged in a square fashion. Downhill is toward the right edge of the drawing. The outer ring of the plot represents 5.52 dBi.

Solid and dotted lines = configuration D(i) Dashed and dot-dashed traces = configuration D(ii)

or partially downhill (the two lines of fire will be 45° to either side of the line of steepest descent).

Information on the forward gain and front-to-back ratio for all eight configurations is listed in **Table 4**. Although the array has four directions of fire, the results for the two downhill quadrants were always extremely similar to each other, as was true for the two uphill azimuths. Thus, only two sets of values are shown for each arrangement of radials.

In many of the configurations that were modeled, the elevation-plane radiation pattern did not have a distinct rear lobe, so it was impossible to accurately determine a specific number for the front-to-back ratio, as noted in the table. An azimuth-plane radiation pattern was also derived for each version (at the takeoff angle where maximum forward gain occurred) and this value of front-to-back ratio was recorded from the plot.

Table 5 includes the computer-predicted driving-point impedances when the antennas are firing in different directions, along with the average variations of the input resistances and reactances, as the beam is switched from partially downhill to partially uphill. Together, these two tables contain much important data. We can see that only configuration A(ii) fails to achieve at least 5 dB of gain in all directions of fire. Version D(ii) has the highest average gain, at 5.36 dBi, followed closely by A(i) at 5.35 dBi, and D(i) at 5.34 dBi. Configuration D(ii) is very consistent as we switch the direction of fire, with no more than 0.14 dB variation in forward gain from one quadrant to another. This version also yields an average front-to-back ratio of well over 20 dB. The main disadvantage with configuration D(ii) is the relatively large change in both input resistance and reactance that occurs when switching directions. One possible solution would be to use two separate phasing/matching networks, one for downhill operation and another for uphill.

Configuration D(i) has the *smallest* overall variations in its driving-point impedances, averaging about 3 Ω for the real parts and just 0.5 Ω for the imaginary parts. It should be possible to construct a single phasing/matching network for this array, if the design calculations use median values of resistance and reactance as the input parameters. This antenna has very good gain when firing both uphill and downhill, with an average of 5.34 dBi, just 0.02 dB less than the best performer. Version D(i) also provides the highest overall front-to-back ratio in the azimuth plane, at more than 30 dB in any direction of fire, while the front-to-back ratio in the elevation plane is at least 25 dB at most take-off angles. Gain variations when switching from uphill to downhill amount to 0.36 dB, nearly the worst of any

Table 5

Input-Impedance Data for Eight Configurations of the Four-Square Array

The elements are oriented in a square configuration, as shown in Fig 17A.

Configuration	*Vertical Monopole Input Impedances (Ω)*	*Average Variation (Ω) in Resistances*	*Reactances*
A(i) Downhill	68.6 +*j* 52.3, 37.8 –*j* 1.7, 35.8 –*j* 8.9, 7.4 –*j* 0.7		
Uphill	66.8 +*j* 45.1, 36.0 –*j* 9.2, 37.7 –*j* 1.7, 9.1 +*j* 6.7	1.80	7.32
B(i) Downhill	61.5 +*j* 48.3, 34.7 –*j* 4.7, 43.4 –*j* 2.3, 11.4 +*j* 4.8		
Uphill	72.4 +*j* 52.7, 39.7 –*j* 2.2, 32.7 –*j* 6.8, 6.3 +*j* 2.2	7.93	3.47
C(i) Downhill	60.7 +*j* 46.4, 34.4 –*j* 5.3, 37.6 –*j* 5.9, 8.6 –*j* 0.7		
Uphill	65.6 +*j* 43.8, 35.6 –*j* 9.4, 32.8 –*j* 3.4, 7.2 +*j* 3.3	3.06	3.34
D(i) Downhill	69.6 +*j* 54.3, 38.2 –*j* 1.0, 41.5 –*j* 5.5, 10.3 +*j* 4.6		
Uphill	73.7 +*j* 54.0, 40.1 –*j* 1.9, 37.4 –*j* 5.3, 8.2 +*j* 5.4	3.05	0.52
A(ii) Downhill	73.5 +*j* 69.7, 47.6 +*j* 6.5, 44.2 –*j* 2.4, 9.2 –*j* 2.4		
Uphill	70.4 +*j* 58.9, 44.6 –*j* 1.8, 47.2 +*j* 8.2, 12.1 +*j* 6.1	3.02	9.52
B(ii) Downhill	62.8 +*j* 59.2, 40.4 +*j* 1.6, 54.6 –*j* 6.6, 12.7 –*j* 9.3		
Uphill	80.6 +*j* 62.6, 48.5 –*j* 6.9, 36.7 –*j* 9.7, 4.0 –*j* 1.1	13.12	5.75
C(ii) Downhill	60.5 +*j* 57.0, 39.1 +*j* 2.9, 45.3 –*j* 2.7, 11.8 –*j* 5.1		
Uphill	68.0 +*j* 58.0, 42.9 –*j* 1.8, 37.6 –*j* 4.1, 8.0 +*j* 0.2	5.70	3.09
D(ii) Downhill	76.5 +*j* 72.6, 48.1 +*j* 4.7, 55.8 –*j* 3.6, 10.0 –*j* 5.6		
Uphill	88.2 +*j* 72.5, 51.8 –*j* 4.0, 42.7 –*j* 4.7, 6.5 +*j* 3.4	8.01	4.78

array in the table, but not unreasonably large.

Fig 18 shows the principal elevation-plane radiation patterns for arrays D(i) and D(ii), when firing either partially downhill or partially uphill. System D(i) has a slight gain advantage when firing downhill, but D(ii) has the edge when the directivity is reversed. At low takeoff angles, D(i) has superior front-to-back ratio in all four quadrants, while the same is true for D(ii) at elevation angles above roughly 50°. Plots for the corresponding azimuth-plane radiation patterns are included as **Fig 19**.

All eight configurations yield a very slight azimuth skewing in their radiation patterns. The nose of the beam does not occur *precisely* at the expected compass angle, but is shifted by 1° or 2° because of the effects of the sloping ground. However, the impact of this azimuth offset upon the performance of the arrays is minor, and can probably be ignored.

Now let's look at the results from case (b) of **Fig 17**, where the four-square array is oriented to resemble a diamond. This arrangement produces radiation in any of four quadrants: straight uphill, straight downhill, or in either direction perpendicular to the line of steepest descent. As usual, eight different permutations must be modeled, using the four choices (A, B, C or D) from Figs 3 and 4, in combination with the two possible radial orientations (i or ii) from Fig 5.

Table 6 lists the gain and front-to-back ratio for each of the eight configurations. These arrays are capable of firing in four different quadrants, of course, but the results for the two directions which are oriented alongside the hill were always quite similar to one another. Therefore, only three sets of data are displayed for each version.

As was true before, some of the elevation-plane radiation patterns did not have a distinct rear lobe, and in these cases the front-to-back ratio had to be estimated. An azimuth-plane radiation pattern was also plotted for each configuration, to determine the front-to-back ratio at the take-off angle where maximum forward gain occurs.

The driving-point impedances for the eight antennas, in various directions of fire, are supplied in **Table 7**. This compilation also includes the average variations in input resistance and reactance values, as the main lobe was switched around the points of the compass.

An examination of these two tables reveals that version D(ii) has the highest average gain at 5.405 dBi, followed by A(ii) with 5.355 dBi, while arrangement B(i) finishes a very close third at 5.350 dBi. These values are so similar that they would probably be indistinguishable on the air. However, the average front-to-back ratio of configuration B(i) is less than 25 dB, clearly inferior to that of both A(ii) and D(ii), which typically attain values of 30 dB or more. When switching from one quadrant

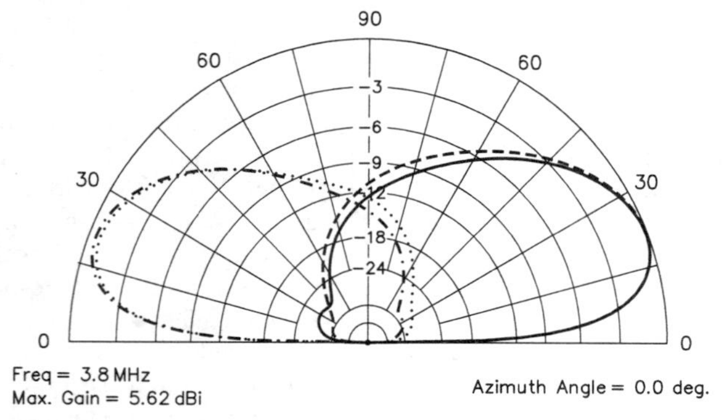

Fig 20—Elevation-plane radiation patterns for two of the four-square configurations discussed in the text. The monopoles are arranged in a diamond fashion. Downhill is toward the right edge of the drawing. The outer ring of the plot represents 5.62 dBi.

Solid and dotted lines = configuration A(ii)
Dashed and dot-dashed traces = configuration D(ii)

Table 6

Power Gain and Front-to-Back Ratio Data for Eight Configurations of the Four-Square Array with Elevated Radials

The elements are oriented in a diamond configuration, as shown in Fig 17B.

Configuration	*Forward Gain (dBi)*	*Change in Gain (dB)*	*Front/Back Ratio (dB) Elevation*	*Azimuth*
A(i) Downhill	5.27		23.95	25.15
Uphill	4.76		22.48	23.46
Sides	5.05	0.51	22.88	24.02
B(i) Downhill	5.01		16.39	18.71
Uphill	5.69		23.98	26.73
Sides	5.35	0.68	21.80	25.84
C(i) Downhill	4.88		20.17	20.84
Uphill	5.10		20*	21.35
Sides	5.18	0.30	22.03	25.19
D(i) Downhill	5.39		18.63	21.28
Uphill	5.34		17.64	21.99
Sides	5.26	0.13	18.80	22.05
A(ii) Downhill	5.52		28.82	29.48
Uphill	5.18		33.81	36.16
Sides	5.36	0.34	30*	30.8
B(ii) Downhill	5.52		25*	27.08
Uphill	4.99		27.40	28.33
Sides	5.27	0.53	25.41	26.3
C(ii) Downhill	5.47		30*	29.67
Uphill	5.05		28.95	30.0
Sides	5.28	0.42	33.37	33.71
D(ii) Downhill	5.62		30*	33.41
Uphill	5.16		30.17	31.12
Sides	5.42	0.46	33.45	33.55

Notes

All of the arrays using radial arrangement (i) have a quarter-wavelength radial extending from the base of each monopole directly toward the exact center of the array. Since these radials partially overlap each other, these four interior radials were deleted and replaced by two longer radials that span the diagonal distance across the center of the array. Thus, these systems have a total of 14 radials; 12 are 0.25 λ and the other two are 0.3535 λ.

Since all four sides of the square are equal to 0.25 λ, and the length of the radials is 0.25 λ, those arrays that use arrangement (ii) for their radials can share several common radials. Thus, these systems have a total of only 12 radials rather than 16; the four redundant radials were omitted.

Several radials in the arrays using configuration (i) cross at right angles, and could touch each other if suspended at the same height. Insulated wires would prevent this occurrence; I addressed this issue in my computer models by raising two of the antennas by 1 inch (to 15 feet, 1 inch above ground) and lowering the other two to 14 feet, 11 inches.

*The actual front-to-back ratio for these configurations is uncertain, since there is no distinct rear lobe in the radiation patterns of these arrays. The value given in the table is an estimate.

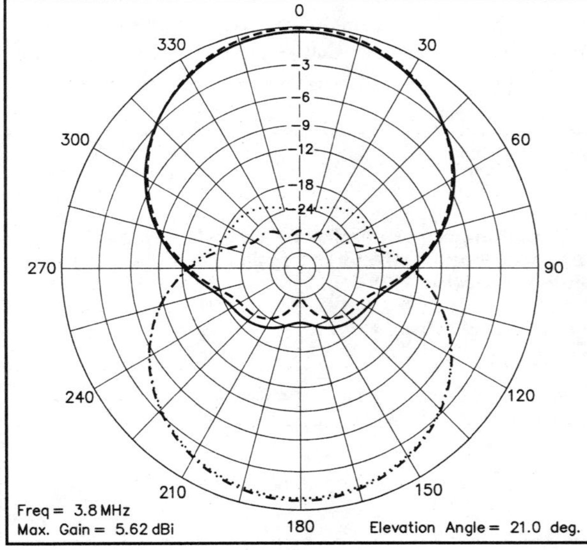

Fig 21—Azimuth-plane radiation patterns for two of the four-square configurations discussed in the text. The monopoles are arranged in a diamond fashion. Downhill is toward the upper edge of the drawing. The outer ring of the plot represents 5.62 dBi.

Solid and dotted lines = configuration A(ii)
Dashed and dot-dashed traces = configuration D(ii)

to another, the gain of arrangement A(ii) is more consistent than D(ii), but only by about 0.12 dB. Comparing the variations in driving-point impedances, A(ii) does well in terms of input resistance, but rather poorly for reactance, while the reverse is true for D(ii). Overall though, version D(ii) has better performance in this area, and a single set of phasing/matching networks should suffice for driving this array.

Fig 20 shows the principal elevation-plane radiation patterns for arrays A(ii) and D(ii), when firing either downhill or uphill. System D(ii) has a very small advantage when firing downhill, but the forward gain of both systems is virtually identical when the directivity is reversed. At low takeoff angles, D(ii) has superior downhill front-to-back ratio, while A(ii) is better when beaming uphill; the reverse is true for both antennas at elevation angles above roughly 45°. Plots for the corresponding azimuth-plane radiation patterns are illustrated in **Fig 21**.

When configurations A(ii) and D(ii) are firing alongside the hill, the resulting elevation- and azimuth-plane patterns are given in **Fig 22A** and **B**, respectively. Patterns for only one side of the hill are displayed, since their size and shape are the same in both directions. We can see that version D(ii) has a bit more forward gain at most elevation angles, plus better rejection of undesired signals off the back of the beam at all takeoff angles, when the antennas are firing alongside the hill.

All of the eight diamond configurations modeled for this section of the analysis produced a slight azimuth skewing of the radiation pattern when the arrays were firing alongside the hill. Here, maximum gain does not occur precisely at the expected azimuth angle, but is shifted from 1° to 3° due to the sloping ground. As before, the repercussions seem to be negligible, and most hams probably could not detect the change.

Four Square over Flat Ground

Finally, for comparison, let's examine the performance of the four-square array when it is installed over flat, level ground. **Fig 23** is a top view showing two ways to orient the radials of the system with respect to the sides of the square. In part (A) we see that the radials are either parallel or perpendicular to the sides of the square. Only 12 radials are needed, since several are shared by each vertical monopole; similarly, just 12 support masts are required.

Part (B) shows the second configuration, where all the radials are at a 45° angle with respect to the sides of the square. One radial from each monopole normally extends directly toward the geometrical center of the array, and would partially overlap the radial coming from the diagonally-opposite corner, if allowed to do so. This is prevented by

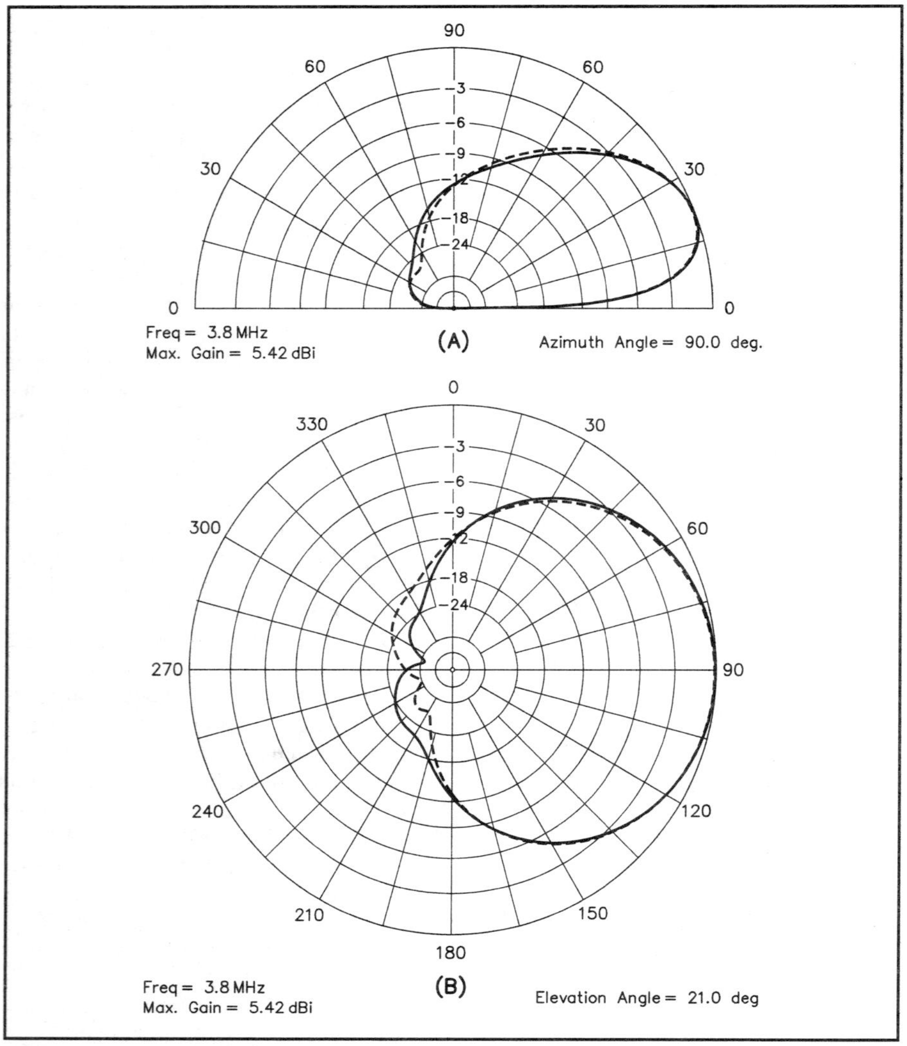

Fig 22—At A, elevation-plane radiation patterns for two of the four-square configurations discussed in the text. The monopoles are arranged in a diamond fashion and firing alongside the hill. At B, the corresponding azimuth-plane radiation patterns for A. The outer ring of the plots represents 5.42 dBi.

Solid line = configuration A(ii)
Dashed line = configuration D(ii)

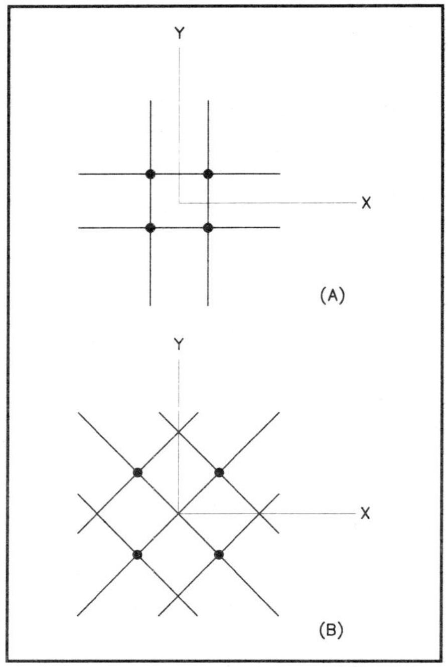

Fig 23—Two ways to configure a four-square array over flat, smooth ground: At A, the radials lie parallel and perpendicular to the sides of the square; at B, radials are all at 45° to the sides of the square.

replacing the four quarter-wave radials that occupy the interior of the system with two long radials that span the diagonals of the square. As a result, the total number of radials is reduced from 16 to 14. Notice however, that if all of these radials were actually at the same height above ground (15 feet), there are several places where two radials would intersect as they cross each other at right angles. To solve this problem, I raised two of the antennas by one additional inch (to 15 feet, 1 inch) while simultaneously lowering the other pair by an inch (to 14 feet, 11 inches).

Fig 24 compares the elevation-plane radiation patterns for the two radial arrangements, while **Fig 25** does the same for the azimuth plots. Configuration (A) has more peak gain (5.39 versus 5.07 dBi, both at a 21° takeoff angle); in fact, (A) seems to have higher gain at all takeoff angles below about 50°. Similarly, version (A) provides about 10 dB of extra rejection off the back at low elevation angles, and is roughly equal in performance at higher angles. Since (A) also requires two less radials and four fewer masts, it is definitely the preferred choice.

As mentioned earlier, Tables 4 through 7 indicate that several of the four-square arrays constructed over sloping ground can

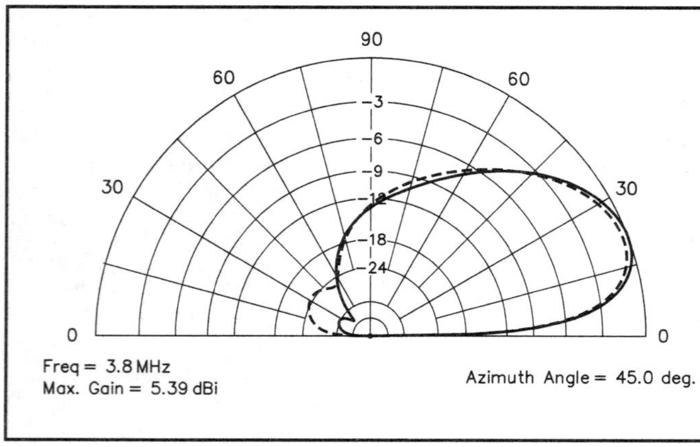

Fig 24—Elevation-plane radiation patterns for the two flat-ground four-square arrays of Fig 23. The outer ring of the plots represents 5.39 dBi.

Solid line = configuration A
Dashed line = configuration B

Table 7

Input-Impedance Data for Eight Configurations of the Four-Square Array with Elevated Radials

The elements are oriented in a diamond configuration, as shown in Fig 17B.

Configuration	*Vertical Monopole Input Impedances (Ω)*	*Average Variation (Ω) in Resistances*	*Reactances*
A(i) Downhill	74.1 +*j* 71.8, 46.0 +*j* 1.8, 46.0 +*j* 1.8, 8.6 −*j* 4.2		
Uphill	69.8 +*j* 56.8, 46.0 +*j* 3.4, 46.0 +*j* 3.4, 12.7 +*j* 7.7		
Sides	72.0 +*j* 64.5, 47.9 +*j* 9.3, 43.7 −*j* 4.1, 10.7 +*j* 2.0	3.17	10.49
B(i) Downhill	58.6 +*j* 59.1, 48.3 −*j* 1.5, 48.3 −*j* 1.5, 11.2 −*j* 13.2		
Uphill	82.0 +*j* 59.0, 39.6 −*j* 7.2, 39.6 −*j* 7.2, 4.0 −*j* 2.1		
Sides	71.9 +*j* 62.5, 36.3 −*j* 5.7, 51.7 −*j* 11.1, 8.7 −*j* 4.2	13.66	7.51
C(i) Downhill	58.1 +*j* 58.6, 47.8 +*j* 0.3, 47.7 +*j* 0.3, 9.9 −*j* 4.6		
Uphill	69.4 +*j* 56.9, 43.8 +*j* 1.2, 43.8 +*j* 1.3, 5.9 −*j* 5.0		
Sides	69.3 +*j* 61.4, 36.4 −*j* 3.9, 43.9 −*j* 4.4, 13.5 −*j* 1.1	8.54	4.8
D(i) Downhill	74.8 +*j* 72.3, 46.4 −*j* 0.4, 46.4 −*j* 0.4, 9.9 −*j* 12.6		
Uphill	83.4 +*j* 59.2, 40.5 −*j* 6.1, 40.5 −*j* 6.1, 12.0 +*j* 11.6		
Sides	74.2 +*j* 65.6, 47.1 +*j* 7.3, 50.5 −*j* 11.3, 5.5 −*j* 0.8	8.05	15.4
A(ii) Downhill	68.9 +*j* 54.1, 36.8 −*j* 5.6, 36.8 −*j* 5.6, 6.9 −*j* 1.9		
Uphill	66.3 +*j* 44.0, 36.9 −*j* 5.8, 36.9 −*j* 5.8, 9.4 +*j* 8.5		
Sides	67.8 +*j* 48.4, 38.1 +*j* 0.1, 35.5 −*j* 10.2, 8.3 +*j* 2.7	1.94	7.75
B(ii) Downhill	59.7 +*j* 47.8, 39.0 −*j* 3.1, 39.0 −*j* 3.1, 13.5 +*j* 5.9		
Uphill	76.0 +*j* 54.7, 34.6 −*j* 5.0, 34.6 −*j* 5.0, 5.9 +*j* 2.4		
Sides	66.2 +*j* 50.0, 32.7 −*j* 6.4, 44.4 −*j* 1.2, 8.0 +*j* 2.9	10.02	4.39
C(ii) Downhill	58.8 +*j* 45.1, 35.2 −*j* 5.9, 35.2 −*j* 5.9, 8.6 −*j* 1.6		
Uphill	64.6 +*j* 42.2, 33.2 −*j* 6.9, 33.2 −*j* 6.9, 6.6 +*j* 3.1		
Sides	61.6 +*j* 43.7, 32.4 −*j* 4.6, 36.2 −*j* 8.3, 7.6 +*j* 0.9	3.41	3.09
D(ii) Downhill	70.4 +*j* 57.1, 41.5 −*j* 2.3, 41.5 −*j* 2.3, 11.0 +*j* 5.3		
Uphill	78.3 +*j* 56.9, 39.0 −*j* 3.2, 39.0 −*j* 3.2, 8.0 +*j* 7.1		
Sides	74.1 +*j* 56.9, 37.9 −*j* 2.3, 43.2 −*j* 3.3, 9.2 +*j* 6.2	4.65	0.99

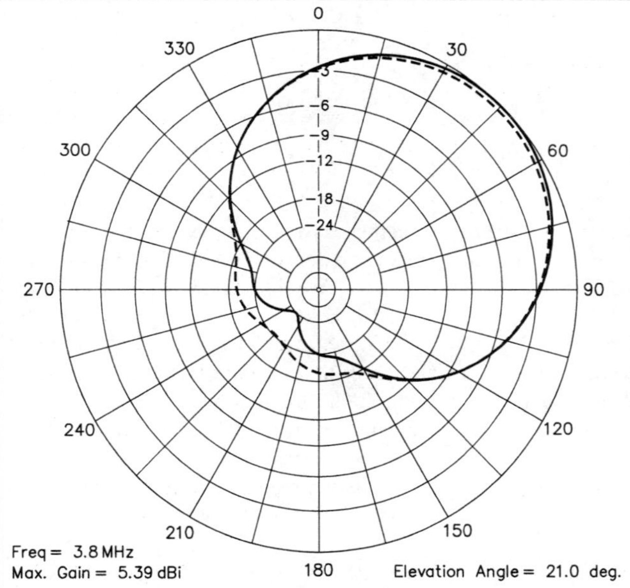

Fig 25—Azimuth-plane radiation patterns for the two flat-ground four-square arrays of Fig 23. The outer ring of the plot represents 5.39 dBi.

Solid line = configuration A
Dashed line = configuration B

achieve higher peak values of forward gain than their level-ground counterparts.

Conclusion

This article has evaluated the performance of elevated vertical-monopole antennas and phased vertical arrays when they are located over sloping ground. Alternative configurations were modeled using the *EZNEC pro* software package, with the results described in both tabular and graphical form. Many of the choices that were presented have values of forward gain that are similar (within a decibel or so), although the front-to-back ratios can vary much more widely. In certain instances, the driving-point impedances change quite a bit as the directivity is switched from one azimuth to another. Some of these variations can be tolerated by the commercial boxes in use today, but other situations may require the construction of several properly designed feed networks to assure correct operation. I hope that the information presented here will help those who are planning to build an elevated vertical antenna in hilly terrain.

Notes

[1]*ELNEC* is available from Roy Lewallen, W7EL, PO Box 6658, Beaverton OR 97007.

[2]Several versions of *EZNEC* are available from Roy Lewallen. See Note 1.

[3]G. J. Burke and A. J. Poggio, "Numerical Electromagnetics Code (*NEC*)—Method of Moments" (San Diego: Naval Ocean Systems Center, January 1981).

[4]J. C. Logan and J. W. Rockway, "The New MiniNEC (Version 3): A Mini-Numerical Electromagnetic Code" (San Diego: Naval Ocean Systems Center, September 1986).

[5]A. Christman, "Elevated Vertical Antenna Systems," *QST*, Aug 1988, pp 35-42.

Short Radials for Ground-Plane Antennas

By Rudy Severns, N6LF
PO Box 589
Cottage Grove, OR 97424
Rudys@ordata.com

Think your elevated radials always have to be full size? N6LF lets you in on some great ideas to lessen the "wingspan" of radials, especially near the beach.

Dean Straw's (N6BV) article[1] in this book describing the 6Y2A operation and beach-front verticals for DXpeditions shows how useful a vertical or vertical array can be—if you can put it over or adjacent to saltwater. For 20 meters and higher in frequency it is practical to use λ/2 verticals with little or no ground plane. For 40 meters and lower in frequency, however, a λ/2 height becomes prohibitive and a λ/4 ground-plane with elevated radials is a more practical form of vertical. Unfortunately, as you go down in frequency the length of the quarter-wave radials becomes very long (approximately 132 feet on 160 meters), and this takes up a lot of area. In addition, most DXpeditions can't place the radials very high off the ground. This results in a number of wires to trip over or strangle on. And if you have several verticals, the beach really becomes an obstacle course!

One way to reduce the problem is to shorten the radials (leaving the vertical part of the antenna as near a quarter-wave as possible) and use either a loading inductor, a top-loading hat or some combination of the two. The question is, "How much do you lose as you shorten the radials?" I took a look at this using *GNEC-4*, a *NEC-4.1*-based modeling program, and the following is what I discovered. Keep in mind of course that all this assumes *NEC* knows what it's talking about!

A 160-Meter Vertical

I have been planning on a beach-front 160-meter vertical for some property I have on Willapa Bay, WA, so I started with that model. While you probably wouldn't try to construct a full-size quarter-wave 160-meter vertical for a DXpedition, the computed results are very similar no matter what band you use with a beach-front vertical. The antenna I am planning will use four elevated radials and will be made of #13 wire. I was planning to use a wooden A-frame made from three Douglas fir trees (as shown in my *QEX* article[2]) to support the antenna. For modeling, the initial lengths of the radials and the vertical were made equal and adjusted for resonance at 1.840 MHz. I then progressively shortened the radials (keeping the vertical height the same) and re-resonated the antenna with a single series inductor feeding all four radials at the feed point.

An inductor Q of 250 was assumed and with a little care this should be readily achievable. In a salt atmosphere you must put the coil in a sealed enclosure, or by morning the Q will be close to zero. I modeled the base of the antenna at 1 foot and at 10 feet, and for comparison used three types of ground: perfect, seawater (ε = 80, the dielectric constant, and σ = 5.0 S/m, the conductivity) and average (ε = 13, σ = 0.005 S/m).

The results are shown in **Figs 1** and **2**. These graphs include both the ground loss and the loss due to the series resistance of the loading inductor. The small wire loss was not included. We can see from Fig 1 the advantage of seawater over average ground: about 4.2 dB more gain for full-length radials. In addition, the peak gain occurs at an elevation angle of 7° for seawater, as opposed to 21° for average ground. As the radial length is reduced the peak gain angle changes very little, but the peak gain goes down. The height of the radials over seawater made very little difference, and the difference between ideal ground and seawater was also very small. The primary difference over seawater is the added loss in the loading inductor.

While Fig 1 shows the peak gain, you can see the variation much better in Fig 2, where the *change* in gain is plotted. Even if you use radials only 40 feet long (0.07 λ!), over seawater the loss is less than 0.2 dB. This is very attractive for DXpeditions. The value of the loading inductor is very nearly the same for all the grounds and heights so that the loss due to the inductor's series resistance is pretty much the same at each radial length. Over average ground, however, the gain reduction is much larger due to in-

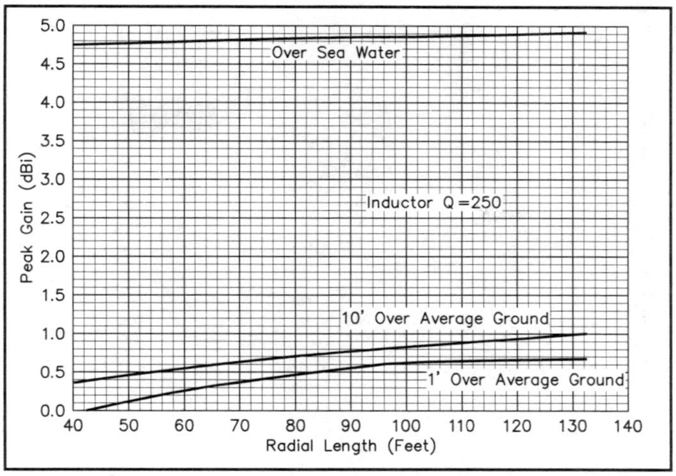

Fig 1—Peak gain for (λ/4-high vertical with four, coil-loaded short radials on 160 meters. The loading coil is assumed to have an unloaded Q of 250. Over seawater, the peak gain doesn't change much, even for quite short radials, while the gain is close to 5 dB less over ground with average conductivity and dielectric constant.

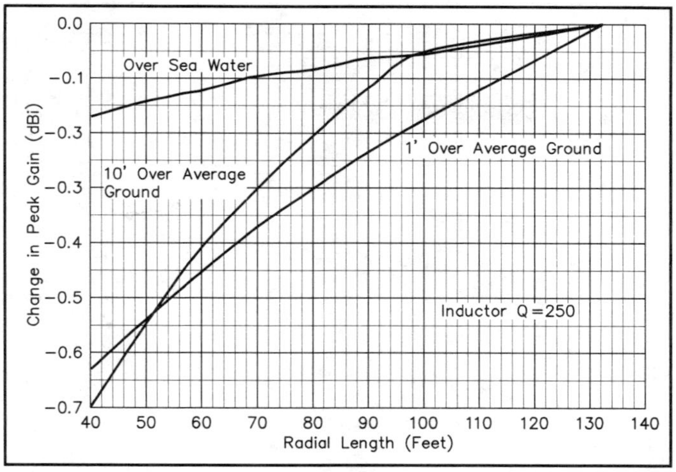

Fig 2—The same data presented in Fig 1, but magnified by showing the change in peak gain versus radial length on 160 meters.

creased ground losses as the radials are shortened.

At some sites the antenna may actually be over seawater, but it is more likely it will be up on the beach adjacent to seawater. How much effect will that have? That depends on two things: the beach's ground characteristics and the distance to the water. If the ground under the antenna is regularly flooded with seawater the conductivity is going to be pretty high. But that may not always be the case, and fairly poor ground characteristics may be encountered—especially on coral islands.

To check this out I modeled the 160-meter antenna site as though it were a circular island located in a sea of saltwater. The island was made up of ground with average conductivity and dielectric constant, and the distance to the saltwater was varied by changing the radius of the island. The results are shown in **Fig 3**. As soon as you move away from the water (that is, you have a larger-diameter island) the peak gain starts to drop and the increased ground loss due to shorter radials shows up.

The message is simple—select a nice salt marsh that is flooded twice a day, or put the antenna out on the reef with water under it! Otherwise, put the antenna as close to the water as practical.

Reducing Losses

We know that the losses due to use of a loading device such as a coil can be reduced by using a higher-Q coil, by moving a loading coil from the base of the antenna to a more optimum location up the vertical radiator, by top-loading or some combination of these.[3,4] For a 1/4-λ vertical with shortened, loaded radials over seawater the ground losses are very small, and even the losses due to base loading of a shortened vertical radiator are small. It is questionable, therefore, whether it is worth the trouble to spend much time trying to minimize the loading loss, except for the case where a vertical's electrical height is much shorter than λ/4.

Unfortunately, short verticals with loading are often used for 80 and 160 meters. On those bands all of the tricks for minimizing losses will have to be used, because shortening the radials as well as the vertical itself can seriously degrade performance, even with a seawater ground.

I looked at modifying the ground plane to see if a more complex radial structure would help. Using eight radials, the difference in ground loss was insignificant. However, the additional radials did reduce the reactance needed to resonate the antenna by almost 1/2. That would reduce the loading coil loss, since a smaller amount of inductance would be needed.

I then looked at tying the ends of the radials together with cross conductors to form a square wagon-wheel shape with four radials. Again, the ground losses were not reduced greatly (≈ 0.2 dB). There appears to be no substitute for long radials if you want that last fraction of a dBi in gain. We really have known this since the 1930s![5,6,7] The reactance, however, was greatly reduced and with the wagon-wheel structure the antenna is resonant—without loading—with a radius of 58 feet, less than half that for normal radials.

Given the fierce corrosion experienced over or near seawater, it would be a good idea to use insulated wire for the radials, some paint on the vertical tubing, conductive joint compound and very careful sealing of all joints and connections.

A Closer Look at Ground Losses

The increase in ground loss with shorter radials is worth a closer look. The additional loss shows up as an increase in feedpoint resistance over that for ideal ground. **Fig 4** is a graph of feedpoint resistance as a function of radial length, without the resistance of the loading inductor. Over seawater the effect of ground loss is very small. It's hard to see it on the graph. Over average ground, however, the effect is very obvious and the loss increases at lower heights.

When generating the data for Fig 3, I noticed that the feedpoint resistance was constant for different values of "island" radius. This is due to the way *NEC* computes impedance, where it takes into account only the first ground characteristic and assumes for this purpose that the ground under the antenna is infinite. For far-field calculations, however, the two ground zones (that is, the ground under the vertical and the seawater surrounding our model island) are taken into account. This means that the ground loss in the model, as reflected in the feedpoint resistance, may be higher than it actually is when close to the water.

NEC can provide a direct calculation of total ground losses using the so-called "RP" card. This card sets the parameters for radiation patterns and can provide a calculation of average gain. An example is given in the Appendix.

Conclusions

If you are lucky enough to be near or on seawater, you can drastically reduce the size of your elevated ground-plane. With a little

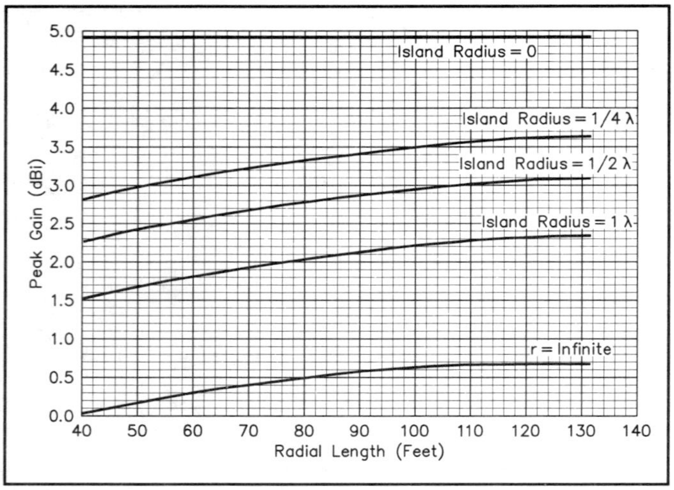

Fig 3—Peak gain for a (λ/4-high vertical with four, coil-loaded short radials on 160 meters, but this time where the antenna is located on a circular island in a saltwater ocean. Three different radial heights are shown over average ground. The radius of the circular island is equal to the length of the shortened radials in this model. Obviously, you should mount your antenna and radials over—or at least as close to—salt water, as you possibly can!

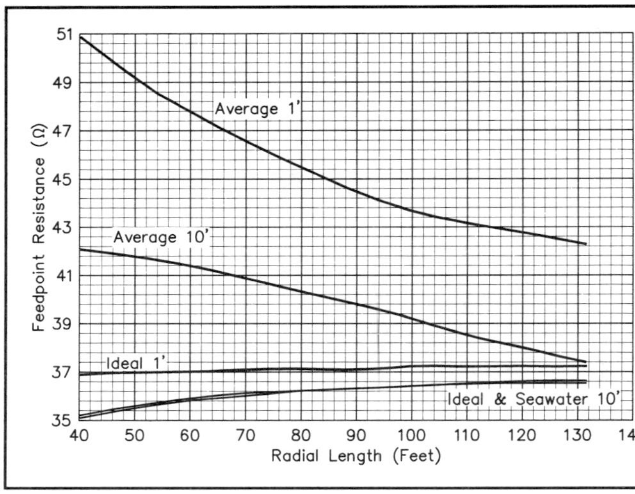

Fig 4—Feed-point resistance as a function of radial length and radial height above ground. Here, the loss resistance of the loading inductor is removed. The effect of ground loss over soil is large compared to that over salt water.

care, loss due to the loading components can be small, on the order of a few tenths of a dB. If you must place the antenna over poorer ground, such as the beach, you should try to get as close to the water as possible and keep the radius of the ground-plane radials greater than λ/8. You can, of course, use a smaller ground plane if you are willing to accept the reduced peak gain.

And here is another important consideration: Even with λ/4 radials, you should decouple the feed line from the antenna with a common-mode choke (balun). As the radials become shorter this becomes even more important, since the voltage between the base of the antenna and ground increases.

References

[1]D. Straw, N6BV, "Antennas Here Are Some Verticals on the Beach...," Elsewhere in this book.

[2]Rudy Severns, N6LF, "Another Way To Look at Vertical Antennas," *QEX*, Mar/Apr 1999.

[3]J. Belrose, VE3BLW, "Short Antennas for Mobile Operation," *QST*, Sep 1953; see also subsequent *ARRL Mobile Manuals.*

[4]B. Brown, W6TWW, "Optimum Design Of Short Coil-Loaded High-Frequency Mobile Antennas," *The ARRL Antenna Compendium, Vol 1* (Newington: ARRL, 1985), pp 108-115.

[5]G. Brown, "The Phase and Magnitude of Earth Currents near Radio Transmitting Antennas," *Proc IRE*, Feb 1935, Vol 23, No. 2, pp 168-182.

[6]Brown, Lewis and Epstein, "Ground Systems as a Factor In Antenna Efficiency," *Proc IRE*, Jun 1937, Vol 25, No. 6, pp 753-787.

[7]J. Devoldere, ON4UN, "Antennas and Techniques for Low-Band DXing," (Newington: ARRL, second edition 1994), Chapter 9, Fig 9-40 and supporting text.

[8]GNEC-4 Users Manual, Nittany Scientific, Airline Highway, Suite 361, Hollister, CA 95023.

[9]Gerald J. Burke, Numerical Electromagnetics Code—"NEC-4, Part I: Users Manual (NEC-4.1)," Lawrence Livermore National Laboratory, Jan 1992, UCRL-MA-109338 Pt I, pp 76-78.

Appendix

The average gain for a lossless antenna in free space is 1.0 (or 0 dBi). For a lossy antenna the average gain will be lower by the amount of the total loss. If you model using ideal conductors and lossless loads, then the reduction in average gain directly reflects the ground loss. Inserting the correct parameters for average gain can be a little tricky, however, until you get used to it. For example, when modeling an antenna over ground, instead of averaging over the surface of a sphere, the averaging is done over a hemisphere. Because the total power is radiated into only half as much space, the gain of a lossless antenna will be 2.0 (or 3.01 dBi).

You have to keep track of these things as you go along. As a check on my "card" entries (following the terminology for the FORTRAN-based NEC-4 software) when starting a new analysis, I make the ground perfect and the conductors lossless. I thus should get an average gain very near 1.0 or 2.0, depending whether I'm modeling in free space or over ground. If all is well, then I insert the real ground constants and proceed with the modeling.

The RP card has the following format:[8,9]

RP I1 I2 I3 I4 F1 F2 F3 F4 F5 F6

where:

I1 = Selects the mode of calculation; I1 = 0 for this example

I2 = Number of values of theta at which the field is to be calculated

I3 = Number of values of phi at which the field is to be calculated

I4 = An integer consisting of 4 digits (XNDA), each of which has a different function

X = Controls antenna output format; X = 1 for this example

N = Causes normalized gain to be printed; N = 0 for this example

D = Selects either power gain or directive gain; D = 0 for this example

A = requests calculation of average gain; A = 2 for this example

F1 = Initial theta angle

F2 = Initial phi angle

F3 = Increment for theta

F4 = Increment for phi

F5 and F6 are not needed for this example.

Greek letter θ (theta) is the angle measured from zenith (directly overhead) downward. For a free-space antenna, θ will vary from 0° to 180° and for an antenna over ground, the range is 0° to 90°. Greek letter ϕ (phi) is the angle moving counter-clockwise (viewed at the antenna from the X axis towards the Y axis) rotating around the Z axis. The range of ϕ is 0° to 360°. The number of values for theta (I2) and phi (I3) will be the range selected divided by the increment (F3 or F4) plus 1. The number of values must be an integer.

The number of increments of theta and phi must be large enough to cover the entire field, unless there are known symmetries that can be exploited to reduce the number of calculation points. For example, a free-space antenna will require a sphere, and an antenna over ground will require a hemisphere. For the case of an elevated-radial, ground-plane antenna with four radials, the field will repeat every 90° of phi. It is thus only necessary to compute one quadrant of the hemisphere. The accuracy of the averaging will depend on the number of points over which the gain is averaged. Fewer points mean less accuracy but much faster computation. The way you check your setup is to calculate the average gain for a lossless system, with perfect ground, no wire loss, etc. Under those perfect conditions the average gain ideally will be 1.0 or 2.0. The difference in the actual calculation is the error due to specification of overly coarse steps in the angles. The error in dB, for antennas over ground, can be expressed as error = 10 log (average gain/2). This gives the error directly in dB. Typically, I accept an error of 0.01 dB.

I usually start with 2° increments and go up or down after checking the lossless gain. The number of field points generated with ½° increments can be quite large and noticeably slows the computation even on a workstation. This great a resolution will seldom be needed but you should always check an ideal version of the antenna model before proceeding with a real ground and lossy antenna.

In some cases where the field does not vary greatly with either θ or ϕ you can use larger increments in one plane to reduce the computation time. For example, with a four-radial ground-plane antenna, the field variation with ϕ, at a fixed θ, is quite small and usually needs only two values for ϕ—, that is, 0° and 90°. This greatly reduces the computation time. However, for a two-radial antenna, the pattern is asymmetrical and you must use smaller increments for ϕ.

A key point is to recognize that the number of points (I2 and I3) must be adjusted to give total coverage of the desired sector (sphere, hemisphere, or quadrant) when the increments (F2 and F3) are selected. Don't forget to include one extra point for the ends.

My RP card looks like this for one quadrant and 1° increments:

RP 0 91 91 1002 0 0 1 1

With four radials and only a small error this can be reduced to:

RP 0 91 2 1002 0 0 1 90

For an antenna over ground with a pattern symmetrical about the X axis:

RP 0 91 181 1002 0 0 1 1

The average gain will appear at the end of the output file in terms of absolute gain. You can convert it to dB by using 10 log (absolute gain) or 10 log (absolute gain/2).

A 10 Through 80-Meter Ground Plane/Vertical Antenna: A Result of New Technology

By Ralph Crumrine, NØKC
1621 E Sunvale Dr
Olathe, KS 66062-2108
e-mail: **CrumrineRJ@aol.com**

CAD software, a fast computer and a novel capacitor design result in a complex multiband antenna structure—on the first try!

I needed a fresh approach to an antenna to go along with my renewed interest in ham radio. Buy or build? Commercial vertical antennas seemed to be common these days, but in one way or another they missed the mark. What new technology could I bring to bear on antenna design and what type of antenna should it be?

Nearly 50 years ago I had built a 40-meter vertical ground plane antenna out of beer cans, a popular thing to do at the time. For a spate of activity 20 years ago I had hung out some dipoles from rooftop to trees. All these simple structures had the disadvantage of single-band operation or high takeoff angles for low dipoles.

What was desired was an antenna that was multiband, with full bandwidth on each band. Electrical efficiency was of prime importance. I ruled out the use of lossy and overly complex matching networks, lossy coaxial tuning stubs, ferrite-cored transformers or resistors.

Because of suburban restrictions, towers and beams were out of the question. Since that old 40-meter beer-can vertical ground plane had worked wonderfully well years ago, I settled quickly on a ground plane and/or vertical as the prototype structure around which to design the new antenna. Furthermore, I figured a ground plane has a built-in maintenance-free antenna rotator!

Antenna Design

Antenna CAD tools[1,2] have been available for some time but only the recent availability of ultra-fast computers has made the application of that software really practical for complex designs. The design job could be done. The question was: How accurately?

If the results of the CAD design called for component values that weren't commercially available how would that be handled? The coils for inductors can be shaped to any value. Capacitors are another thing.

Binary adjustment of small-signal chip capacitors is a staple in the microwave receiver industry. Taking a cue from these parts I decided to take on as a side project the development of a binary-adjustable, printed-circuit capacitor along those lines. With such a capacitor, I could create whatever value came out of the computer design. Having a continuum of capacitor and inductor values allowed me both degrees of freedom in the design of an LC parallel circuit—both center frequency and impedance level. See sidebar *Making a High-Voltage Fixed Capacitor*.

Achieving full bandwidth in an efficient manner for each band was one design goal. However, after many months of design, plus trial and error at the computer keyboard, I determined that I couldn't reach this goal for all bands from 10 through 80 meters. So I discarded the so-called WARC bands: 12, 17 and 30 meters. An antenna for those bands would have to be a follow-on project. Further, I decided to concentrate on the 10 through 40-meter bands in the form of a ground-plane antenna and then merge this into the mounting base to produce an 80-meter vertical antenna that used conventional on-ground radials. More on that later.

I started the design by establishing a reasonable mechanical size for the antenna. Since tubing comes in 12-foot lengths, I made the vertical sections, VS1, VS2 and VS3 (see schematic drawing in **Fig 1**) out of two telescoping sections of tubing. I also made each radial, RDL1 through RDL4, from 12-foot sections of smaller-diameter tubing. I found that an advantage of a large-radius radial system for the antenna is design stability. If I paid careful attention to mechanical dimensions and accurate implementation of electrical parts from the computer design, my goal was that the antenna would perform to specifications without any adjustment.

I determined coverage on the 20 and 40-meter bands on the prototype antenna structure using an LC tank circuit (L1-C1 in Fig 1). My first attempt was to tune this to the geometric center of the two bands, although after manipulation using the computer I ended up with a parallel resonance at 10.93 MHz. I placed the tank at the approximate midpoint of the vertical element. At frequencies below LC resonance, the circuit looks inductive, stretching the length of the antenna to resonate on 40 meters. Above the LC resonant frequency, the tank circuit is a capacitive reactance, which shortens the effective antenna length to make it resonant on 20 meters. One advantage to this approach is that the LC tank circuit is not operated at resonance on any band, and hence has lower voltages across it for all bands.

I defined the part of the antenna structure

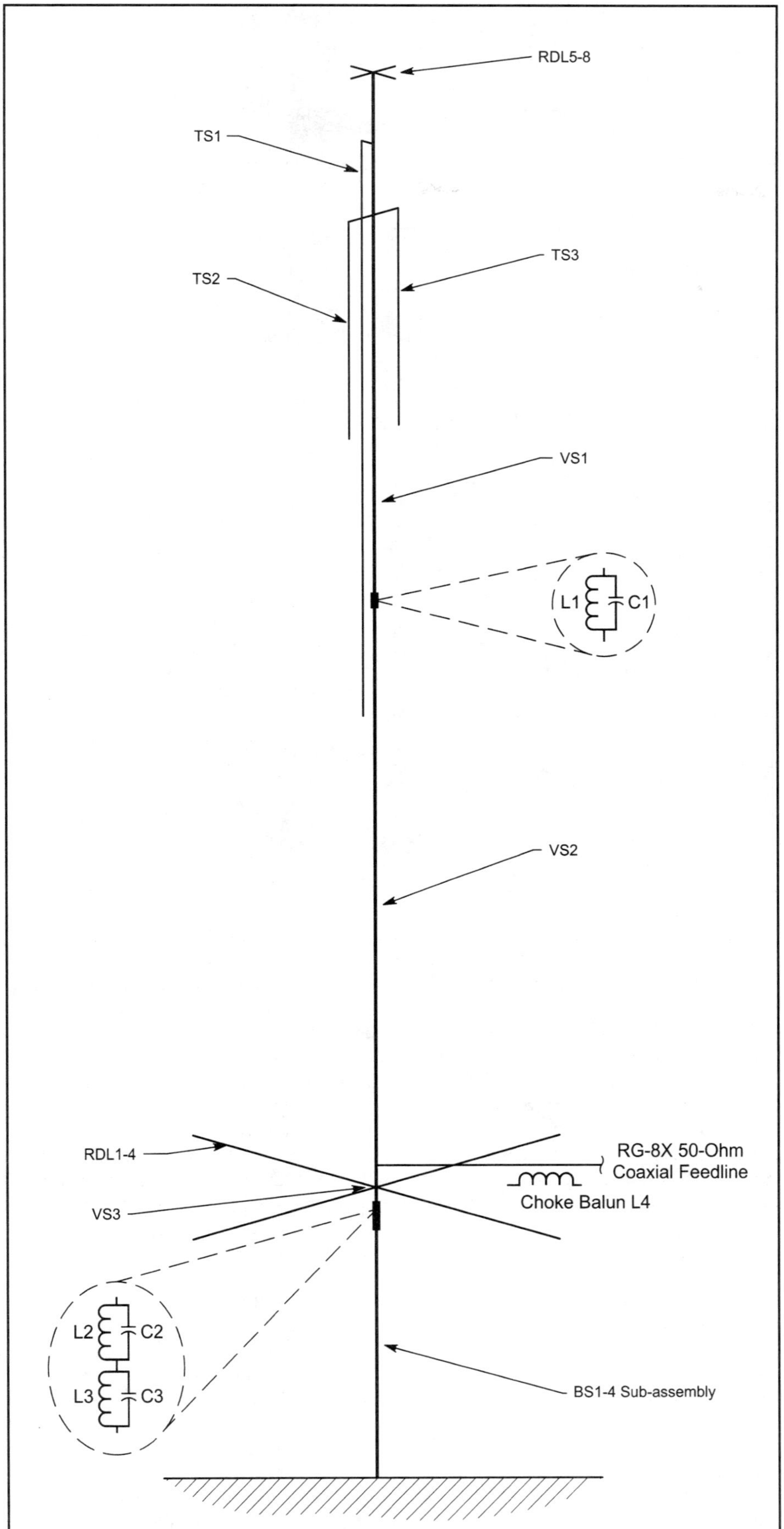

Fig 1—Schematic representation of 10 through 40-meter ground-plane antenna, with 80-meter capability.

that functions as a 15-meter antenna by mounting a ¼-wave 15-meter transmission line stub (TS1 in Fig 1) on the main vertical element, open-circuited at the bottom. Since a ¼-wave stub shorted at one end will be an infinite impedance at the open end, the resulting high impedance isolates the upper part of the overall antenna structure for operation on 15 meters.

The full length of the antenna structure is approximately two ½-wavelengths on 10 meters. To adjust the resonant length to exactly 10 meters and to effect an impedance transformation at the feed point, I added two approximately ⅛-wave-long stubs TS2 and TS3 in parallel, as shown in Fig 1. Think of this as a very broad stub that brackets the center section of the antenna.

All of these pieces of a multiband antenna interact with each other, so this is where the power of the computer comes into play. After many computer design trials (several weeks or more), I established final locations, component values and mechanical dimensions for all of the parts. Finally, to allow another degree of freedom in computer optimizing the design, I added a top hat (RDL5 through 8).

The feed point is at the junction of the radials and the vertical radiator. I ran several computer studies to investigate bandwidth. First, I examined bandwidth as a function of the diameter of the antenna parts. I fixed the main vertical section at 1.5 inches in diameter because going to a larger diameter produced very little improvement in bandwidth. Likewise, I studied how the number of radials affected bandwidth. There is a nice improvement in bandwidth between three and four radials, but beyond four radials the improvement in bandwidth becomes negligible.

I optimized the antenna mounted with its base eight feet high. The antenna unit is insulated from the mounting base just below the radials with a phenolic tube at the bottom of VS3 in Fig 1. With the addition of two parallel LC networks in series, just below the radials and across the insulating gap between the antenna and the mounting base, the structure becomes a loaded 80-meter vertical antenna. This requires a traditional ground-mounted radial system[3,4] for 80 meters. The installation of the network for 80 meters has a small effect on the 40-meter bandwidth and no noticeable effect on any of the higher bands.

The eight-foot base mounting height is not an accident. Besides being a good dimension to accommodate the 80-meter band, it allows you to walk around the yard, mow or garden under the antenna without getting poked in the eye by a radial.

Mechanical Design and Construction

I made the vertical sections, VS1, VS2

and VS3, from 1½-inch OD aluminum tubing. The radials, RDL1 through 4, are 12-foot lengths of ½-inch OD aluminum tubing. The final assembly of the antenna is shown in the photo in **Fig 2**. The main vertical sections, VS1 and VS2, of the antenna are joined (insulated) with a 13-inch long section of 1³/₈-inch OD thick-wall phenolic tubing, MS1. L1 and C1 are placed in parallel across this insulated gap. Again, see Fig 1. [Detailed engineering drawings for this antenna and *EZNEC* files are on the ARRLWeb site at **www.arrl.org/8608**—*Ed.*]

I tapped the phenolic tubing for 8-32 screws. The radials, RDL1 through 4, connect to the bottom vertical section, VS3, using a collar, CO1, I machined out of a section of aluminum round-bar stock. A 28-inch piece of phenolic tubing, MS2, passes through VS3 and joins this to the middle vertical section, VS2, at the top and the mounting base, BS1, at the bottom.

The bottom insulating section of the antenna, MS2, fits into the base mount, BS1, which is a seven-foot section of 1¼-inch Schedule-40 galvanized pipe. The bottom end of this pipe has a 1¼ to 1½-inch reducer, plus a six-inch section of 1½-inch Schedule 40 pipe. I attached a floor flange, BS4, to the bottom of the 1½-inch pipe.

This mounting arrangement requires two persons to raise the antenna; one to hold the antenna erect and one to bolt up the flange. If you were to weld a hinge to the bottom of the mounting base instead of the floor flange, then erection could probably be done by one person.

Fig 2—View of installed antenna showing vertical radiator, radials, guy attachment to vertical section and radials, feed-line attachment with strain relief, and 80-meter network at the very bottom.

The antenna is designed to be guyed, using four ¹/₈-inch OD nylon polyester cords attached to a guy ring, GR1, just below the midpoint of the vertical section. Since the radials are long and flexible, they require support along their length. The guy cords are thus secured to the radials at the point where they pass the radial on the way to the ground. The radial to guy-attaching bracket grasps the radial and has a hole large enough to permit a loop of guy rope to be passed through it. I placed a piece of wooden dowel in the loop and when the loop is pulled taut, the dowel is held firmly by the loop. This provides an adjustment to hold the radial horizontally and in the right location along the length of each guy rope. Again, see the photo in Fig 2.

The vertical sections of the stubs, TS1, TS2 and TS3, are made of ³/₈-inch OD aluminum tubing. They are spaced away from the center vertical section with ½-inch OD polycarbonate insulating rods and 0.040-inch aluminum brackets. The top hat, RDL5 through 8, is formed by crossed pieces of ³/₈-inch OD tubing. They are securely held in place by a 6-32 screw where the rods cross in the middle of the tubing cross section. See **Table 1** for a complete list of mechanical materials.

Electrical Construction

Electrical construction is straightforward. Using a grid dipper I adjusted the coil length of L1 to put the resonance of the L1-C1 circuit at the design frequency, monitoring the exact frequency with a calibrated receiver. I tuned L1-C1 on the bench and then with a dial caliper measured the final length of the coil. I then installed the coil on the antenna, keeping the same measured length.

The 80-meter network is within easy reach from a stepladder. See **Fig 3**. After erection and testing of the 10 through 40-meter antenna, I put the 80-meter network in place. The procedure for adjusting the 80-meter network is identical to that described above, with the exception that there are two resonators involved in series, L2-C2 and L3-C3. Each LC section should be tuned separately. Mount L2 and L3 to minimize magnetic coupling between them.

The transmission-line balun, L4, is formed by threading 12 to 14 turns of the coax feed line through a ferrite toroidal core. When the turns are pulled tight on the core, leave a 3 to 4 foot pendant length. This is then attached to the antenna's feed point. Dress the coax away from the antenna, in the plane of the radials. Some means of strain relief is appropriate at the antenna to keep the coax from breaking away from its attachment point. I used a short piece of nylon cord wrapped in a self-tightening binding (Chinese finger trap) on the coax and tied to the antenna just above the feed point.

Electrical Components

As I mentioned earlier, the development of a new capacitor ran in parallel with the antenna design. The prototype of this antenna was installed with a new planar Teflon® capacitor of my own design wherever I required capacitors. A close-up picture of these capacitors is shown in **Fig 4**. They have the advantage of being adjustable over a ±15% range, in four-bit binary-coded increments. The prototype capacitors are similar in quality to a 850S ceramic "doorknob" capacitor, but in their present form can only handle about 500 to

Table 1
List of Materials (Mechanical)

Item	*Material*	*Function*	*Source/Remarks*
BS1	1¼" × 7' Schedule 40 steel pipe	Top base section	local hardware store
BS2	1½" × 1¼" Schedule 40 reducer	Top to mid section coupler	local hardware store
BS3	1½" × 6" Schedule 40 steel pipe	Lower base section	local hardware store
BS4	1½" Schedule 40 floor flange	Bottom base adapter	local hardware store
CO1	2½" OD, 1½" ID × 1½" long	Radial attaching collar	See note 2
GR1	Fab from 0.040" alum. sheet	Guy rope attaching ring	See note 3
MS1	1³/₈" OD × 0.75" ID × 13" phen. tube	Mid vertical mast section splicer	U.S. Plastics, Lima, Ohio See note 5.
MS2	1³/₈" OD × 0.75" ID × 28" phen. tube	Bottom vert. mast section splicer	U.S. Plastics, Lima, Ohio See note 5.
RA 1-4	Fab from 0.040" alum. sheet	Radial to guy attaching bracket	See note 3
RDL 1-4	½" OD × 0.058" wall × 12'	Radials (4 total)	See note 1
RDL 5-8	³/₈" OD × 0.058" wall × 9¾"	Top hat (4 radials from 2 pcs)	See note 1
SP1	Fab from 0.040" alum. sheet	15 Meter stub bracket	See note 3
SP 2-3	Fab from 0.040" alum. sheet	10 Meter stub bracket	See note 3
SS1	½" OD × 11³/₈" Polycarb. rod	15 Meter stub spacer	U.S. Plastics, Lima, Ohio See note 4.
SS2	½" OD × 9⁵/₈" Polycarb rod	10 Meter stub spacer	U.S. Plastics, Lima, Ohio See note 4
TS1	³/₈" OD × 0.058" wall × 11' 4⁵/₈"	15 Meter tuning stub	See note 1
TS2-3	³/₈" OD × 0.058" wall × 4' 8³/₄"	10 Meter tuning stub	See note 1
VS1	1½" OD × 0.049" wall × 12'	Top vertical section	See note 1
VS2	1½" OD × 0.049" wall × 10' 9⁷/₈"	Middle vert. section	See note 1
VS3	1½" OD × 0.049" wall × 12"	Bottom vert. section	See note 1
1 ea	¼-20 × 2" bolt and nut	Base insulator insert limit	local hardware store
100 ft	¹/₈" Dia. polyester rope, UV stabilized		local hardware store
4 ea	⁵/₁₆" Dia. × 1" wooden dowel rod	Guy rope adjusters (RDL1-4)	local hardware store
4 ea	6-32 × ½" Phillips Pan head screw, SS	Radial bracket attachment	local hardware store
4 ea	6-32 nuts, SS	Radial bracket attachment	local hardware store
1 ea	6-32 × 1" Phillips Pan Head screw, SS	Top hat assembly	local hardware store
24 ea	8-32 × ½" Phillips Pan Head screw, SS	Vertical mast assembly	local hardware store
4 ea	8-32 × ½" Allen head set screw, SS	Radial to collar attaching	local hardware store
4 ea	10-24 × ½" Allen head set screw, SS	Collar to mast attaching	local hardware store

Note 1: Aluminum tubing from local distributor or Texas Towers (1-800-272-3467). Note that Texas Towers stock is 0.058" wall thickness and 1¹/₂" OD, so 1³/₈" insulating splicer may be a tight fit.
Note 2: Collar machined from 2½" OD 6061-T6 aluminum round bar per drawing supplied here or inquire about price and availability from NØKC.
Note 3: Sheet metal parts fabricated from 0.040" thick 3003 aluminum sheet metal stock per drawings supplied here or inquire about price and availability from NØKC.
Note 4: U.S. Plastics (1-800-537-9724) ½" OD polycarbonate round rod, stock #43284; 2' minimum order
Note 5: U.S. Plastics ¾" ID ⁵/₁₆" wall phenolic tubing, stock #47106; single piece required

700 W of peak power. Not so bad for a first try, but not good enough to handle the legal power limit without some redesign.

The experimenter building this antenna can choose between the Teflon® capacitors I used (which are available from me) or 850S ceramic doorknob capacitors, available from RF Parts Co (1-800-RF-PARTS) or Fair Radio (1-419-223-2196).

The fixed values and tolerances of the commercial doorknob capacitors will likely result in some minor adjustments to your installed antenna. To reproduce the results obtained here—and to accommodate any changes desired, such as a greatly different mounting elevation—you should have an adjustable capacitor and a capacitance meter. Many modern digital multi-meters have a capacitance measuring range that can measure to a tolerance of 1%. Note that the capacitance values given in **Table 2** are actual values for the capacitor component itself—that is, after stray capacitances have been accounted for.

I determined that the ferrite choke balun ought to have a value of at least 50 to 100 µH of inductance and should be located about three feet away from the feed point. The requirement is satisfied with 12 to 14 turns of RG-8X transmission line wound through a FT-240-43 core. These turns can be about 1¾ inches in diameter on the core; not so tight that they distort the impedance of the transmission line and not so long that they waste transmission-line length.

I designed the coils using software provided by G4FGQ.[5] Results correlated well—except for predicted stray capacitance—with design information provided in the book *Reference Data for Radio Engineers*.[6] I deemed coil Qs in excess of 400 to be satisfactory. The Q of the coil is preserved by side mounting rather than coaxial mounting.

Electrical Performance—Computed and Measured

I compared the calculated gains and elevation patterns for the new design to standard ground-plane antennas (a vertical antenna for 80 meters). My antenna has a little less gain, –2.80 dBi on 80 meters compared to a full-sized standard vertical, but at 20 meters it is a tie and at 15 and 10 meters there is a small advantage in gain over the reference ground plane antenna. Elevation (takeoff) angles are nearly identical to the standard antennas.

One of the goals of this project was to see just how accurately the software could model this type of antenna. The results presented here are for the prototype as constructed and installed using the computed component values and dimensions without adjustments. The antenna has measured feed-point SWRs shown in **Fig 5**. I made

Fig 3—View of the top of the base-mounting section and antenna base, showing the 80-meter network.

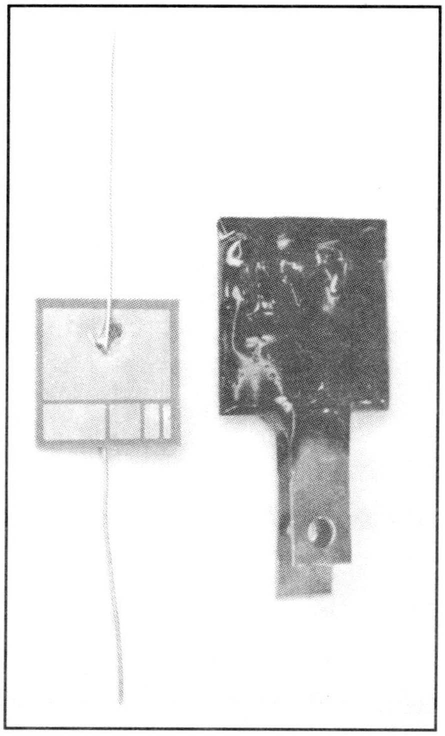

Fig 4—Power planar Teflon capacitors. Shown here is an uncoated substrate with test leads connected and a finished capacitor with heavy-duty, heat-dissipating copper strap leads and epoxy coating. Note the binary-sized capacitive sections that can be jumpered to the main area for adjustment.

these SWR measurements at the transmitter end of an 85-foot, RG-8X feed line for the prototype antenna. I made small design corrections to center the 15 and 10-meter bands in the next-generation hardware—that is, the hardware described in this article.

The 80-meter VSWR curve, while higher than desired, is acceptable because it is within the reach of the antenna tuner of the Kenwood TS440S-AT transceiver I use. Also a worst case SWR of 4:1 is within the tuning range of a well-designed linear amplifier with a Pi-L output network. An analysis[9] of the RG-8X coax shows the worst case heat rise to be within acceptable limits for full legal power when operating CW or SSB, and I computed losses in the cable to be only 0.4 dB. To some extent, the 80-meter SWR values are influenced by the ground radial system, of course.

You may want to mount this antenna at an elevation other than eight feet. Several computer runs indicate that center frequencies will change about ± 1% as the antenna base elevation is varied from one to 15 feet. This effect will be felt most on the 40-meter band. Either you will have to recalculate the principal elements of the antenna or else use some trial-and-error experimentation to accommodate the frequency shift at the extremes of base elevations mentioned here. If the base elevation is changed, you will have to recalculate the 80-meter network for operation on that band.

RF Exposure Considerations

With the current FCC rules regarding RF energy exposure[7,8] the design of a new antenna isn't complete without an investigation of near-electromagnetic fields to establish safe distances from the antenna. Computer modeling is an acceptable method under the rules. I calculated the closest-approach distance at a 1500-W level at 14 feet from the center of the antenna. At 100 W, the distance is 5 feet.

Table 2
List of Materials (Electrical)

Item	*Description*	*Source/Remarks*
C1	72.5 pF; HF power capacitor	See text
L1	2.80 µH; 8 T, #10AWG Cu wire, 2.0" Dia., 1.50" long	Hand fab, note 1,2
L4	60 µH; 12 T, RG8/X on Amidon FT-240-43 core	Loose 1¾" Dia. turns
The following parts are required only if operation on 80M is desired:		
C2	218 pF; HF power capacitor	See text
C3	85.0 pF; HF power capacitor. See text	
L2	2.5 µH; 8 T, #10AWG Cu wire, 2.0" Dia., 1.75" long	Hand fab, note 1,2
L3	1.48 µH; 7 T, #10AWG Cu wire, 1.45" Dia, 1.13" long	Hand fab, note 1,2
4ea	8-32 × ½ Phillips Pan head screws, SS	Local hardware store
4ea	8-32 Nuts, SS	Local hardware store
12ea	8-32 Washers, SS	Local hardware store
1ea	4.5" × 6.5" Phenolic vectorboard	Digi-Key #V1043-ND; Tel. 1-800-344-4539
2ea	1½" Radiator hose clamp, SS	Auto parts store

Note 1: Bend the coil leads at a sharp right angle from the coil such that the leads lie in a plane under the coil body. In one lead form an eyelet that will accept an 8-32 screw. Form an elongated eyelet in the other coil lead that accommodates the appropriate center to center mounting spacing and allows some inductance adjustment.

Note 2: Final adjustment of L1, L2 and L3 is done with a Grid Dip meter. With capacitors C1, C2 and C3 in hand, adjust the length of each inductor to resonate with its respective capacitor to a resonance frequency as shown:

L1C1 10.93 MHz
L2C2 6.82 MHz
L3C3 14.20 MHz

Monitor the Grid Dip meter signal with a calibrated receiver to obtain the necessary frequency accuracy for these adjustments.

Making a High-Voltage Fixed Capacitor

PC boards made with PTFE insulation (commonly known as Teflon) can be used to make high-voltage, high RF-current capacitors that have some very desirable characteristics. While Teflon doesn't have a very high dielectric constant, it can withstand high temperatures (think of Teflon-coated skillets), doesn't absorb moisture and is stable under UV. Furthermore, it has very low dielectric losses, even at microwave frequencies, and it can withstand high voltages.

The general expression for a parallel-plate capacitor is given by:

$$C = 0.225\,\varepsilon_r \left[(N-1)\,A / t\right] \qquad \text{Eq 1}$$

where N is 2 (representing the top and bottom surfaces of PC-board plate area) and the dielectric constant ε_r for Teflon is 2.1. A is the area in one side of the plate in square inches and t is the thickness of the dielectric between the plates in inches. Here, we'll use a 5 mil PCB substrate thickness and rearrange Eq 1 to calculate the area for a desired amount of capacitance.

$$A = 0.01058 \text{ square inches per pF} \qquad \text{Eq 2}$$

Note that the shape of the "plate" can be anything—a square, a rectangle, a circle, etc. For 100 pF, we need 1.058 square inches of copper, minus a 1% to 3% area reduction depending on the area-to-perimeter ratio to allow for fringing effects. A practical limit to this technique is a capacitor on the order of 300 pF, or about 1¾ inches on a side.

In Fig 4 you can see a capacitor made with "binary" adjustability to give a total of 15 binary weighted possible adjustments, with a 2-pF smallest "bit" of extra capacitance. I solder heavy copper straps (20 mils thick by ³⁄₈-inch wide) onto the PCB top and bottom. These straps conduct heat from the Teflon dielectric and also make a low-inductance connection. Remember, a square inch of Teflon might be asked to dissipate 3 or 4 W of RF power due to its dissipation factor (rated at 0.00015) for typical values of RF current in these kinds of applications!

Fig 4 also shows a capacitor after tweaking for a desired level of capacitance. It has been coated with epoxy to seal it against the weather. These capacitors broke down around the edges at about 1200 V rms of RF, so they are safe at applications up to the same sorts of levels that 850S-type doorknob capacitors can handle. Sometime in the near future I expect to improve the capacitor design and I will write an article about that. Stay tuned!

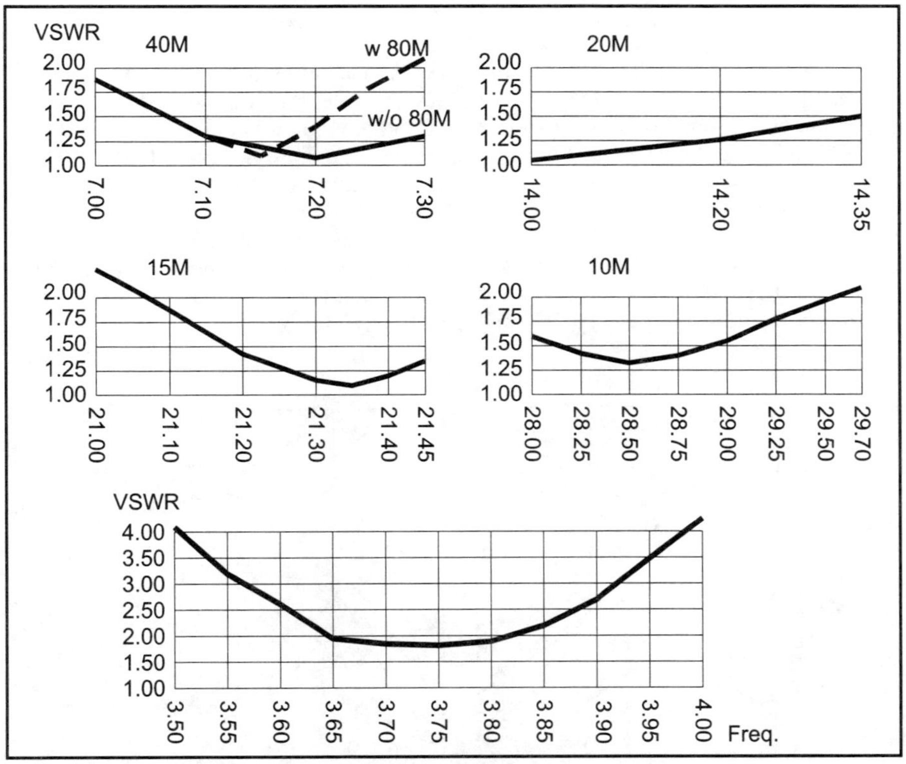

Fig 5—SWR measurement for the 80 through 10-meter bands.

Summary

This antenna resulted in worst-case center frequency errors of 1% or less for the 10 and 15-meter bands when comparing installed with computed results. With the exception of the 80-meter band, the SWRs are close to or better than 2:1 across the full bandwidth of each amateur band, 10 through 40 meters. My prototype antenna went up and has stayed up, without changes, after being built to the numbers crunched by the computer. As a computer exercise, I only had to make slight design adjustments to more perfectly center the SWR across the bands. This is shown in *EZNEC* file 1040GPD3.EZ.

I have used this antenna now for several years. It has withstood the "ice storm of the century" here in Kansas City. I have worked many DX contacts, CW and SSB, and have achieved all-mode DXCC. The local ground-wave tests that I've made indicated a remarkable advantage over a dipole, as expected.

References and Notes

[1] *EZNEC* software, Roy Lewallen, W7EL, PO Box 6658, Beaverton, OR 97007, **http://eznec.com/**.

[2] *NEC-Win Basic, NEC-Win Plus* antenna CAD software, Nittany Scientific, 1733 West 12600 South, Suite 420, Riverton, UT 84065, Phone/Fax: (801) 446-1426, **http://www.nittany-scientific.com/**.

[3] Rudy Severns, N6LF, "Verticals, Ground Systems and Some History," *QST*, Jul 2000, pp 38-44.

[4] R.J. Edwards, G4FGQ, Radial Systems CAD shareware, **http://www.btinternet.com/~g4fgq.regp/**.

[5] R.J. Edwards, G4FGQ, Solenoid Inductor CAD shareware, **http://www.btinternet.com/~g4fgq.regp/**.

[6] *Reference Data for Radio Engineers, Fifth Edition,* Howard W. Sams & Co, Inc., 1974.

[7] Ed Hare, W1RFI, "The FCC's New RF-Exposure Regulations," *QST*, Jan 1997, **http://www.arrl.org/news/rfsafety/qst9710.html**.

[8] Ed Hare, W1RFI, "FCC RF-Exposure—the Station Evaluation,", *QST*, Jan 1998, **http://www.arrl.org/news/rfsafety/eval/index.html**.

[9] R. J. Edwards, G4FGQ, Coaxial Ratings CAD software, **http://www.btinternet.com/~g4fgq.regp/**.

A MEMORIAL

This article is dedicated to the memory of Maurice "Buz" Baer, WØJAS. We called him "Uncle Wojazz." Buz was devoted to promoting Amateur Radio through teaching the art and science of radio, by construction and operation. His professional life was devoted to teaching. He was my friend and mentor.

A Practical 30-Meter Vertical

By Ken Elsberry, WD4ERM
503 Blaine Cr
Quitman, GA 31643

Here's how to put up a 30-meter wire vertical and how to tame the ground radials.

This antenna is a quarter-wave vertical, ground mounted and using ground radials. Over the years I have used several arrangements for dipole and inverted-V designs for 30 meters, but I have had better overall results with this vertical. This is the only band I have found where the vertical has an advantage over simple horizontal antennas, although I have not had the opportunity to check it against a 30-meter beam.

Noise seems to be noticeably less on the vertical than on other antennas at my location. This is unusual, as it is normally the other way around. Sometimes if the received signal is even less on the S meter without noise, it is indeed louder to the ear, or at least easier to copy.

I used the traditional formula (234 divided by the frequency in MHz) to arrive at the length of the radiator, 23.15 feet. I made the radials 5% longer, at 24.3 feet. I used eight radials in the three different versions of these antennas I have built. However, I don't think you can have too many radials, so use as many as you can, with a minimum of at least eight.

I have always believed it is best to keep construction as simple as practical, so I decided to use a solid #10 vertical wire as the radiator. I started out with an eight-foot ground rod close to the base of a pine tree. I leave four to six inches of the ground rod sticking out of the ground and attach a simple ground clamp to the top of the ground rod. This holds a screw-in type insulator used for electric fences. See **Fig 1** for details. The radiator wire loops around the insulator and at the other end, some 23 feet away, I attach some form of small insulator, usually a ceramic egg-type or plastic "dog bone."

From the top insulator I attach a weather-resistant $^3/_{16}$-inch polyester rope and pull it over a convenient limb located directly over the ground rod and bottom insulator. To make the wire radiator stand parallel to the tree, I pull the rope down tight, tying it off to a spike driven into the tree or to one of the clamps on the ground rod, as shown in **Fig 2**.

I next attach radials to the ground rod using another ground clamp below the first one, extending the radials in a spoke-like pattern from the rod. You can either bury the radials or pin them down tight to the ground and let grass grow over them. Either way will work fine electrically. I have used a yard edger to cut grooves in which to bury radials.

Lately, I have started using so-called "plumber's pipe hooks" with the 90° end crooks cut off. **Fig 3** shows two of these side-by-side: the original at the upper left and the modified version with the cut-off crooks on the lower right. I found mine at Home Depot, by PHP (Prairie Home Products, Inc.). These are ½ inch by 6 inch wire pipe hooks and sold for $2.41 in a package of 20. I find that pinning the radials with these giant hairpins about 3 to 5 feet apart works fine for me.

Fig 1—Closeup view of ground rod and clamp used to hold mounting screw for fence-post type insulator. The vertical wire insulator is attached to the top of the insulator.

Fig 2—Completed 30-meter vertical, with ground radials clamped to ground rod with a second ground clamp. The pull-down polyester rope is at the left, tied to the second clamp. The RG-8X coax is the white cable at the right.

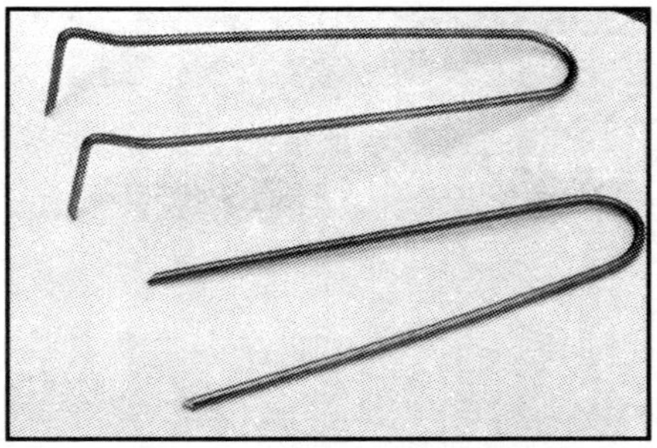

Fig 3—Before and after view of "pipe hook" used by WD4ERM. The one at the upper left is the original shape, and the one on the bottom right has had its 90° crooks cut off.

I attached the center conductor of RG-8X coax to the radiator, with the braid going to the ground rod. Before you solder the connection to the center conductor, make sure you measure the SWR and make any necessary adjustments to the length of the radiator. You will probably have to shorten it somewhat to achieve resonance. My vertical had better than a 1.2:1 SWR over the 30-meter band.

This vertical is certainly simple to build, since most hams can get the materials easily. It is also inconspicuous, especially with the green-insulated wires I used for the radials, which barely show in the grass!

An HF Multiband Vertical Antenna

By Gordon Moogk, VE3DBP
645 Lakeside Trail
Point Clark, ON N0G 2R0
Canada

VE3DBP describes his unique multiband vertical that uses no traps.

Until recently most multiband antennas have used traps of one form or another. Several years ago, Gary Breed, K9AY, described a different approach that uses multiple elements instead of traps.[1]

Fig 1 shows a vertical antenna covering all bands from 40 thru 6 meters using K9AY's approach. The antenna requires a ground counterpoise to work properly and the more radials in the counterpoise the better. In this antenna I used 16 radials of varying lengths of #12 copper wire.

The radials are not visible in **Fig 2**, since they are under the shingles on the roof of the garage and down the inside walls of the garage. The driven element is a 40-meter ¼ λ vertical radiator and is fed with 50-Ω coaxial cable at the base. The additional elements are not fed directly, but instead are tied to the counterpoise and parasitically coupled to the driven element.

The 40-meter element is constructed of telescoping aluminum conduit 1.5 inches in diameter at the bottom and 1 inch at the top. It has two sets of guys made of ¼-inch nylon or other weather-resistant rope. The bottom insulator is a nylon hose coupling, which is available at any hardware store.

The additional vertical elements are for 30, 20, 17, 12 and 10 meters. These elements are spaced evenly around and are supported with insulated brackets to the 40-meter element. The parasitic elements are made of 1 inch or less diameter aluminum tubing. All element lengths are calculated using the handbook formula 246/f MHz.

Tuning Things Up

Once the 40-meter vertical and the counterpoise are in place, check the tuning of the antenna on 40 and 15 meters. You should find a good match on both 40 and 15 meters. You can then start adding the additional elements for the bands you wish to cover, one at a time. Spacing at the bottom of these elements is critical, and I used a screw and locking nut device (see **Fig 3**) to allow precise adjustments.

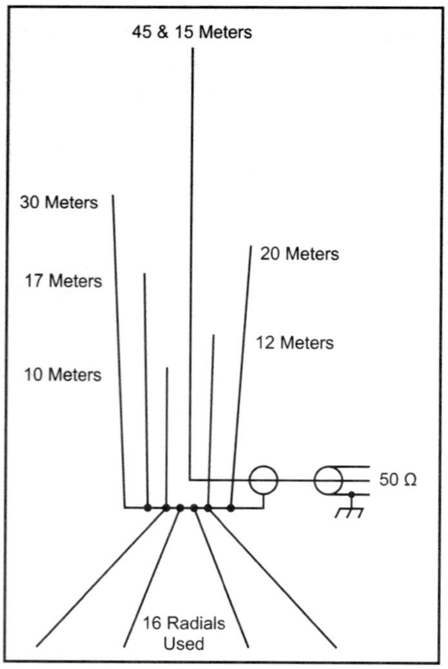

Fig 1—The configuration shown here also provides a good match on the bottom end of the 6-meter band. I've used it successfully during auroral and other band openings. It is probably operating on the 5th harmonic of the 30-meter element.

Fig 2—The completed antenna mounted on my roof. Insulated guys are used to brace the assembly in high winds.

Fig 3—A low-impedance connection at the bottom of the vertical is required to tie the radials and the parasitic elements together. While I used square aluminum tubing, a sheet of copper or aluminum would probably work just as well. Note the threaded-rod and L-bracket fine-tuning adjustment system for each parasitic element. The threaded rod, bracket, nuts and washers are made of aluminum to avoid corrosion problems.

A spacing of about 6 to 7 inches between the bottom of the 40-meter radiator and the added elements would be a good starting point. An insulated spacer tied to the 40-meter vertical supports the new elements. The spacing at the top of the elements being added is not as critical and varies from about 16 to 30 inches. Using an antenna analyzer or VSWR bridge, adjust the spacing at the bottom initially and then the top until a 1:1 match is obtained. You might also have to make slight adjustments to the element length. Use the same procedure to add additional elements. The end result is an antenna that will work effectively on all bands without an antenna tuner.

Fig 4—The Plexiglas insulating brackets from the parasitic elements to the center 40-meter element. I've used this antenna successfully on all bands, although the bandwidth on 10 meters is limited to about 200 kHz.

Notes and References

[1]G. Breed, K9AY, "The Coupled-Resonator Principle: A Flexible Method for Multiband Antennas," *The ARRL Antenna Compendium, Vol 6* (Newington: ARRL, 1996), pp 109-112.

FEEDBACK

Please use this form to give us your comments on this book and what you'd like to see in future editions, or e-mail us at **pubsfdbk@arrl.org** (publications feedback). If you use e-mail, please include your name, call, e-mail address and the book title, edition and printing in the body of your message. Also indicate whether or not you are an ARRL member.

Where did you purchase this book?

☐ From ARRL directly ☐ From an ARRL dealer

Is there a dealer who carries ARRL publications within:

☐ 5 miles ☐ 15 miles ☐ 30 miles of your location? ☐ Not sure.

License class:

☐ Novice ☐ Technician ☐ Technician with code ☐ General ☐ Advanced ☐ Amateur Extra

---

Name ______________________ ARRL member? ☐ Yes ☐ No

Call Sign ______________________

Daytime Phone ( ) ______________________ Age ______________________

Address ______________________

City, State/Province, ZIP/Postal Code ______________________

If licensed, how long? ______________________ e-mail address: ______________________

Other hobbies ______________________

____

Occupation ______________________

For ARRL use only	MVAC
Edition	1 2 3 4 5 6 7 8 9 10 11 12
Printing	1 2 3 4 5 6 7 8 9 10 11 12

From ______________________________

____

____

Please affix postage. Post Office will not deliver without postage.

EDITOR, MORE VERTICAL ANTENNA CLASSICS
ARRL—THE NATIONAL ASSOCIATION FOR AMATEUR RADIO
225 MAIN STREET
NEWINGTON CT 06111-1494

- - - - - - - - - - - - - - - - - please fold and tape - - - - - - - - - - - - - - - - -